Key Concepts in Critical Theory

ECOLOGY

EDITED BY

Carolyn Merchant

Humanity
Books

an imprint of Prometheus Books
59 John Glenn Drive, Amherst, New York 14228-2197

Published 1999 by Humanity Books, an imprint of Prometheus Books

03 02   9 8 7 6

Library of Congress Cataloging-in-Publication Data

Ecology / edited by Carolyn Merchant.
       p.   cm. — (Key concepts in critical theory)
    Originally published: Atlantic Highlands, NJ : Humanities Press International,
Inc., 1994
    Includes bibliographical references and index.
    ISBN 1–57392–600–0
    1. Human ecology—Philosophy. 2. Social ecology—Philosophy. 3. Ecology.
I. Merchant, Carolyn. II. Series.
GF21.E25   1994
304.2—dc20   94–3090
                                            CIP

Printed in the United States of America on acid-free paper

ECOLOGY

Key Concepts in Critical Theory

Series Editor
Roger S. Gottlieb

JUSTICE
Edited by Milton Fisk

GENDER
Edited by Carol C. Gould

DEMOCRACY
Edited by Philip Green

RACISM
Edited by Leonard Harris

ECOLOGY
Edited by Carolyn Merchant

EXPLOITATION
Edited by Kai Nielsen and Robert Ware

ALIENATION AND SOCIAL CRITICISM
Edited by Richard Schmitt and Thomas E. Moody

CONTENTS

PART VI: SPIRITUAL ECOLOGY

PART VII: POSTMODERN SCIENCE

SERIES EDITOR'S PREFACE

THE VISION OF A rational, just, and fulfilling social life, present in Western thought from the time of the Judaic prophets and Plato's *Republic*, has since the French Revolution been embodied in systematic *critical theories* whose adherents seek a fundamental political, economic, and cultural transformation of society.

These critical theories—varieties of Marxism, socialism, anarchism, feminism, gay/lesbian liberation, ecological perspectives, discourses by antiracist, anti-imperialist, and national liberation movements, and utopian/critical strains of religious communities—have a common bond that separates them from liberal and conservative thought. They are joined by the goal of sweeping social change; the rejection of existing patterns of authority, power, and privilege; and a desire to include within the realms of recognition and respect the previously marginalized and oppressed.

Yet each tradition of Critical Theory also has its distinct features: specific concerns, programs, and locations within a geometry of difference and critique. Because of their intellectual specificity and the conflicts among the different social groups they represent, these theories have often been at odds with one another, differing over basic questions concerning the ultimate cause and best response to injustice, the dynamics of social change, the optimum structure of a liberated society, the identity of the social agent who will direct the revolutionary change, and in whose interests the revolutionary change will be made.

In struggling against what is to some extent a common enemy, in overlapping and (at times) allying in the pursuit of radical social change, critical theories to a great extent share a common conceptual vocabulary. It is the purpose of this series to explore that vocabulary, revealing what is common and what is distinct, in the broad spectrum of radical perspectives.

For instance, although both Marxists and feminists may use the word "exploitation," it is not clear that they really are describing the same phenomenon. In the Marxist paradigm the concept identifies the surplus labor appropriated by the capitalist as a result of the wage-labor relation. Feminists have used the same term to refer as well to the unequal amounts of housework, emotional nurturance, and child raising performed by women in the nuclear family. We see some similarity in the notion of group inequality (capitalists/workers, husbands/wives) and of unequal exchange. But we also see critical differences: a previously "public" concept extended to the private realm; one first centered in the economy of goods now moved into the life of emotional relations. Or, for another example, when deep ecologists speak of "alienation" they may be exposing the contradictory and destructive relations of humans *to* nature. For socialists and anarchists, by contrast, "alienation" basically refers only to relations among human beings. Here we find a profound contrast between what is and is not included in the basic arena of politically significant relationships.

What can we learn from exploring the various ways different radical perspectives utilize the same terminology?

Most important, we see that these key concepts have histories and that the theories of which they are a part and the social movements whose spirit they embody take shape through a process of political struggle as well as of intellectual reflection. As a corollary, we can note that the creative tension and dissonance among the different uses of these concepts stem not only from the endless play of textual interpretation (the different understandings of classic texts, attempts to refute counterexamples or remove inconsistencies, rereadings of history, reactions to new theories), but also from the continual movement of social groups. Oppression, domination, resistance, passion, and hope are crystallized here. The feminist expansion of the concept of exploitation could only grow out of the women's movement. The rejection of a purely anthropocentric (human-centered, solely humanistic) interpretation of alienation is a fruit of people's resistance to civilization's lethal treatment of the biosphere.

Finally, in my own view at least, surveys of the differing applications of these key concepts of Critical Theory provide compelling reasons to see how complementary, rather than exclusive, the many radical perspectives are. Shaped by history and embodying the spirit of the radical movements that created them, these varying applications each have in them some of the truth we need in order to face the darkness of the current social world and the ominous threats to the earth.

ROGER S. GOTTLIEB

ACKNOWLEDGMENTS

The following publishers and authors have kindly granted permission to reprint or quote from their articles:

Karl Marx and Friedrich Engels, "Marx and Engels on Ecology" from *Marx and Engels on Ecology*, ed. Howard Parsons (Westport, Conn.: Greenwood Press [an imprint of Greenwood Publishing Group, Inc., Westport Conn.], 1977), 129–85, selections. © 1977 by Howard Parsons. Reprinted with permission.

Max Horkheimer and Theodor Adorno, "The Concept of Enlightenment," from *Dialectic of Enlightenment* by Max Horkheimer and Theodor Adorno, trans. John Cumming (New York: Herder & Herder, 1972), 3–28, 42, excerpts. English translation copyright © 1972 by Herder & Herder, Inc. Reprinted by permission of The Continuum Publishing Company.

Herbert Marcuse, "Ecology and Revolution," from *Liberation* 16 (September 1972): 10–12.

William Leiss, "The Domination of Nature," from William Leiss, "Technology and Domination," in *The Domination of Nature* (New York: George Braziller, 1972), 145–55, 161–65. Copyright © 1972 by William Leiss. Reprinted by permission of George Braziller, Inc.

Robyn Eckersley, "The Failed Promise of Critical Theory." Reprinted from *Environmentalism and Political Theory* by Robyn Eckersley (Albany: State University of New York Press, 1992), 97–106, by permission of the State University of New York Press. Copyright © 1992 by the State University of New York.

Alan Miller, "Economics and the Environment," from *A Planet to Choose: Value Studies in Political Ecology* (New York: Pilgrim Press, 1978), 23–35, 49. Reprinted by permission of the publisher, The Pilgrim Press, Cleveland, Ohio. Copyright © 1978.

Barry Commoner, "Poverty and Population," from Barry Commoner, "How Poverty Breeds Overpopulation (and Not the Other Way Around)," *Ramparts* (1974): 21–25, 58–59, excerpts. Revised 1990.

Herman Daly, "Steady-State Economics," from *Steady State Economics* by Herman E. Daly (San Francisco: W. H. Freeman, 1977), 2–18, excerpts. Copyright © 1977 by W. H. Freeman and Company. Reprinted with permission.

Len Doyal and Ian Gough, "Human Needs and Social Change," from *The Living Economy: A New Economics in the Making*, ed. Paul Ekins (New York: Routledge, 1986), 69–72, 78–79.

Brian Tokar, "Creating a Green Future," reprinted from *The Green Alternative: Creating an Ecological Future*, rev. ed., by Brian Tokar (San Pedro: R. & E. Miles, [1987] 1993), 141–48. Copyright © 1992, R. & E. Miles, by permission.

Arne Naess, "Deep Ecology," reprinted from Arne Naess, "The Shallow and the Deep, Long-Range Ecology Movement: A Summary," *Inquiry* 16 (1973): 95–100, by permission of Scandinavian University Press, Oslo, Norway.

Bill Devall, "The Deep Ecology Movement," from *Natural Resources Journal* 20 (April 1980): 299–313.

George Sessions, "Ecocentrism and the Anthropocentric Detour," from *ReVision* 13, no. 3 (Winter, 1991): 109–15. Reprinted with permission of the Helen Dwight Reid Educational Foundation. Published by Heldref Publications, 1319 18th Street, N.W., Washington, D.C. 20036–1802. Copyright 1991.

Murray Bookchin, "The Concept of Social Ecology," from: *CoEvolution Quarterly* (Winter 1981): 15–22.

James O'Connor, "Socialism and Ecology," from *Capitalism, Nature, Socialism* 2, no. 3 (1991): 1–12. "Socialism and Ecology" has been reprinted in *Our Generation* (Canada), *Il Manifesto* (Italy), *Ecologia Politica* (Spain), *Nature and Society* (Greece), *Making Sense* (Ireland), and *Salud y Cambio* (Chile). A revised version will appear in James O'Connor, *Capitalism and Nature: Essays in the Political Economy and Politics of Ecology*, to be published by Guilford Publications in association with *Capitalism, Nature, Socialism*.

Françoise d'Eaubonne, "The Time for Ecofeminism," trans. Ruth Hottell, from *Le Fémminisme ou la Mort* (Paris: Pierre Horay, 1974), 213–52.

Ynestra King, "Feminism and the Revolt of Nature," from *Heresies*, no. 13 (1981): 12–16.

Val Plumwood, "Ecosocial Feminism as a General Theory of Oppression," from *Ecopolitics V Proceedings*, ed. Ronnie Harding (Kensington, NSW, Australia: Centre for Liberal & General Studies, University of New South Wales, 1992), 63–72.

Elizabeth Carlassare, "Essentialism in Ecofeminist Discourse." A modified version appears in *Capitalism, Nature, Socialism* 5, no. 3 (September 1994): 1–18.

Freya Mathews, "Ecofeminism and Deep Ecology," from "Relating to Nature," in *Ecopolitics V Proceedings*, ed. Ronnie Harding (Kensington, NSW, Australia: Centre for Liberal & General Studies, University of New South Wales, 1992), 489–96. A longer version appears in *The Trumpeter* 11, no. 2 (Fall 1994): 159–66.

Peter Wenz, "The Importance of Environmental Justice," reprinted from *Environmental Justice* by Peter Wenz (Albany: State University of New York Press, 1988), xi–xii, 6–21, excerpts, by permission of the State University of New York Press. Copyright (c) 1988 by the State University of New York.

Robert Bullard, "Environmental Racism and the Environmental Justice Movement," from *Confronting Environmental Racism: Voices from the Grassroots*, ed. Robert Bullard (Boston: South End Press, 1993), 15–24, 38–39.

Winona LaDuke, "From Resistance to Regeneration," from *The Nonviolent Activist* (September–October 1992): 3–6.

Vandana Shiva, "Development, Ecology, and Women," from *Staying Alive: Women, Ecology, and Development* by Vandana Shiva (London: Zed Books, 1988), 1–9, 13.

Ramachandra Guha, "Radical Environmentalism: A Third-World Critique," from Ramachandra Guha, "Radical Environmentalism and Wilderness Preservation: A Third World Critique," *Environmental Ethics* 11 (Spring 1989): 71–80.

Joanna Macy, "Toward a Healing of Self and World," from Joanna Macy, "Deep Ecology Work: Toward the Healing of Self and World," *Human Potential Magazine* 17, no. 1 (Spring 1992): 10–13, 29–31.

Charlene Spretnak, "The Spiritual Dimension of Green Politics," reprinted from *The Spiritual Dimension of Green Politics* by Charlene Spretnak (Santa Fe, N.M.: Bear & Co., 1986), 52–69. Copyright 1986, Bear & Co. Inc., P.O. Box 2860, Santa Fe, N.M. 87504.

Carol Christ, "Why Women Need the Goddess," from *Womanspirit Rising: A Feminist Reader in Religion* by Carol P. Christ and Judith Plaskow (San Francisco: Harper and Row, 1979), 273–89. Reprinted by permission of HarperCollins Publishers, Inc. Also printed in Carol P. Christ *Laughter of Aphrodite: Reflections on a Journey to the Goddess* (San Francisco: Harper and Row, 1987), 117–32. First published in *Heresies No. 5: The Great Goddess*, 1978.

John Cobb, Jr., "Ecology and Process Theology," from "Process Theology and an Ecological Model," *Pacific Theological Review* 15, no. 2 (Winter 1982): 24–27, 28.

Paula Gunn Allen, "The Woman I Love Is a Planet," from *Reweaving the*

World: The Emergence of Ecofeminism, ed. Irene Diamond and Gloria Orenstein (San Francisco: Sierra Club Books, 1990), 52–57.

Fritjof Capra, "Systems Theory and the New Paradigm," reprinted from "Physics and the Current Change of Paradigms," in *The World View of Contemporary Physics: Does It Need a New Metaphysics?*, ed. Richard F. Kitchener (Albany: State University of New York Press, 1988), 144–52, by permission of the State University of New York Press. Copyright © 1988 by the State University of New York.

David Bohm, "Postmodern Science and a Postmodern World," reprinted from *The Reenchantment of Science: Postmodern Proposals*, ed. David Ray Griffin (Albany: State University of New York Press, 1988), 57–58, 60–66, 68, by permission of the State University of New York Press. Copyright © 1988 by the State University of New York.

James Lovelock, "Gaia," from *Planet Earth* (Fall 1986): 3–22.

Edward Lorenz, "Predictability: Does the Flap of a Butterfly's Wings in Brazil Set Off a Tornado in Texas?," from *The Essence of Chaos* by Edward Lorenz (Seattle: University of Washington Press, 1993), 181–84, copyright © 1993, University of Washington Press. Reprinted with permission of the University of Washington Press.

Ilya Prigogine, "Science in a World of Limited Predictability," from "The Rediscovery of Time: Science in a World of Limited Predictability," paper presented to the International Congress on Spirit and Nature, Hanover, Germany, 21–27 May, 1988, excerpts. Copyright © Stiftung Niedersachsen, Hanover, Germany.

The First National People of Color Environmental Leadership Summit, "Principles of Environmental Justice," Washington, D.C. 24–27 October 1991, adopted 27 October 1991.

I thank Jessea Greenman for assistance in preparing notes and obtaining permissions, Celeste Newbrough for compiling the index, and Kathy Delfosse for editorial advice and assistance. Funding for research and permission costs was contributed by the Agricultural Experiment Station and the Committee on Research, University of California, Berkeley.—C.M.

INTRODUCTION

CAROLYN MERCHANT

DOMINATION IS ONE OF our century's most fruitful concepts for understanding human-human and human-nature relationships. The theme of domination and its reversal through liberation unites critical theorists and environmental philosophers whose work spans the twentieth century. When the domination of nonhuman nature is integrated with the domination of human beings and the call for environmental justice, Critical Theory instills the environmental movement with ethical fervor. This book brings together the Frankfurt school's analysis of domination with the insights of today's deep, social, and socialist ecologists, ecofeminists, people of color, spiritual ecologists, and postmodern scientists. But the project of analyzing and overcoming domination is not dealt with uniformly by all parties. There is much disagreement over why and how nature and humans are linked and what to do about changing those linkages.

The problem of domination was explored in depth by Max Horkheimer and Theodor Adorno, in exile from Germany during World War II, in New York and in Los Angeles, California. Together they wrote *Dialectic of Enlightenment*, published in 1945, and Horkheimer followed with *The Eclipse of Reason* in 1947. From the 1920s, the Institute for Social Research in Frankfurt had attempted to develop a multidisciplinary theory of society and culture that would bring together critiques and alternatives to mainstream social theory, science, and technology and address social problems. They expanded on Marxism by extending its analysis of political economy to the interconnections between and among the spheres of nature, economy, society, politics, psychology, and culture. They characterized their approach as a comprehensive, totalizing theory of modernization and its alternatives and emphasized Hegel's theory of dialectical interactions among the various aspects of society. They were deeply concerned about the problems they associated with modernity—the period from the Renaissance and Reformation to the era of state capitalism in the twentieth century—and the concept of enlightenment that epitomized the ideology of the modern world. Horkheimer and Adorno exposed the Greco-Roman roots of individualism, science, and the domination of nature that reached a crescendo in the eighteenth century's Age of Enlightenment. Rather than seeing the progressive

1

aspects of modernity in which science, technology, and capitalism increasingly improve on the human condition, they emphasized modernity's dehumanizing tendencies, its destruction of the environment, its potential for totalitarian politics, and its inability to control technology.[1]

Critical Theory drew its initial inspiration from Marxism. Karl Marx and Friedrich Engels's nineteenth-century critique of the inequalities created by industrialization's separation of capital from labor, entrepreneurs from working people, mind from body, and humanity from nature went to the heart of the problem of social justice. Under capitalism, justice and equality for every person could never be achieved because of the structural constraints built into capital's need to expand its sphere continually, by using nature and humans as resources. Nevertheless, Marx and Engels viewed capitalism as only a stage in the progression to an equitable socialist society in which the basic needs of all people for food, clothing, shelter, and energy would be fulfilled. But both capitalism and socialism would achieve these human gains over the "necessities of nature," through the domination of nonhuman nature by science and technology. While Marx and Engels displayed an extraordinary understanding of and sensitivity toward the "ecological" costs of capitalism, as revealed in the selection of their work edited by Howard Parsons, they nevertheless bought into the Enlightenment's myth of progress via the domination of nature.[2] It was this myth that Horkheimer and Adorno sought to expose.

Horkheimer and Adorno employed Hegelian dialectics to analyze society as a totality, drawing on the humanism of the early Marx. They saw culture as continually changing and developing in an open-ended transforming process. They emphasized the relative autonomy of the cultural superstructure rather than the later Marx's determinism of the economic base on the legal-political superstructure. Horkheimer had set the agenda when he became the director of the Institute for Social Research in 1931. He saw the institute's task as delineating the interconnections between economic life, the psychological development of individuals, and culture, including science, technology, spiritual life, ethics, law, and even entertainment and sports. In addition to Horkheimer and Adorno, various members of the institute, such as Herbert Marcuse, Karl Wittfogel, Erich Fromm, and Leo Lowenthal, carried out critiques of philosophy, bureaucratic administration, ideology, sociology, psychology, literature, and popular music as they contributed to the dominant ideology. They saw social theories as reproducing the dominant practices of capitalist society. Scientific theories and cultural concepts were representations of the material world rather than absolute truths. In opposition to the absolutes of idealism, they defended a materialist approach to nature and reality, emphasizing the material conditions of human needs. Material life created the human subject, while the subjects in turn transformed their own specific historical and material conditions.

To the Frankfurt theorists, the material conditions in an inegalitarian class society create suffering. Yet the unequal social conditions that lead to unequal pain are not natural or inevitable. They can be changed. Moral outrage over suffering leads not only to sympathy and compassion, but also to efforts to transform the particular social conditions that give rise to pain. Freedom from pain and suffering can be achieved in a just society, and it is Critical Theory's goal to envision that society and its attainment through the fulfillment of human needs and potentials.

CRITICAL THEORY AND THE DOMINATION OF NATURE

Working in New York City and Los Angeles, California in the 1940s, Horkheimer and Adorno turned their attention to the problem of the domination of nature and human beings. In the ancient world the emergence of a sense of self as distinct from the external natural world entailed a denial of internal nature in the human being. "In class history," said Horkheimer and Adorno, "the enmity of the self to sacrifice implied a sacrifice of the self, inasmuch as it was paid for by a denial of nature in man for the sake of domination over non-human nature and over other men. . . . With the denial of nature in man not merely the telos of the outward control of nature, but the telos of man's own life is distorted and befogged."[3] Odysseus in the ancient world and Francis Bacon at the onset of modernity epitomized the break with the enchanted past of myth and mimesis (imitation). Mimesis is participation in nature through identification with it. Odysseus represents the struggle to overcome the imitation of nature and immersion in the pleasures of animal life and tribal society. Through the emergence of his own identity as an individual self, he is able to break the hold of the mythic past and control his animal instincts, his men, his wife, and other women. He becomes alienated from his own emotions, bodily pleasures, other human beings, and nature itself.

Tribal societies pursued their needs through the imitation of nature. Human beings became as much like the animals they hunted as possible. Power over nature, hence self-preservation, was achieved through imitative magic. Enlightenment thinking disenchants nature by removing that magic and turning the subject into an object, and that process of objectification distances subject from object. In the early modern era the domination of nature reached a new level in the thought of Francis Bacon. In the early seventeenth century, Bacon advocated extending the dominion of man over the entire universe through an experimental science that put nature on the rack and tortured "her" to reveal "her" secrets. Power over nature was subsequently extended to mathematics and physics by René Descartes and Isaac Newton.

Horkheimer and Adorno used critical philosophy to expose the underlying instrumental reasoning behind both scientific thought and capitalist society. The disenchantment of external nature was achieved by despiritualizing a human being's internal nature: "The subjective spirit which cancels the animation of nature can master a despiritualized nature only by imitating its rigidity and despiritualizing itself in turn."[4] Bourgeois society lived according to quantification, calculation, profits, exchange, and the logic of identity. But so deeply imbedded does this way of thinking become that it is presumed to be reality by mainstream society. So powerful is the mystique of reason as instrument in the control of nature and human bodies that it banishes other modes of participating in the world to the periphery of society. Logic and mathematics along with the calculus of capital become privileged modes of thought, defining the very meaning of truth. The identity "a = a" in logic and between the two sides of an equation in mathematics allows instrumental reason to describe the natural world. Describing the world through logic and mathematics in turn leads to prediction and hence to the possibility of controlling nature. Instrumental reason and enlightenment are thus synonymous with domination.

The domination of external nature by internal nature exacts a cost. The unrestrained use of nature destroys its own conditions for continuation, as the inexorable expansion of capital undercuts its own natural resource base. Similarly, the repression of human emotions and animal pleasures leads not to human happiness, but to anguish. But the tighter the rein, the greater the potential for rebellion. The revolt of nature is thus contained within the enlightenment project. Internal nature rebels psychically, spiritually, and bodily. External nature revolts ecologically. Here Critical Theory and the ecology movement intersect.

Horkheimer and Adorno entitled their book *Dialectic of Enlightenment* to emphasize the flux between mimesis and enlightenment. Mimesis, like enlightenment, depended on identity. Identifying with another subject through imitation allowed for self-preservation in tribal societies. Naming of plants and animals entailed the identification of characteristics that remain constant through change. Names in turn bring power in ritual magic. Yet at the same time, mimesis retains its original sense of participation in nature. On the other hand, enlightenment is also myth. Its historical specificity, its contextuality within modernity, and its limited truth domain destroy its claims to absolute knowledge. The myth of enlightenment is the guiding story of modernity.

The domination of internal nature makes possible the domination of external nature, which in turn leads to the domination of human beings. The relationship between the domination of nature and the domination of human beings was explicitly stated by Adorno in a 1955 essay on the German historian Oswald Spengler: "The confrontation of man with nature, which first produces the tendency to dominate nature, which in turn results in the domination of men by other men, is nowhere to be seen in the *Decline of the West*."[5] The domination of external nature therefore precedes, and is a condition for, the domination of human beings in society. Horkheimer and Adorno's analysis of the dialectics of various forms of domination—internal, external, human, and nonhuman—thus sets up a problematic for future work on the roots of the ecological crisis. It opens room for discussion of the many aspects of the relationships between domination and the life-world—women, minorities, technology, economy, justice, spirit, and science—the topics considered in this book.

Another member of the Institute for Social Research, who took up residence at the University of California, San Diego, following the World War II exodus was Herbert Marcuse. His *Eros and Civilization* (1955), *One-Dimensional Man* (1964), *Counterrevolution and Revolt* (1972) applied Critical Theory to social reproduction by examining the role of mass media, the control of information, and the decline of the family in maintaining the culture of needs necessary to the success of capitalism. Although Marcuse was not an ecological philosopher, he did address the issue of ecology in a special symposium on "Ecology and Revolution," published in *Liberation* magazine in 1972.[6] He saw ecology as a revolutionary force of life, against which the counterrevolutionary forces of ecocide were destroying "the sources and resources of life itself" in the service of monopoly capital. Writing in the context of the Vietnam War, Marcuse connected the genocide and ecocide in North Vietnam, which at the time were being carried out through modern science and technology's showcase weapons—the "electronic battlefield" and chemical defoliants—with the wider war against nature. Similarly, the same process that transforms people into objects in a market society also transforms nature into commodities, leaving little natural beauty, tranquility, or untouched space. The ecology movement exposes this war against nature and attacks the space capital carves out for its own. The movement attempts to defend what is left of untouched nature and argues that the entire production-consumption model of "war, waste, and gadgets" must be halted in the name of the very survival of life itself.

The same year that Marcuse published his brief article, William Leiss, who had obtained his doctoral degree in philosophy from the University of California, San Diego in 1969, published a major treatise on *The Domination of Nature* (1972). Leiss explored the legacy of Horkheimer and Adorno, who had returned to Frankfurt after the war. Elaborating on the role of Francis Bacon in setting the modern agenda of power over nature through science and technology, Leiss analyzed technology's role in mastering both the external world of nature and the human being. In the modern world, nature has already been subjected to widespread exploitation, and for many in the Western world, material well-being has been achieved. Yet the production of an endless parade of technological improvements maintains the subjection of people's internal natures by chaining them to the manufacturing process. As aggression against external nature accelerates, the technological connection renders people's internal natures increasingly passive. At the same time, the potential for social conflict increases because of growing differentials in wealth and the specialized geographical locations of essential resources such as oil. "The cunning of unreason takes its revenge," writes Leiss. "In the process of globalized competition men become the servants of the very instruments fashioned for their own mastery over nature." The result is the revolt of nature, both internal and external, in a "blind irrational outbreak of human nature" and the concommitant collapse of ecological systems.[7]

Yet Critical Theory's astute analysis of the human domination of nature, according to Australian political theorist Robyn Eckersley, fails to offer any real alternative to an anthropocentric approach. Critical Theory and the green movement have not been as mutually supportive as their common German origins and interests in new social movements might suggest. Horkheimer and Adorno did not develop their promising insights into the liberatory aspects of critical reason, the ecological dimensions of the revenge of nature, or the voice of nonhuman nature as speech for all that is mute. Marcuse's analysis of the ecology movement as a revolutionary force and his concept of nature as an opposing partner did not move toward an ecocentric ethic that gave value or rights to nonhuman organisms and inanimate entities. Jürgen Habermas, who in 1964 took over Horkheimer's chair in philosophy at Frankfurt, retained his alignment with the Social Democrats rather than supporting the German Greens in the 1980s, and he viewed the radical ecology movement as neoromantic rather than emancipatory. According to Eckersley, political theory needs to move toward an ecocentric approach to the human-nature relationship. "Ecocentric theorists," she writes, "are concerned to develop an ecologically informed approach that is able to value (for their own sake) not just individual living organisms, but also ecological entities at different levels of aggregation, such as populations, species, ecosystems, and the ecosphere (Gaia)."[8]

ENVIRONMENTAL ECONOMICS AND POLITICS

The ecology movement of the 1960s and 1970s extended the critique of the domination of nature and human beings by industrial capitalism begun by Marx, Engels, and the Frankfurt theorists. Political economists looked at the relationships between First-World capitalism and Third-World colonialism through Immanual Wallerstein's model of core and peripheral economies.[9] Although the industrial revolutions in eighteenth-century Europe and nineteenth-century America stimulated economic production, raised living standards, reduced death rates, and led to smaller family sizes (the demographic transition), they did so at the expense of Third-World peoples and resources. Political ecologist Alan Miller discusses the linkages between the core economies of Europe, North America, and Japan (after World War II) and the resource extraction loops of the Third World in which soils, forests, and mines were used to produce export crops such as coffee, sugar, hemp, beef, copper, and aluminum. Wealthy Third-World landowners imported luxury items from the Northern Hemisphere, raising their own standard of living, in turn pushing the vast majority of people deeper into poverty on marginal lands and in urban slums.

Deeply interconnected with colonialism and the political economy of the Third World are issues of expanding population. As Barry Commoner argues, population growth rates decline when countries undergo the demographic transition: declining death rates, resulting from rising standards of living, are followed by declining birthrates, as large numbers of children are no longer needed for rural labor and old-age security. But unlike industrial development in the First World, Third-World dependency relations have prevented a rapid reduction in population growth rates. Much depends, therefore, on how development occurs and whether that development will be environmentally sustainable and socially just.[10]

The goal, in the view of World Bank economist Herman Daly, is to reverse the environmental policies of international development agencies, many of which have led to environmental and social deterioration, and to move toward a steady-state or very-low-growth global economy in which population, resources, and artifacts remain at a constant level. A steady-state economy would be maintained by the lowest possible energy costs in production, by constant recycling of nonrenewable resources, by replacing renewable resources at the same rate they are used, and by minimizing pollution. Quality of life, culture, knowledge, and psychic satisfaction could continue to grow and develop; physical materials and energy would be conserved. Yet a rapid transition to a steady-state society could disproportionately affect Third-World peoples and the poor. Social justice is therefore a key factor in the transition to a sustainable world and the liberation of nature.

While Daly sets out conditions for a sustainable economy, sociologists Len Doyal and Ian Gough delineate conditions for a socially just society. "It is clear," they argue, "that the Baconian vision of nature as an unending storehouse and unfillable cistern is over. The new reality is that awareness of the delicacy of the biosphere must now go hand in hand with democratically planned production for human need."[11] Fulfillment of basic needs entails individual human health and autonomy. For individual needs to be fulfilled, societies have to be able to reproduce themselves. They have to produce the basic needs of material life—food, clothing, shelter, and energy—and to provide the conditions for physical and emotional health. Children must be socialized, educated, and taught to communicate in order to become independent adults, and there must also be some means of maintaining social order. Social justice involves a guarantee of basic liberties, such as free speech and the right to assemble, the removal of social inequalities that prevent equitable distribution of social goods, and equality of opportunity. How these basic needs are to be satisfied in a sustainable world entails the relationship between the state and human liberty and the problem of the liberation of nature.

One avenue for bringing about a sustainable society and a socially just world is through green politics. Green movement theorist Brian Tokar envisions the development of community-based institutions and greens in local government as a way of implementing the mandate to "think globally, act locally." Creating a green future starts at the grass-roots level with individual empowerment, coalition building, and living within the means of the bioregion. The green program for change is based on four pillars: ecology, grass-roots democracy, social justice, and nonviolence. As green groups and parties expand throughout the world, the goal of restructuring politics itself along green lines represents a vision for the future.

DEEP, SOCIAL, AND SOCIALIST ECOLOGY

Although green economics and politics provide a practical means for moving toward global sustainability, overcoming the deeply engrained anthropocentrism at the root of the domination of nature requires a new philosophy and ethics. Deep ecologists argue that mainstream environmentalism is limited and incremental in scope and that what is needed before real change can occur is a transformation in consciousness. In 1973 Norwegian philosopher Arne Naess published a foundational article, "The Shallow and the Deep, Long-Range Ecology Movement,"[12] and the idea of Deep Ecology was taken up by California sociologist Bill Devall and philosopher George Sessions, who promoted and developed it in a series of newsletters and articles. They argued that the anthropocentric core of mainstream Western philosophy must be overturned and replaced with a new metaphysics, psychology, ethics, and science. They looked for alternatives within the Western traditions, Eastern philosophy, and the insights of indigenous peoples. A major compilation of sources was published by Devall and Sessions under the title *Deep Ecology: Living as if Nature Mattered* (1985), and Naess revised the platform in *Ecology, Community, and Lifestyle,* (1989).[13] In 1990, Australian philosopher Warwick Fox linked Deep Ecology with transpersonal psychology through the ideas of identification with the nonhuman world and the notion of an expanded self that was capable of moving beyond the atomized, isolated ego. These works were followed in 1991 by Australian philosopher Freya Mathews's development of the foundations of Deep Ecology in relation to the principles of interconnectedness, intrinsic value, and self-realizing systems in her book *The Ecological Self.*[14] Fundamental to the deep ecological approach is the need to overcome a narrow anthropocentrism dominant in Western culture. Yet Deep Ecology's focus on anthropocentrism also masks the role of capitalism and political economy in the domination of nature *and* human beings.

An alternative to Deep Ecology's placement of the blame for the ecological crisis at the doorstep of anthropocentrism is social ecology.[15] Defined and defended by social philosopher Murray Bookchin, social ecology, like

Critical Theory, grounds its analysis in domination.[16] But for Bookchin, as opposed to deep ecologists, critical theorists, and Marxists, the domination of human beings is historically and causally prior to the domination of nature. Bookchin writes, "By the early sixties, my views could be summarized in a fairly crisp formulation: the very notion of the domination of nature by man stems from the very real domination of human by human."[17] Whereas early tribal societies were basically egalitarian and lived within nature, the increasing prestige of male elders created social hierarchies and inequalities that led to power over other human beings, especially women, and ultimately over nature. The growth of ancient city-states, medieval walled towns, and state capitalism depended on increasingly entrenched hierarchies of elders over other tribal members, men over women, and elites over laborers and slaves. This social domination led to the domination of people over nature. The goal of social ecology is to remove hierarchy and domination from society, including the domination of people over nature. Bookchin's ecological anarchism envisions an ecological society to be achieved through reliance on the resources and energy of the local bioregion, face-to-face grass-roots democracy within libertarian municipalities linked together in a confederation, and the dissolution of the state as a source of authority and control.

Socialist ecology, as conceptualized by Marxist economist James O'Connor, contrasts with Bookchin's social ecology in that it is grounded not in the concept of domination but in political economy.[18] Only the Marxist categories of labor exploitation, production, the profit rate, capital circulation and accumulation, and so on can adequately account for the degradation of nature under capitalism. Socialist ecology is distinguished from state socialism as represented by the failed Soviet Union and eastern-bloc countries, whose industrial-growth models resulted in environmental disaster. Instead it looks toward new forms of eco-socialism brought about by green social movements with commitments to democracy, internationalism, and ways to overcome the dualism of local versus state control and administration. But like deep and social ecology, it recognizes the autonomy of nonhuman nature, ecological diversity, and the science of ecology as the basic science of survival for the twenty-first century.

ECOFEMINISM

Deep Ecology's efforts to place the blame for ecological deterioration on the domination of nature by human beings (anthropocentrism) meets resistance from ecofeminists who see the domination of both nature and women by men (androcentrism) as the root cause of the modern crisis. French feminist Françoise d'Eaubonne set up Ecologie-Féminisme in 1972 as part of the project of "launching a new action: *ecofeminism*" and in 1974 published a chapter entitled "The Time for Ecofeminism" in her book *Feminism or Death*.[19]

In that foundational chapter, translated here for the first time by French feminist scholar Ruth Hottell, d'Eaubonne states that women in the "Feminist Front," separated from the movement and founded the information center called the "Ecology-Feminism Center." The new action, christened "ecofeminism," attempted a synthesis "between two struggles previously thought to be separated, feminism and ecology." The goal was to "remake the planet around a totally new model," for it was "in danger of dying, and we along with it." It called for a mutation of the world that would allow the human species to escape from death and to continue to have a future. Writing as a militant radical feminist, d'Eaubonne placed the problem of the death of the planet squarely on the shoulders of men. The slogan of the Ecology-Feminism Center was "to tear the planet away from the male today in order to restore it for humanity of tomorrow. . . . If the male society persists there will be no tomorrow for humanity."[20]

Presenting a litany of planetary ills, ranging from the global population explosion to worldwide pollution and American consumption, urban crowding, and violence, d'Eaubonne argued that both capitalism and socialism were scenes of ecological disasters. The most immediate death threats to the planet were overpopulation (a glut of births) and the destruction of natural resources (a glut of products). Although many men attempted to label "overpopulation" a "Third-World problem," the real cause of the sickness was patriarchal power.[21]

D'Eaubonne followed the analysis of nineteenth- and early-twentieth-century proponents of ancient matriarchal societies such as Johann Bachofen, Friedrich Engels, Robert Briffault, and August Bebel, who saw "the worldwide defeat of the female sex" some five thousand years ago as the beginning of an age of patriarchal power. It was the male System created five thousand years ago, not capitalism or socialism, that gave men the power to sow both the earth (fertility) and women (fecundity). The iron age of the second sex began, women were caged, and the earth was appropriated by males. The male society "built by males and *for* males" that took over running the planet did so in terms of competition, aggression, and sexual hierarchy, "allocated in such a way to be exercised by men over women." Patriarchal power produced agricultural overexploitation and industrial overexpansion. "The Earth, symbol and former preserve of the Great Mothers," d'Eaubonne notes, "has had a harder life and has resisted longer; today, her conqueror has reduced her to agony. This is the price of phallocracy."[22] Today husbands who control women's bodies and implant them with their seed, doctors who examine them, and male priests who call for large families are the bearers of phallocratic power over women's wombs.[23]

D'Eaubonne saw ecofeminism as a new humanism that put forth the goals of the "feminine masses" in an egalitarian administration of a reborn world. A society in the feminine would not mean power in the hands of women but no power at all. The human being would be treated as a human being, not as a male or female. Women's personal interests join those of the entire human community, while individual male interests are separate from the general interests of the community. The preservation of the earth was a question not just of change or improvement, but of life or death. The problem, she said, paraphrasing Marx, is "to change the world . . . *so that there can still be a world*." But only the feminine which is concerned with all levels of society and nature can accomplish "the ecological revolution." She concluded her foundational essay with the telling words: "And the planet placed in the feminine will flourish for all."[24]

In the United States, the term "ecofeminism" was used at Murray Bookchin's Institute for Social Ecology in Vermont in about 1976 to identify courses as ecological, namely, eco-technology, eco-agriculture, and ecofeminism. The course on ecofeminism was taught by Ynestra King, who used the concept in 1980 as a major theme for the conference "Women and Life on Earth: Ecofeminism in the '80s," held in Amherst, Massachusetts. King published "Feminism and the Revolt of Nature" in 1981 in a special issue of *Heresies* on "Feminism and Ecology." Her article and forthcoming book, "Feminism and the Reenchantment of Nature," both reflect the Frankfurt school's conceptual framework of the disenchantment of the world and the revolt of nature. The promise of ecological feminism lies in its liberatory potential to "pose a rational reenchantment that brings together spiritual and material, being and knowing."[25]

King conceptualized ecological feminism as a tranformative feminism drawing on the insights of both radical cultural feminism and socialist feminism. Radical cultural feminists such as Mary Daly in *Gyn/Ecology* (1978) and Susan Griffin in *Woman and Nature* (1978) linked the domination of women and nature under patriarchy. Men use both to defy death and attain immortality. Woman's oppression is rooted in her biological difference from men, who use women to secure their own immortality through childbearing. Nature's oppression is rooted in its biological otherness from men, who secure immortality as rational creators of human culture. For radical feminists, women and nature can be liberated only through a feminist separatist movement that fights their exploitation through the overthrow of patriarchy. Socialist feminists, however, ground their analysis not in biological difference, but in the historically constructed material conditions of production and reproduction as a base for the changing superstructure of culture and consciousness. Underlying both positions, King argues, is a false separation of nature from culture. Instead, a transformative feminism offers an understanding of the dialectic between nature and culture that is the key to overcoming the domination of both women and nature. Such a position is needed if an ecological culture that reconnects nature and culture is to emerge.

Australian philosopher Val Plumwood extends the analysis of domination initiated by d'Eaubonne and King by comparing the debates between deep ecologists, social ecologists, and ecofeminists. Each group of eco-philosophers makes valid points, but in so doing tries to reduce the others to its own critique. Thus Deep Ecology is correct to challenge the human-centeredness of social ecology, but social ecology is also right in its analysis that hierarchical differences within human society affect the character of environmental problems. Thus an alternative, cooperative approach is needed. Ecofeminism, with its emphasis on relations, has the potential to see connections among various forms of oppression, such as those affecting women, minorities, the colonized, animals, and nature. Recognition of the weblike character of various forms of domination suggests a cooperative strategy of web repair. The ecofeminist approach focuses on relations and interconnections among the various ecology movements and leads to the possibility of a liberatory theory and practice.

A major problem for ecofeminist theory is essentialism. Do women (and men) have innate, unchanging characteristics (or essences), or are all male and female qualities historically contingent? Do women's reproductive biological functions (ovulation, menstruation, and the potential for pregnancy, childbearing, and lactation) constitute their essence? The essentialist perception of women as closer to nature, as a result of their biological functions of reproduction, has historically been used in the service of domination to limit their social roles to childbearers, child rearers, caretakers, and house-keepers.[26] Furthermore, do women have a special relationship to nature that men cannot share? If women declare that they are different from men and as ecofeminists set themselves up as caretakers of nature, it would seem that they cement their own oppression and thwart their hopes for liberation and equality. These issues are addressed by Elizabeth Carlassare.

Carlassare argues that the essentialist debate in ecofeminism reflects the concerns raised by social and socialist ecofeminists that cultural ecofeminism is essentialist. Social/ist ecofeminists employ the label "essentialist" to privilege their own materialist/constructivist ways of knowing over cultural feminism's spiritual/intuitive ways of knowing. But, she argues, some cultural ecofeminist texts actually explore the historical and cultural process by which essences are constructed, and some social/ist ecofeminist texts have essentialist ideas not too far beneath their surface, inasmuch as gender is constructed out of "sexed anatomical raw material." Instead of one group of ecofeminists at-tempting to dominate and marginalize another group through use of the essentialist label, the variety of voices within the ecofeminist movement should be taken as a sign of its vitality. Ecofeminists can unite in a political struggle against the oppression of women and nature without having to assume a unified, yet totalizing, epistemology. "Unity in diversity" can be read as "difference does not have to mean domination."[27]

Whereas Elizabeth Carlassare's ultimate goal is to create unity within the diversity of the ecofeminist movement, Freya Mathews, in her article on "Ecofeminism and Deep Ecology," wishes to reconcile ecofeminism with Deep

Ecology. The differences between the two movements have been detailed by a number of scholars, including Ariel Salleh, Val Plumwood, Marti Kheel, Warwick Fox, Michael Zimmerman, and Jim Cheney.[28] Deep Ecology's holistic view of the world as an extended self-writ-large contrasts with ecofeminism's emphasis of the world as a community of beings with which one has compassionate, caring relationships. For ecofeminism, each being in the community is respected as distinct, rather than as part of the cosmic whole with which one should identify and into which one can merge. Mathews concludes that through its ethic of compassion, ecofeminism can humanize Deep Ecology, which has become embittered toward human rapaciousness, seeing humans as a species bent on destroying other species. But Deep Ecology can deepen ecofeminism by asking it to see the whole as an internally connected moral order, not just a family of individuals for whom one intimately cares.

Ecofeminism, therefore, as developed by d'Eaubonne, King, and Plumwood, not only deepens Critical Theory's analysis of the domination of nature by connecting it to the historically constructed domination of women, but it also indicates pathways to liberation. Plumwood and Mathews suggest ways of cooperating with social and Deep Ecology, while King and Carlassare promote cooperation within feminism and ecofeminism.

ENVIRONMENTAL JUSTICE

Reversing the domination of nature and human beings requires environmental justice. Environmental justice entails the fulfillment of basic needs through the equitable distribution and use of natural and social resources and freedom from the effects of environmental misuse, scarcity, and pollution. According to philosopher Peter Wenz, environmental justice is a problem of distributive justice. In some cases, such as the Garrett Hardin's tragedy of the commons, in which herdsmen overgraze a pasture, lumber companies deplete a timber supply, or chemical companies pollute a river, the solution is one of coordinated or mandated restraint. In other cases, where a basic need, such as water or food, is in short supply, the scarce resource must be equitably distributed through tribal or governmental law and policy. Democratic societies that are relatively nonrepressive and open need to be perceived as being socially just or civil rebellion may occur. People who share the benefits, responsibilities, and burdens of living in an open society may have to make sacrifices to maintain those advantages. Environmental laws and policies, therefore, must embody principles of environmental justice that a majority of people agree are reasonable.

Problems of environmental justice arise, however, when past decisions and practices disproportionately affect certain groups of people. Robert Bullard argues that in the United States, minorities, women, and the poor are often the victims of environmental racism. Internal colonies of minorities have experienced racism in the form of polluted air and water, toxins from landfills and incinerators, and hazardous waste facilities. Pollution goes hand in hand with poverty, deteriorating buildings, poor schools, and inadequate health care and is reinforced by government neglect. The mainstream environmental movement has ignored urban ecology, choosing to focus on wilderness preservation, pollution abatement, and population control. For environmental justice to occur, the effects of decade- or century-old decisions must be ameliorated and new socially just policies put in place. People of color have come together in a powerful movement to reverse environmental racism and promote environmental justice through grass-roots organizations, conferences, protests, and demonstrations. The 1987 United Church of Christ "Report on Toxic Wastes and Race in the United States"; the Mothers of East Los Angeles coalition; the Race, Poverty, and Environment Newsletter; and the 1991 First National People of Color Environmental Leadership Summit, with its "Principles of Environmental Justice," are but a few recent manifestations of a powerful new movement for environmental justice.

Among the North American groups longest and most virulently affected by environmental racism are Native Americans, who, according to activist Winona LaDuke, lived according to an ethical accountability to natural law. Because they relied on the animate world for sustenance, in compliance with natural law, they took only what they needed and gave thanks. This reciprocity between Native Americans and nature created sustainable communities that contrast with the capitalist-industrial model of accumulation. The impact of industrialization on Native American communities has been to create toxic and nuclear waste dumps on reservations, especially when urban communities resist them. Uranium and coal mining employed Indians on their own lands, but left them sick with lung and skin cancers. The shift from nuclear to hydropower (such as the James Bay hydro project) affected many northern United States and Canadian tribes. To turn the world around, says LaDuke, we need the leadership of indigenous and marginalized peoples. These are the people who know how to survive and who can bring a vision to the future.

Environmental justice for Third-World peoples involves a wholesale questioning and restructuring of Western-style development projects. Since World War II, argues Indian physicist and philosopher Vandana Shiva, development has actually been maldevelopment. Development, which was supposed to have been a postcolonial project, is rooted in the domination of women, tribal peoples, and nature by patriarchy and capitalism. Cash cropping undermines traditional subsistence and land rights, destroys soil, water, and forests, and renders people and nature passive. Maldevelopment subverted the traditional feminine principle in nature and resulted in the feminization of poverty. It named nature's own reproductive systems nonproductive unless they produced surplus capital and profits. "Maldevelopment," says

Shiva, "is the violation of the integrity of organic, interconnected and interdependent systems, that sets in motion a process of exploitation, inequality, injustice and violence." Recovering the feminine principle in nature that produces diversity and women's traditional connections to the ecosystem that produce life would mean rethinking development in the name of environmental justice.

The political consequences for the Third World, not only of Western-style development but also of the American version of Deep Ecology "are very grave indeed," writes Indian environmental historian Ramachandra Guha. The anthropocentric domination of nature that Deep Ecology implicates in the destruction of wilderness and its reversal by cordoning off huge tracts of lands from use of human beings, subverts the very real needs for survival by Third-World peoples. Setting aside wilderness in densely populated India has transferred vital subsistence resources from the poor to the rich. The real causes of resource depletion are overconsumption and militarization by the industrialized world. Deep Ecology appropriates and distorts the spiritual traditions of Eastern thinkers in its effort to become a new universal philosophy. It is actually a nature preservationist philosophy that in the hands of groups such as Earth First! uses militant methods of saving so-called wild nature (a concept that ignores the extensive use of lands by indigenous peoples for thousands of years). The claims of deep ecologists do not extend beyond the unique situation of American frontier expansion into lightly populated Native American lands and are counter to the real needs for environmental justice in the Third World.

The environmental justice movement thus extends Critical Theory's analysis of domination to encompass minorities and Third-World peoples who have been the victims of colonial and capitalist expansion. Redistribution of natural and social resources that fulfill basic needs, redressing past injustices, and rethinking the very meanings of concepts such as development, wilderness, and nature are essential first steps toward liberation.

SPIRITUAL ECOLOGY

In a famous article published in 1967, historian Lynn White, Jr., argued that the domination of nature stemmed from the Judeo-Christian mandate expressed in Genesis 1:28 to "increase, multiply, replenish the earth and subdue it." "Christianity," he declared, "is the most anthropocentric religion the world has ever seen. . . . By destroying pagan animism, Christianity made it possible to exploit nature in a mood of indifference to the feelings of natural objects."[29] Since then, hundreds of articles debating the validity of his thesis have been written and numerous alternative interpretations and practices offered. Mainstream religions of all denominations have re-read and reinterpreted their traditional relationships to nature in an effort to offer alternatives to the Genesis mandate. Ecologically based forms of spirituality that draw inspiration from Eastern religions, Native American philosophies, and pagan traditions offer connections to the natural world in order to engender appreciation, care, and environmental action on behalf of the planet.

Joanna Macy draws on Deep Ecology, systems theory, and Bodhisattva Buddhism to develop a sense of a self interconnected with the natural world. Deep Ecology's expanded self (John Seed's, "I am the rain forest protecting myself"), Gregory Bateson's systems theory (the self as individual plus its environment—the entire pattern that connects), and Buddhism's sense of interconnectedness (the dependent co-arising of phenomena) help us break out of the imprisonment of the self-contained ego. The new ecological self is a self connected to the world, bringing new resources of courage, ingenuity, and endurance to combat despair and engage in healing our body— the world.[30]

Charlene Spretnak relates the efforts of mainstream religions to find ecological alternatives to the Judeo-Christian mandate to subdue the earth by looking within their own traditions for alternative interpretations and passages. The stewardship ethic of humans as caretakers of the rest of creation, ecologically based worship services, respect for women and the female principle in creation, care and concern for homeless and poor people, and development of worker-owned and community-based economic alternatives in depressed areas are but a few examples. Creation spirituality, which emphasizes the interconnectedness of all creation, revivals of the old religion (paganism and wicca), and learning respect for the land from indigenous peoples are other examples from outside the mainstream. These spiritual approaches are consistent with principles of green politics, such as ecology, nonviolence, postpatriarchal principles, and grass-roots democracy.[31]

"Do women need the Goddess?" asks Carol Christ. Yes, answered Christ unequivocally in 1979. "Absolutely not," counters Pope John Paul II in a 1993 condemnation of "'Nature Worship' by some feminist U.S. Catholics." "The Christian faith itself," said the Pope, "is in danger of being undermined. Sometimes forms of nature worship and the celebration of myths and symbols take the place of the worship of the God revealed in Jesus Christ."[32] According to Carol Christ, religious symbols are part of powerful and compelling patriarchal systems that create moods and motivations in people. The domination of women by men in society can be legitimated by religions that worship a male God. Goddess symbols affirm female power as well as women's cultural and social heritage of spirituality and bonding, as well as the celebration of women, the life-cycle of nature, and the female body. The goddess for some women is the female aspect of God, whereas for others she is a powerful presence within nature, and for still others the vibrant sexuality of nature's reproductive capacities.

John Cobb offers another ecological approach to spirituality, via process theology, based on the philosophy of Alfred North Whitehead's *Process and*

Reality (1928). This philosophy views the world as an organism comprising individual organisms existing in relationship to the environment. Every organism is constituted by its set of relations with the rest of the world. The ecological model is a relational science that is consistent with relativity and quantum field theory. An organism is a series of events that Whitehead called "occasions"; the world is a vast field of occasions in enduring patterns. God for process theologians is a name for the ecological model in which all the relations are complete and perfect. This god is not a domineering god imposing his will on a separate creation, but an open, receptive, and responsive god. Process theologians and their followers are deeply involved in issues of ecology, feminism, peace, social justice, liberation, and freedom.

Native American writer Paula Gunn Allen writes of the planet as a woman. The physical planet includes human beings and their spiritual, emotional, and social lives, and all other life, as well as the climate, rocks, and minerals. The planet, which includes ourselves and all these aspects of physicality, is a mother becoming a grandmother as she emerges into a new consciousness. She is in a crisis caused, apparently, by human failings, but is actually entering an initiation into a sacred planet. As the planet, our mother, undergoes her initiation rites, we change with her. Part of her fears the initiation and wants to remain the same, part of her fears she will fail the initiation tests, and part of her longs for the fulfillment of completion. With her we make ourselves ready, with her we rejoice, with her we enter a new sacred life.

The forms of spirituality offered by Macy, Spretnak, Christ, Cobb, and Gunn Allen help to overcome the Genesis 1:28 version of nature domination through alternative earth-based forms of spirituality. In so doing, human domination of the earth can be transformed into the more participatory, mimetic modes of relating to nature identified by the Frankfurt theorists.

POSTMODERN SCIENCE

From the point of view of science, as well as of religion, our ways of relating to the planet are likewise undergoing a significant transformation. The Enlightenment ethic of the domination of nature fostered by mechanistic science's reduction of the world to dead atoms moved by external forces is being replaced by a postmodern, ecological world view based on interconnectedness, process, and open systems. Fritjof Capra contends that physics is in the midst of a paradigm shift to a new set of assumptions about reality and new ways of representing the world that will replace those of the modern era. This transformation in physics mirrors a much larger cultural and social transformation resulting from dislocations such as the environmental, nuclear, and poverty crises. The problems facing science and society today reflect the inadequacy of the structures of modernism—

mechanistic physics, industrialization, and the inequalities of class, race, and gender—that are fundamental to the domination of nature and human beings. Solving them requires a new social paradigm—"a [new] constellation of concepts, values, perceptions, and practices shared by a community, which form a particular vision of reality that is the basis of the way the community organizes itself."[33] Capra proposes that the ecological systems view of life is the new paradigm emerging to replace the mechanistic world view. This includes an emphasis on the whole over the parts, on process over structure, on the relative knowability of the external world, the idea of networks of knowledge and information, and the recognition of the necessity of approximation. The assumptions of the systems approach entail a new ethic that is life-affirming rather than life-destroying, and it recognizes the interconnectedness of all things and the human place in the network.

Postmodern science challenges the limitations of modern science first through early twentieth-century developments in relativity and quantum mechanics and then through postmechanistic physics. Physicist David Bohm has developed a theory of unbroken wholeness as the ground of matter, energy, and life that he argues resolves many of the contradictory assumptions underlying relativity theory and quantum mechanics. Bohm's theory is based on the underlying holomovement—a flow of energy in multidimensional space-time out of which unfolds the three-dimensional world described by Newtonian physics. The mechanistic world is implicit within, or enfolded into, a higher order of reality. This visible world is the explicate order on which most successful work in physics has been done. "In my technical writings," says Bohm, "I have sought to show that the mathematical laws of quantum theory can be understood as describing the holomovement, in which the whole is enfolded in each region, and the region is unfolded into the whole."[34] The holomovement is life implicit and consciousness implicit. This view militates against the fragmentation of the mechanistic world view and suggests an integration between matter and consciousness, value and fact, ethics and science.

James Lovelock's Gaia hypothesis pushes the ecological world view further by asking the question, Is the earth a living organism? Gaia is the Greek word for the earth goddess (see Spretnak), and its use implies that the earth, a seemingly inert object made of rocks and water, is alive. How so? Lovelock's answer is that life on earth constantly maintains atmospheric and hydrological conditions comfortable for its own continuation. It does so through feedback processes because the environment is an integral part of every living thing. The atmosphere, which is composed of nonliving gases, is nevertheless part of Gaia's life, just as is the shell on the back of a snail or the fur on a cat. Only when things are studied in isolation from their surroundings do they appear to be inanimate. Systems theory explains what mechanistic science cannot. The Gaia hypothesis means that the system as a whole is powerful enough to withstand climatic disasters and human dep-redations. "But, of course," Lovelock concludes, "if we transgress in our pollutions and our forest clearance, Gaia can move to a new stable state and one that's no longer comfortable for us. So living with Gaia is not so different from a human relationship."[35]

The postmodern, ecological world view, unlike the modern mechanistic one, is based on the impossibility of completely predicting the behavior of the natural world. Chaos theory suggests that most environmental and bio-logical systems, such as weather, noise, population, and ecological patterns, cannot be described accurately by the linear equations of mechanistic sci-ence and may be governed by nonlinear chaotic relationships.[36] To atmo-spheric physicist Edward Lorenz is attributed the famous metaphor of the butterfly effect, influential in the early work in chaos theory, that describes sensitive dependence on initial conditions. This approach questions the ability of science to make predictions in all but the most unusual situations, that is, the limited number of closed, isolated systems successfully described by mechanistic science.

In his 1972 paper presented to the annual meeting of the American As-sociation for the Advancement of Science, Lorenz asked whether the flap of a butterfly's wings in Brazil could set off a tornado in Texas. His point was that we cannot predict the results of small effects, such as a butterfly, on the weather because the atmosphere is unstable with respect to perturbations of small amplitude. We don't know how many small effects there are (such as butterflies), or even where they are located. We can't even set up a controlled experiment to find out if the atmosphere is unstable because we can never know what might have happened if we hadn't disturbed it. Most important is the problem of the "inevitable approximations which must be introduced in formulating . . . [the governing physical] principles as proce-dures which the human brain or the computer can carry out." The rapid doubling of errors precludes great accuracy in real world forecasting. The best that we can hope for is to make the "best forecasts which the atmo-sphere is willing to have us make."[37]

The problem of predictability is pushed further by Ilya Prigogine in his work with Isabelle Stengers on *Order Out of Chaos* and the concept of self-organization.[38] Classical thermodynamic processes discovered in the nine-teenth century, such as equilibrium and near-equilibrium cases that describe the steam engine and the refrigerator, suggest that the universe is running down and becoming more disorderly and chaotic. Yet in Prigogine's far-from-equilibrium thermodynamics—in situations found in hydrodynamics, many chemical processes, and evolution—a new reorganization can occur in which order can emerge out of chaos. In such situations, irreversibility and nonlinearity can lead to self-organization. Irreversibility and nonlinearity increase the role of fluctuations and lead to bifurcations (divisions) in which the system can go in several directions, that is, the nonlinear equations have several different solutions. The outcome cannot be predicted with certainty. Unstable dynamic systems, such as the weather systems described by Lorenz, behave differently than stable systems, such as the planetary systems described by Newton. Prigogine suggests that the unpredictability of systems holds implications for the domination of nature and leads to alternative strategies for human interactions with natural systems:

> The important element is that unstable systems are not controllable. . . . The classical view of the laws of nature, of our relation with nature, was domination. That we can control everything. If we change our initial conditions, the trajectories change slightly. . . . But that is not the general situation. . . .
>   We see in nature the appearance of spontaneous processes which we cannot control in the strict sense in which it was imagined to be possible in classical mechanics. . . . The world in which we are living is highly unstable. What I want to emphasize, however, is that this knowledge of instability may lead to other types of strategies, may lead to other types of interacting systems.[39]

What is remarkable about a number of the advances in modern science at the end of the twentieth century, Prigogine continues, is their appearance "at the very moment humanity is going through an age of transition, when instability, irreversibility, fluctuation, amplification, are found in every human activity. . . . What is so interesting is that there is a kind of overall cultural atmosphere, be it in science, be it in human science, which is developing at the end of this century."[40]

CONCLUSION

The work of postmodern scientists on unpredictability implies that human beings must give up the possibility of totally dominating and controlling nature. Because ecological and social systems are open, interacting, and unpredictable, we must allow for the possibility of surprise. Global weather patterns that include hurricanes and tornados; geological changes, such as earthquakes and volcanos; and ecological and evolutionary processes cannot be predicted with sufficient certainty to give human beings complete control over nonhuman nature. We must leave room within our planning of industry, agriculture, forestry, and water usage and in our construction of dams, factories, and housing developments for nature's unpredictable events. We cannot dam every wild river, cut every old-growth forest, irrigate every desert, or build homes in every flood plain.

The reenchantment of nature called forth by the Frankfurt school's analysis of domination implies a partnership with nonhuman nature. Nature is an equal subject, not an object to be controlled. A partnership ethic means that a human community is in a sustainable ecological relationship with its surrounding natural community. Human beings are neither inferior to nature and dominated by it, as in premodern societies, nor superior to it through their science and technology, as in modern societies. Rather human beings and nonhuman nature are equal partners in survival. The continuance of life as we know it requires an ethic of restraint, a holding back in implementing and producing some of the things potentially possible through science and technology (nuclear bombs and power, for example). It requires an active effort to restore the earth through replanting prairies, forests, and meadows. It requires reparations to the earth and social justice for human beings oppressed by the colonization of their lands, bodies, and hearts. As the First National People of Color Environmental Summit put it in their "Principles of Social Justice" in 1991:

> Environmental justice affirms the sacredness of Mother Earth, ecological unity and the interdependence of all species, and the right to be free from ecological destruction. . . . Environmental justice requires that we, as individuals, make personal and consumer choices to consume as little of Mother Earth's resources and to produce as little waste as possible; and make the conscious decision to challenge and reprioritize our lifestyles to insure the health of the natural world for present and future generations.[41]

Notes

1. The following discussion of critical theory and the Frankfurt school draws on Douglas Kellner, *Critical Theory, Marxism, and Modernity* (Baltimore: Johns Hopkins University Press, 1989), esp. chaps. 1–5; Martin Jay, *The Dialectical Imagination: A History of the Frankfurt School and The Institute of Social Research, 1923–1970* (Boston: Little, Brown, 1973); Martin Jay, *Marxism and Totality: The Adventures of a Concept from Lukács to Habermas* (Berkeley and Los Angeles: University of California Press, 1984), esp. chaps. 6–8; Martin Jay, *Force Fields: Between Intellectual History and Cultural Critique* (New York: Routledge, 1993), esp. chaps. 1, 2, 8, 9, 10; Mark Poster, *Critical Theory and Poststructuralism: In Search of a Context* (Ithaca, N.Y.: Cornell, 1989); Morris Berman, *The Reenchantment of the World* (Ithaca: Cornell, 1981).
2. The extent to which Marx was "ecological" is debatable. See Albert Schmidt, *The Concept of Nature in Marx*, trans. Ben Fowkes (London: New Left Review, 1971); Howard Parsons, *Marx and Engels on Ecology* (Westport, Conn.: Greenwood Press, 1977); Donald C. Lee, "On the Marxian View of the Relationship between Man and Nature," *Environmental Ethics* 2 (Spring 1980): 3–16 and "Toward a Marxian Ecological Ethic: A Response to Two Critics," *Environmental Ethics* 4 (Winter 1982): 339–43; Val Routley [Plumwood], "On Karl Marx as an Environmental Hero," *Environmental Ethics* 3 (Fall 1981): 237–44; Charles Tolman, "Karl Marx, Alienation, and the Mastery of Nature," *Environmental Ethics* 3 (Spring 1981): 63–74; Hwa Yol Jung, "Marxism, Ecology, and Technology," *Environmental Ethics* 5 (Summer 1983): 169–71; Reiner Grundmann, *Marxism and Ecology* (New York: Oxford University Press, 1991).
3. Max Horkheimer and Theodor Adorno, *Dialectic of Enlightenment*, trans. John Cummings (New York: Herder and Herder, 1972), 54.

4. Horkheimer and Adorno, *Dialectic of Enlightenment*, 57.

5. Theodor Adorno, "Spengler after the Decline," in *Prisms*, trans. Samuel and Shierry Weber (from the German edition of 1955) (1967; Cambridge: MIT Press, 1981), 67. I thank Murray Bookchin for this reference.

6. Marcuse's relationship to ecology has been debated. See Andrew Light, "Rereading Bookchin and Marcuse as Environmentalist Materialists," *Capitalism, Nature, Socialism* 4, no. 1 (March 1993); Murray Bookchin, "Response to Andrew Light's 'Rereading Bookchin and Marcuse as Environmentalist Materialists'," *Capitalism, Nature, Socialism* 4, no. 2 (June 1993): 101–20; Tim Luke, "Marcuse and Ecology," in *Marcuse Revised*, ed. John Bokina and Timothy Luke (Lawrence: University of Kansas Press, 1993).

7. William Leiss, *The Domination of Nature* (New York: George Braziller, 1972), 158, 164. On William Leiss, see Koula Mellos, "Leiss's Critical Theory of Human Needs," *Perspectives on Ecology: A Critical Essay* (New York: St. Martin's Press, 1988), 129–43.

8. Robyn Eckersley, *Environmentalism and Political Theory* (Albany: State University of New York Press, 1992), 47.

9. Immanual Wallerstein, *The Modern World-System* (New York: Academic Press, 1974–80).

10. On the debate over population, see Paul Ehrlich, *The Population Explosion* (New York: Simon and Schuster, 1990); "The Population Bomb: An Explosive Issue for the Environmental Movement?" *Utne Reader* 27 (May/June 1988): 78–88; "Is Aids Good for the Earth?" *Utne Reader* 36 (Nov./Dec. 1987): 14; David Harvey, "Population, Resources, and the Ideology of Science," *Economic Geography* 50, no. 3 (July 1974): 256–77; Murray Bookchin, "The Population Myth—I," *Green Perspectives* 8 (July 1988); Murray Bookchin, "The Population Myth—Part II," *Green Perspectives* 15 (Apr. 1989).

11. Len Doyal and Ian Gough, "Human Needs and Strategies for Social Change," in *The Living Economy: A New Economics in the Making*, ed. Paul Ekins (New York: Routledge, 1986), 79. See below, chap. 9, III.

12. Arne Naess, "The Shallow and the Deep, Long-Range Ecology Movement: A Summary" *Inquiry* 16 (1973): 95–100. See below, chap. II, 120–24.

13. Bill Devall and George Sessions, *Deep Ecology: Living as if Nature Mattered* (Salt Lake City: Peregrine Smith Books, 1985); Arne Naess, *Ecology, Community, and Lifestyle*, trans. David Rothenburg (Cambridge: Cambridge University Press, 1989). Naess, working with Sessions, revised the 7 initial "tendencies" (1972) to the more conservative 8-point "platform," eliminating class and adding population reduction.

14. Warwick Fox, *Toward a Transpersonal Ecology: Developing New Foundations for Environmentalism* (Boston: Shambala, 1990); Freya Mathews, *The Ecological Self* (London: Routledge, 1991).

15. On the debate between social ecology and deep ecology, see Murray Bookchin, "Social Ecology versus Deep Ecology: A Challenge for the Ecology Movement" (1987), reprinted in *Socialist Review* 18, no. 3 (July–Sept. 1988): 9–29; Kirkpatrick Sale, "Deep Ecology and Its Critics," *Nation*, 14 May 1988, 670–75; Murray Bookchin, "As if People Mattered," a response to Kirkpatrick Sale's "Deep Ecology and its Critics," *Nation*, 10 October 1988; Ynestra King, "Letter to the Editor," *Nation*, 12 December 1987; George Bradford, "How Deep Is Deep Ecology," *Fifth Estate* 22, no. 3 (1978): 3–30; Stephan Elkins, "The Politics of Mystical Ecology," *Telos* 82 (Winter 1989–90): 52–70; Tim Luke, "The Dreams of Deep Ecology," *Telos* 76 (Summer 1988): 65–92; Robyn Eckersley, "Divining Evolution: The Ecological Ethics of Murray Bookchin," *Environmental Ethics* 11 (Summer

1989): 99–116; Murray Bookchin, "Recovering Evolution: A Reply to Eckersley and Fox," *Environmental Ethics* 12 (Fall 1990): 253–73.

16. Murray Bookchin, *Our Synthethic Environment* (originally published under pseudonym Lewis Herber, New York: Knopf, 1962; New York: Harper Colophon, 1974); Murray Bookchin, *Post-Scarcity Anarchism* (San Francisco: Ramparts Books, 1971); Murray Bookchin, *Toward an Ecological Society* (Montreal: Black Rose Books, 1981); Murray Bookchin, *The Ecology of Freedom: The Emergence and Dissolution of Hierarchy* (Palo Alto: Cheshire Books, 1982); Murray Bookchin, *The Modern Crisis* (Philadelphia: New Society Publishers, 1986); Murray Bookchin, *Remaking Society: Pathways to a Green Future* (Montreal: Black Rose Books, 1989); Murray Bookchin, *The Philosophy of Social Ecology: Essays on Dialectical Naturalism* (Montreal: Black Rose Books, 1990).

17. Bookchin, *Ecology of Freedom*, 1. Bookchin reverses the causal sequence between the domination of nature and the domination of man as stated by Adorno in *Prisms* (see note 5) and contrasts his position with that of Marx and the Frankfurt school: "However much they opposed domination, neither Adorno or Horkheimer singled out hierarchy as an underlying problematic in their writings. Indeed, their residual Marxian premises led to a historical fatalism that saw any liberatory enterprise (beyond art, perhaps) as hopelessly tainted by the *need* to dominate nature and *consequently* 'man.' This position stands completely at odds with my own view that the *notion*—and no more than an *unrealizable* notion—of dominating nature stems from the domination of human by human. . . . The Frankfurt School, no less than Marxism, in effect, placed the onus for domination on a 'blind,' 'mute,' 'cruel,' and 'stingy,' nature, not (let me emphasize) only society. My own writings . . . argue that the domination of nature first arose within *society* as part of its institutionalization into gerontocracies that placed the young in varying degrees of servitude to the old and in patriarchies that placed women in varying degrees of servitude to men—not in any endeavour to 'control' nature or natural forces." (Murray Bookchin, "Thinking Ecologically: A Dialectical Approach," in *The Philosophy of Social Ecology: Essays on Dialectical Naturalism* [Montreal: Black Rose Books, 1990], 188–89.) Bookchin also contrasts the classical Marxist and liberal formulation of the desirability of the domination of nature by humans with that of certain ecological theorists who desire to continue the domination of humans by nonhuman nature. "Classically, the counterpart of the 'domination of nature by man' has been the 'domination of man by nature'" (Ibid., 163).

18. For an introduction to socialist ecology, see the journal *Capitalism, Nature, Socialism*, especially James O'Connor, *Capitalism, Nature, Socialism* 1, no. 1 (Fall 1988): 11–38; Alexander Cockburn, "Socialist Ecology: What It Means, Why No Other Will Do" *Z Magazine* 2, no. 2 (Feb. 1989): 15–21; Alexander Cockburn, "Whose Better Nature? Socialism, Capitalism, and the Environment," *Z Magazine* 2, no. 6 (June 1989): 27–32. On the debate between social and socialist ecology, see Murray Bookchin, "Letters to Z," *Z Magazine* 2, no. 4 (Apr. 1989): 3.

19. On Françoise d'Eaubonne's founding of the "Ecologie-Féminisme" center in 1972, see the chronology in Françoise d'Eaubonne, "Feminism or Death," in *New French Feminisms: An Anthology*, ed. Elaine Marks and Isabelle de Courtivron (Amherst: University of Massachusetts Press, 1980), 25. On the center's "launching of a new action: *ecofeminism*," see her chapter, "The Time for Ecofeminism," translated here by Ruth Hottell, pp. 174–97. The original French version is Françoise d'Eaubonne, *Le Féminisme ou la Mort* (Paris: Pierre Horay, 1974), 215–52.

20. D'Eaubonne, "The Time for Ecofeminism," quotations below, 175, 176, 193.
21. For quotations from d'Eaubonne's "The Time for Ecofeminism" on the labeling of population as a Third-World problem, see p. 186.
22. Johann Jakob Bachofen, Myth, Religion, and Mother Right, trans. Rudolf Marx (1863; Princeton: Princeton University Press, 1973); Friedrich Engels, "Origin of the Family, Private Property, and the State" (1884), in Selected Works (New York: International Publishers, 1968); Robert Briffault, The Mothers, 3 vols. (1927; abridged ed., New York: Atheneum, 1977); August Bebel, Woman in the Past, Present, and Future (San Francisco: G. B. Benham, 1897). On the male system and its power over women, see below, 177; on the price of phallocracy being paid by the earth, see below, 188.
23. D'Eaubonne, "The Time for Ecofeminism," 188.
24. D'Eaubonne, "The Time for Ecofeminism," quotations 193, 194.
25. Ynestra King, "Feminism and the Revolt of Nature," below, 202.
26. Sherry Ortner, "Is Female to Male as Nature is to Culture?" in Woman, Culture, and Society, ed. Michelle Rosaldo and Louise Lamphere (Stanford, Calif.: Stanford University Press, 1974), 67–87.
27. See Elizabeth Carlassare, below, chap. 19, 232.
28. On the debates between ecofeminism and Deep Ecology, see Ariel Kay Salleh, "Deeper than Deep Ecology: The Ecofeminist Connection," Environmental Ethics 6, no. 4 (Winter 1984): 339–45; Marti Kheel, "Ecofeminism and Deep Ecology: Reflections on Identity and Difference," in Covenant for a New Creation, ed. Carol Robb and Carl Casebolt (New York: Orbis Books, 1991); Jim Cheney, "Eco-Feminism and Deep Ecology," Environmental Ethics 9, no. 2 (Summer 1987): 115–45; Warwick Fox, "The Deep Ecology-Ecofeminism Debate and Its Parallels," Environmental Ethics 11 (Spring 1989): 5–25; Michael Zimmerman, "Feminism, Deep Ecology, and Environmental Ethics," Environmental Ethics 9, no. 1 (Spring 1987): 21–44; Ariel Salleh, "The Ecofeminism/Deep Ecology Debate," Environmental Ethics 14, no. 3 (Fall 1992): 195–216; Ariel Salleh, "Class, Race, and Gender Discourse in the Ecofeminism/Deep Ecology Debate," Environmental Ethics 15, no. 3 (Fall 1993): 225–44. On the debates between ecofeminism and social/socialist ecology, see Janet Biehl, Rethinking Ecofeminist Politics (Boston: South End Press, 1991); Laura Schere, "Feminism, Ecology, and Left Green Politics," Left Green Notes (August–September 1991); Janet Biehl, "Ecofeminism and the Left Greens: A Response to Laura Schere," Left Green Notes (Nov.–Dec. 1991); Discussion between Lori-Ann Thrupp, Daniel Faber, and James O'Connor (on Faber and O'Connor's "The Struggle for Nature"), Capitalism, Nature, Socialism 1, no. 3 (Nov. 1989): 169–74; Discussion between Ariel Salleh, Martin O'Connor, James O'Connor, and Daniel Faber, Capitalism, Nature, Socialism 2, no. 2 (February 1991); Mary Mellor, "Eco-Feminism and Eco-Socialism," Capitalism, Nature, Socialism 3 no. 2 (Spring 1993): 43–62.
29. Lynn White, Jr., "The Historical Roots of Our Ecologic Crisis," Science 155 (10 Mar. 1967): 1203–7, as reprinted in Ian G. Barbour, ed., Western Man and Environmental Ethics (Reading, Mass.: Addison-Wesley, 1973), 18–30, quotations on 25.
30. Joanna Macy, "Toward a Healing of Self and World," human potential magazine 17, no. 1 (Spring 1992): 10–13, 29–33; Joanna Macy, Despair and Personal Power in the Nuclear Age (Philadelphia: New Society Publishers, 1983); Joanna Macy, World as Lover: World as Self (Berkeley, Calif.: Parallax Press, 1991); Macy, Mutual Causality in Buddhism and General Systems Theory: The Dharma of Natural Systems (Albany: State University of New York Press, 1991); John Seed, Joanna Macy, Pat Flemming, Arne Naess, Thinking Like a Mountain: Towards

a Council of all Beings (Philadelphia: New Society Publishers, 1988).

31. Charlene Spretnak, *The Spiritual Dimension of Green Politics* (Santa Fe, N.M.: Bear and Co., 1986); Charlene Spretnak, *States of Grace* (San Francisco: Harper and Row, 1991). The infusion of spirituality into green politics is controversial. For critiques, see Janet Biehl, "Goddess Mythology in Ecological Politics," *New Politics* 2 (Winter 1989): 84–105; Murray Bookchin, "Will Ecology Become 'the Dismal Science'?" *Progressive*, 20 Dec. 1991, 18–21.

32. Alan Cowell, "Pope Issues Censure of 'Nature Worship' by Some Feminists," *New York Times*, 3 July 1993, pp. 1, 4.

33. Fritjof Capra, "Systems Theory and the New Paradigm," below, 335.

34. David Bohm, "Postmodern Science and a Postmodern World," quotation on 349. David Bohm, *Wholeness and the Implicate Order* (Boston: Routledge and Kegan Paul, 1980). See also John P. Briggs and F. David Peat, *The Looking Glass Universe: The Emergence of Wholeness* (New York: Simon and Schuster, 1984).

35. James Lovelock, "Gaia," 359. See also, Lovelock, *Gaia: A New Look at Life on Earth* (New York: Oxford University Press, 1979); Lovelock, *The Ages of Gaia: A Biography of Our Living Earth* (New York: W. W. Norton, 1988). For a critique of Lovelock's Gaia hypothesis, see James Kirchner, "The Gaia Hypothesis: Can it Be Tested?" *Reviews of Geophysics* 27, no. 2 (May 1989): 223–35.

36. James Gleick, *Chaos: The Making of a New Science* (New York: Viking, 1987); M. Mitchell Waldrop, *Complexity: The Emerging Science at the Edge of Order and Chaos* (New York: Simon and Schuster, 1992).

37. Edward N. Lorenz, "Predictability: Does the Flap of a Butterfly's Wings in Brazil Set off a Tornado in Texas?," 362; Lorenz, "Craaford Prize Lecture," *Tellus* (1984): 36A, 98–110.

38. Ilya Prigogine and Isabelle Stengers, *Order Out of Chaos: Man's New Dialogue with Nature* (New York; Bantam, 1984). On Prigogine's work, see Erich Jantsch, *The Self-Organizing Universe* (New York: Pergamon Press, 1980).

39. Ilya Prigogine, "Science in a World of Limited Predictability," quotations below, 367–68.

40. Prigogine, "Science in a World of Limited Predictability," quotation on 369.

41. First National People of Color Environmental Leadership Summit, "Principles of Environmental Justice," adopted 27 Oct. 1991, principles 1 and 17, pp. 371, 372.

PART I

Critical Theory and the Domination of Nature

Marx and Engels on Ecology

KARL MARX AND FRIEDRICH ENGELS
Edited by Howard L. Parsons

I. DIALECTICS

Man's collective and immediate material dealings with nature bring
him into a dialectical relation with it, into a dynamic and potentially
developing interaction with it. Such a relation, when critically analyzed,
reveals nature as continuous motion, interconnection, and transformation. Nature is a ceaseless series of unities of opposites, which are
mutually creative, mutually destructive, and mutually transforming.
—H.L.P.

DIALECTICS VERSUS METAPHYSICS

When we reflect on nature, or the history of mankind, or our own intellectual activity, the first picture presented to us is of an endless maze of relations and interactions, in which nothing remains what, where and as it
was, but everything moves, changes, comes into being and passes out of
existence. This primitive, naive, yet intrinsically correct conception of the
world was that of ancient Greek philosophy, and was first clearly formulated by Heraclitus: everything is and also is not, for everything is in *flux*,
is constantly changing, constantly coming into being and passing away. But
this conception, correctly as it covers the general character of the picture
of phenomena as a whole, is yet inadequate to explain the details of which
this total picture is composed; and so long as we do not understand these,
we also have no clear idea of the picture as a whole. In order to understand
these details, we must detach them from their natural or historical connections, and examine each one separately, as to its nature, its special causes
and effects, etc. This is primarily the task of natural science and historical
research; branches of science which the Greeks of the classical period, on

From Howard L. Parsons, ed., *Marx and Engels on Ecology* (Westport, Conn.: Greenwood
Press, 1977), pp. 129–85, selections.

very good grounds, relegated to a merely subordinate position, because they had first of all to collect materials for these to work upon. The beginnings of the exact investigation of nature were first developed by the Greeks of the Alexandrian period, and later on, in the Middle Ages, were further developed by the Arabs. Real natural science, however, dates only from the second half of the fifteenth century, and from then on it has advanced with constantly increasing rapidity.

The analysis of Nature into its individual parts, the grouping of the different natural processes and natural objects in definite classes, the study of the internal anatomy of organic bodies in their manifold forms—these were the fundamental conditions of the gigantic strides in our knowledge of Nature which have been made during the last four hundred years. But this method of investigation has also left us as a legacy the habit of observing natural objects and natural processes in their isolation, detached from the whole vast interconnection of things; and therefore not in their motion, but in their repose; not as essentially changing, but as fixed constants; not in their life, but in their death. And when, as was the case with Bacon and Locke, this way of looking at things was transferred from natural science to philosophy, it produced the specific narrow-mindedness of the last centuries, the metaphysical mode of thought. . . .

Friedrich Engels, *Herr Eugen Dühring's Revolution in Science*, 26–29.[1]

II. THE INTERDEPENDENCE OF MAN AS A LIVING BEING WITH NATURE

Man depends on nature for vital substances (such as food, water, and oxygen) and processes (such as photosynthesis), and nature in turn is affected by man's activities, such as the use of fire and the domestication of plants and animals. Marx described man's interdependence with nature at both the biochemical and the psychological levels.—H.L.P.

NATURE IS MAN'S BODY, ON WHICH HE LIVES

The life of the species, both in man and in animals, consists physically in the fact that man (like the animal) lives on inorganic nature; and the more universal man is compared with an animal, the more universal is the sphere of inorganic nature on which he lives. Just as plants, animals, stones, air, light, etc., constitute theoretically a part of human consciousness, partly as objects of natural science, partly as objects of art—his spiritual inorganic nature, spiritual nourishment which he must first prepare to make palatable and digestible—so also in the realm of practice they constitute a part of human life and human activity. Physically man lives only on these products of nature, whether they appear in the form of food, heating, clothes, a dwelling, etc. The universality of man appears in practice precisely in the universality which makes all nature his *inorganic* body—both inasmuch as nature is (1) his direct means of life, and (2) the material, the object, and the instrument of his life activity. Nature is man's *inorganic body*—nature, that is, in so far as it is not itself the human body. Man *lives* on nature—means that nature is his *body*, with which he must remain in continuous interchange if he is not to die. That man's physical and spiritual life is linked to nature means simply that nature is linked to itself, for man is a part of nature.

Karl Marx, *The Economic and Philosophic Manuscripts of 1844*, 112.

MAN'S ESSENTIAL INTERDEPENDENCE WITH NATURE

. . . Man is directly a *natural being*. As a natural being and as a living natural being he is on the one hand endowed with *natural powers of life*—he is an *active* natural being. These forces exist in him as tendencies and abilities—as *instincts*. On the other hand, as a natural corporeal, sensuous, objective being he is a *suffering*, conditioned and limited creature, like animals and plants. That is to say, the *objects* of his instincts exist outside him, as *objects* independent of him; yet these objects are *objects* that he *needs*—essential *objects*, indispensable to the manifestation and confirmation of his essential powers. To say that man is a *corporeal*, living, real, sensuous, objective being full of natural vigor is to say that he has *real, sensuous, objects* as the objects of his life, or that he can only express his life in real, sensuous objects. To *be* objective, natural and sensuous, and at the same time to have object, nature and sense outside oneself, or oneself to be object, nature and sense for a third party, is one and the same thing. *Hunger* is a natural *need*; it therefore needs a *nature* outside itself, an *object* outside itself, in order to satisfy itself, to be stilled. Hunger is an acknowledged need of my body for an *object* existing outside it, indispensable to its integration and to the expression of its essential being. The sun is the *object* of the plant—an indispensable object to it, confirming its life—just as the plant is an object of the sun, being an *expression* of the life-awakening power of the sun, of the sun's *objective* essential power. . . .

Karl Marx, *The Economic and Philosophic Manuscripts of 1844*, 180–82.

MATTER AS MOTION, CREATIVITY, AND SENSUOUS QUALITY

The real founder of *English materialism* and all *modern experimental* science was *Bacon*. For him natural science was true science and *physics* based on perception was the most excellent part of natural science. *Anaxagoras* with his *homoeomeria* and *Democritus* with his atoms are often the authorities he refers to. According to his teaching the *senses* are infallible and are the *source* of all knowledge. Science is *experimental* and consists in applying a *rational method* to the data provided by the senses. Induction, analysis, comparison, observation and experiment are the principal requisites of rational method. The first and most important of the inherent qualities of *matter* is *motion*, not only *mechanical* and *mathematical* movement, but still more *impulse, vital life-spirit, tension*, or, to use Jacob Bohme's expression, the throes [*Qual*] of matter. The primary forms of matter are the living, individualizing *forces of being* inherent in it and producing the distinction between the species.

In *Bacon*, its first creator, materialism contained latent and still in a naive way the germs of all-round development. Matter smiled at man with poetical sensuous brightness. The aphoristic doctrine itself, on the other hand, was full of the inconsistencies of theology.

Karl Marx and Friedrich Engels,
The Holy Family, 172.

III. MAN'S INTERDEPENDENCE WITH NATURE AS A BEING THAT MAKES A LIVING

As distinct from other living beings, man not only "lives on inorganic nature," but he also makes a living of nature, interacting with it by means of his brain, hand, and tools in order to subsist. Thus, far more than any other species, the human species puts its "stamp on nature." Marx and Engels addressed themselves to the question of how the human species is differentiated from other animals, because on the one hand they wished to refute the idealists who repudiated man's animal nature and because on the other hand they wanted to demonstrate the great possibilities for fulfillment in man's evolutionary relation to nature.—H.L.P.

BY PRODUCTIVELY INTERACTING WITH NATURE, MAN SUBSISTS

The first premise of all human history is, of course, the existence of living human individuals. Thus the first fact to be established is the physical organization of these individuals and their consequent relation to the rest of nature. Of course, we cannot here go either into the actual physical nature of man, or into the natural conditions in which man finds himself—geological, oro-hydrographical, climatic and so on. The writing of history must always set out from these natural bases and their modification in the course of history through the action of man.

Men can be distinguished from animals by consciousness, by religion or anything else you like. They themselves begin to distinguish themselves from animals as soon as they begin to *produce* their means of subsistence, a step which is conditioned by their physical organization. By producing their means of subsistence men are indirectly producing their actual material life.

The way in which men produce their means of subsistence depends first of all on the nature of the actual means they find in existence and have to reproduce. This mode of production must not be considered simply as being the reproduction of the physical existence of the individuals. Rather it is a definite form of activity of these individuals, a definite form of expressing their life, a definite *mode of life* on their part. As individuals express their life, so they are. What they are, therefore, coincides with their production, both with *what* they produce and with *how* they produce. The nature of individuals thus depends on the material conditions determining their production.

Karl Marx and Friedrich Engels,
The German Ideology, 7.

MAN HAS IMPRESSED HIS STAMP ON NATURE THROUGH HAND AND BRAIN

When after thousands of years of struggle the differentiation of hand from foot, and erect gait, were finally established, man became distinct from the ape and the basis was laid for the development of articulate speech and the mighty development of the brain that has since made the gulf between man and the ape an unbridgeable one. The specialization of the hand—this implies the *tool*, and the tool implies specific human activity, the trans- forming reaction of man on nature, production. Animals in the narrower sense also have tools, but only as limbs of their bodies: the ant, the bee, the beaver; animals also produce, but their productive effect on surrounding nature in relation to the latter amounts to nothing at all. Man alone has succeeded in impressing his stamp on nature, not only by so altering the aspect and climate of his dwelling-place, and even the plants and animals themselves, that the consequences of his activity can disappear only with the general extinction of the terrestrial globe. And he has accomplished this primarily and essentially by means of *the hand*. Even the steam-engine, so far his most powerful tool for the transformation of nature, depends, because it is a tool, in the last resort on the hand. But step by step with the development of the hand went that of the brain; first of all came con- sciousness of the conditions for separate practically useful actions, and later, among the more favoured peoples and arising from that consciousness, in- sight into the natural laws governing them. And with the rapidly growing knowledge of the laws of nature the means for reacting on nature also grew; the hand alone would never have achieved the steam-engine if, along with and parallel to the hand, and partly owing to it, the brain of man had not correspondingly developed.

With man we enter *history*. Animals also have a history, that of their descent and gradual evolution to their present position. This history, how- ever, is made for them, and in so far as they themselves take part in it, this occurs without their knowledge and desire. On the other hand, the more that human beings become removed from animals in the narrower sense of the word, the more they make their history themselves, consciously, the less becomes the influence of unforeseen effects and uncontrolled forces on this history, and the more accurately does the historical result correspond to the aim laid down in advance.

Friedrich Engels, *Dialectics of
Nature*, 46–48.

MAN'S REACTION ON NATURE

Natural science, like philosophy, has hitherto entirely neglected the influence of men's activity on their thought; both know only nature on the one hand and thought on the other. But it is precisely *the alteration of nature by men*, not solely nature as such, which is the most essential and immediate basis of human thought, and it is in the measure that man has learned to change nature that his intelligence has increased. The naturalistic conception of history, as found, for instance, to a greater or lesser extent in Draper and other scientists, as if nature exclusively reacts on man, and natural conditions everywhere exclusively determined his historical development, is therefore one-sided and forgets that man also reacts on nature, changing it and creating new conditions of existence for himself. There is devilishly little left of "nature" as it was in Germany at the time when the Germanic peoples immigrated into it. The earth's surface, climate, vegetation, fauna, and the human beings themselves have infinitely changed, and all this owing to human activity, while the changes of nature in Germany which have occurred in this period of time without human interference are incalculably small.

Friedrich Engels, *Dialectics of Nature*, 306.

THE LAWS OF HUMAN LIFE ARE DIFFERENT FROM THE LAWS OF ANIMAL LIFE

The struggle for life. Until Darwin, what was stressed by his present adherents was precisely the harmonious co-operative working of organic nature, how the plant kingdom supplies animals with nourishment and oxygen, and animals supply plants with manure, ammonia, and carbonic acid. Hardly was Darwin recognized before these same people saw everywhere nothing but *struggle*. Both views are justified within narrow limits, but both are equally one-sided and prejudiced. The interaction of bodies in non-living nature includes both harmony and collisions, that of living bodies conscious and unconscious co-operation as well as conscious and unconscious struggle. Hence, even in regard to nature, it is not permissible one-sidedly to inscribe only "struggle" on one's banners. But it is absolutely childish to desire to sum up the whole manifold wealth of historical evolution and complexity in the meagre and one-sided phrase "struggle for existence." That says less than nothing.

The whole Darwinian theory of the struggle for existence is simply the transference from society to organic nature of Hobbes' theory of *bellum omnium contra omnes* and of the bourgeois economic theory of competition, as well as the Malthusian theory of population. When once this feat has been accomplished (the unconditional justification for which, especially as regards the Malthusian theory, is still very questionable), it is very easy to transfer these theories back again from natural history to the history of society, and altogether too naive to maintain that thereby these assertions have been proved as eternal natural laws of society.

Let us accept for a moment the phrase "struggle for existence," for argument's sake. The most that the animal can achieve is to *collect*; man *produces*, he prepares the means of life, in the widest sense of the words, which without him nature would not have produced. This makes impossible any unqualified transference of the laws of life in animal societies to human society. Production soon brings it about that the so-called struggle for existence no longer turns on pure means of existence, but on means of enjoyment and development. Here—where the means of development are socially produced—the categories taken from the animal kingdom are already totally inapplicable. Finally, under the capitalist mode of production, production reaches such a high level that society can no longer consume the means of life, enjoyment and development that have been produced, because for the great mass of producers access to these means is artificially and forcibly barred; and therefore every ten years a crisis restores the equilibrium by destroying not only the means of life, enjoyment and development that have been produced, but also a great part of the productive forces themselves. Hence the so-called struggle for existence assumes the form: to *protect* the products and productive forces produced by bourgeois capitalist society against the destructive, ravaging effect of this capitalist social order, by taking control of social production and distribution out of the hands of the ruling capitalist class, which has become incapable of this function, and transferring it to the producing masses—and that is the socialist revolution.

The conception of history as a series of class struggles is already much richer in content and deeper than merely reducing it to weakly distinguished phases of the struggle for existence.

Friedrich Engels, *Dialectics of
Nature*, 404–5.

THE PASSAGE FROM ANIMAL NECESSITY TO HUMAN FREEDOM

The seizure of the means of production by society puts an end to commodity production, and therewith to the domination of the product over the producer. Anarchy in social production is replaced by conscious organization on a planned basis. The struggle for individual existence comes to an end. And at this point, in a certain sense, man finally cuts himself off from the animal world, leaves the conditions of animal existence behind him and enters conditions which are really human. The conditions of existence forming man's environment, which up to now have dominated man, at this point pass under the dominion and control of man, who now for the first time becomes the real conscious master of Nature, because and in so far as he has become master of his own social organization. The laws of his own social activity, which have hitherto confronted him as external, dominating laws of Nature, will then be applied by man with complete understanding, and hence will be dominated by man. Men's own social organization which has hitherto stood in opposition to them as if arbitrarily decreed by Nature and history, will then become the voluntary act of men themselves. The objective, external forces which have hitherto dominated history, will then pass under the control of men themselves. It is only from this point that men, with full consciousness, will fashion their own history; it is only from this point that the social causes set in motion by men will have, predominantly and in constantly increasing measure, the effects willed by men. It is humanity's leap from the realm of necessity into the realm of freedom.

To carry through his world-emancipating act is the historical mission of the modern proletariat. And it is the task of scientific socialism, the theoretical expression of the proletarian movement, to establish the historical conditions and, with these, the nature of this act, and thus to bring to the consciousness of the now oppressed class the conditions and nature of the act which it is its destiny to accomplish.

Friedrich Engels, *Herr Eugen Dühring's Revolution in Science* 309–10.

MAN'S BRAIN AND THOUGHTS ARE THE PRODUCTS OF NATURE AND ARE IN CORRESPONDENCE WITH IT

But if the further question is raised: what then are thought and consciousness, and whence they come, it becomes apparent that they are products of the human brain and that man himself is a product of Nature, which has been developed in and along with its environment: whence it is self-evident that the products of the human brain, being in the last analysis also products of Nature, do not contradict the rest of Nature but are in correspondence with it.

Friedrich Engels, *Herr Eugen Dühring's Revolution in Science*, 42–43.

IV. PRECAPITALIST RELATIONS OF MAN TO NATURE

Marx's investigation of the history of man's relation to nature through communal labor led to his concentration on the modes of production and distribution in precapitalist societies, as contrasted with what occurs in capitalist societies.—H.L.P.

COMMUNAL LABOR APPROPRIATES NATURE AS COMMON PROPERTY

In the first form of this landed property, an initial, naturally arisen spontaneous [*naturwüchsiges*] community appears as first presupposition. Family, and the family extended as a clan [*Stamm*], or through intermarriage between families, or combination of clans. Since we may assume that *pastoralism*, or more generally a *migratory* form of life, was the first form of the mode of existence, not that the clan settles in a specific site, but that it grazes off what it finds—humankind is not settlement-prone by nature (except possibly in a natural environment so especially fertile that they sit like monkeys on a tree; else roaming like the animals)—then the *clan community*, the natural community, appears not as a *result* of, but as a *presupposition for the communal appropriation* (temporary) *and utilization of the land*. When they finally do settle down, the extent to which this original community is modified will depend on various external, climatic, geographic, physical etc. conditions as well as on their particular natural predisposition—their clan character. This naturally arisen clan community, or, if one will, pastoral society, is the first presupposition—the communality [*Gemeinschaftlichkeit*] of blood, language customs—for the *appropriation of the objective conditions* of their life, and of their life's reproducing and objectifying activity (activity as herdsmen, hunters, tillers, etc.). The earth is the great workshop, the arsenal which furnishes both means and material of labour, as well as the seat, the *base* of the community. They relate naively to it as the *property of the community*, of the community producing and reproducing itself in living labour. Each individual conducts himself only as a link, as a member of this community as *proprietor* or *possessor*.

Karl Marx, *Grundrisse*, 472.

APPROPRIATION OF THE EARTH BY THE COMMUNE

The main point here is this: In all these forms—in which landed property and agriculture form the basis of the economic order, and where the economic aim is hence the production of use values, i.e. the *reproduction of the individual* within the specific relation to the commune in which he is its basis—there is to be found: (1) Appropriation not through labour, but presupposed to labour; appropriation of the natural conditions of labour, of the *earth* as the original instrument of labour as well as its workshop and repository of raw materials. The individual relates simply to the objective conditions of labour as being his; [relates] to them as the inorganic nature of his subjectivity in which the latter realizes itself; the chief objective condition of labour does not itself appear as a *product* of labour, but is already there as *nature*; on one side the living individual, on the other the earth, as the objective condition of his reproduction; (2) but this *relation* to land and soil, to the earth, as the property of the labouring individual— who thus appears from the outset not merely as labouring individual, in this abstraction, but who has an *objective mode of existence* in his ownership of the land, an existence *presupposed* to his activity, and not merely as a result of it, a presupposition of his activity just like his skin, his sense organs, which of course he also reproduces and develops etc. in the life process, but which are nevertheless presuppositions of this process of his reproduction—is instantly mediated by the naturally arisen, spontaneous, more or less historically developed and modified presence of the individual as *member of a commune*—his naturally arisen presence as member of a tribe etc. An isolated individual could no more have property in land and soil than he could speak. He could, of course, live off it as substance, as do the animals. The relation to the earth as property is always mediated through the occupation of the land and soil, peacefully or violently, by the tribe, the commune, in some more or less naturally arisen or already historically developed form.

Karl Marx, Grundrisse, 485.

V. CAPITALIST POLLUTION AND THE RUINATION OF NATURE

The development of primitive societies into class societies and eventually into capitalist society has produced a transformation in man's relation to nature. Under capitalism this relation is determined primarily by the ruling class of capitalists. Marx and Engels undertook a thorough examination of this new relation. Like the relation that they bear to workers, the relation of capitalists to nature is marked by exploitation, pollution, and ruination. Marx and Engels observed that the capitalists appropriate the resources of the earth without any cost to themselves, that they transform the earth into an "object of huckstering," that under capitalism the original unity of man and nature is breached as well as the unity of manufacture and agriculture, that man's precipitous alteration of nature issues in unforeseen and harmful consequences, and that the capitalists' wastage and exhaustion of the soil, deforestation, disruption of nature's cycle of matter, greedy policy toward nature, and neglect of man's welfare are ruinous to both nature and man.—H.L.P.

THE PRODUCTIVE FORCES OF MAN, SCIENCE, AND NATURE COST CAPITAL NOTHING

We saw that the productive forces resulting from co-operation and division of labour cost capital nothing. They are natural forces of social labour. So also physical forces, like steam, water, &c., when appropriated to productive processes, cost nothing. But just as a man requires lungs to breathe with, so he requires something that is work of man's hand, in order to consume physical forces productively. A water-wheel is necessary to exploit the force of water, and a steam-engine to exploit the elasticity of steam. Once discovered, the law of the deviation of the magnetic needle in the field of an electric current, or the law of the magnetization of iron, around which an electric current circulates, cost never a penny.[2] But the exploitation of these laws for the purposes of telegraphy, &c., necessitates a costly and extensive apparatus. The tool, as we have seen, is not exterminated by the machine. From being a dwarf implement of the human organism, it expands and multiplies into the implement of a mechanism created by man. Capital now sets the labourer to work, not with a manual tool, but with a machine which itself handles the tools. Although, therefore, it is clear at the first glance that, by incorporating both stupendous physical forces, and the natural sciences, with the process of production, Modern Industry raises the productiveness of labour to an extraordinary degree, it is by no means equally clear, that this increased productive force is not, on the other hand, purchased by an increased expenditure of labour. Machinery, like every other component of constant capital, creates no new value, but yields up its own value to the product that it serves to beget. In so far as the machine has value, and, in consequence, parts with value to the product, it forms an element in the value of that product. Instead of being cheapened, the product is made dearer in proportion to the value of the machine. And it is clear as noon-day, that machines and systems of machinery, the characteristic instruments of labour of Modern Industry, are incomparably more loaded with value than the implements used in handicrafts and manufactures.

In the first place, it must be observed that the machinery, while always entering as a whole into the labour-process, enters into the value-begetting process only by bits. It never adds more value than it loses, on an average, by wear and tear. Hence there is a great difference between the value of a machine, and the value transferred in a given time by that machine to the product. The longer the life of the machine in the labour-process, the greater is that difference. It is true, no doubt, as we have already seen, that every instrument of labour enters as a whole into the labour-process, and only piece-meal, proportionally to its average daily loss by wear and tear, into the value-begetting process. But this difference between the instrument as a whole and its daily wear and tear, is much greater in a machine than in a tool, because the machine, being made from more durable material, has a longer life; because its employment, being regulated by strictly scientific laws, allows of greater economy in the wear and tear of its parts, and in the materials it consumes; and lastly, because its field of production is incomparably larger than that of a tool.

Karl Marx, *Capital* 1:386–88.

In the extractive industries, mines, &c., the raw materials form no part of the capital advanced. The subject of labour is in this case not a product of previous labour, but is furnished by Nature gratis, as in the case of metals, minerals, coal, stone, &c.

Karl Marx, *Capital* 1:603.

Natural elements entering as agents into production, and which cost nothing, no matter what role they play in production, do not enter as components of capital, but as a free gift of Nature to capital, that is, as a free gift of Nature's productive power to labour, which, however, appears as the productiveness of capital, as all other productivity under the capitalist mode of production.

Karl Marx, *Capital* 3:745.

CAPITALIST HUCKSTERING OF PEOPLE AND NATURE
. . . To make earth an object of huckstering—the earth which is our one

and all, the first condition of our existence—was the last step toward making oneself an object of huckstering. It was and is to this very day an immorality surpassed only by the immorality of self-alienation. And the original appropriation—the monopolization of the earth by a few, the exclusion of the rest from that which is the condition of their life—yields nothing in immorality to the subsequent huckstering of the earth.

> Friedrich Engels, *Outlines of a Critique of Political Economy*, in *The Economic and Philosophic Manuscripts of 1844*, by Karl Marx, 210.

CAPITALISM RENDS THE UNITY OF AGRICULTURE AND INDUSTRY, DISTURBS MAN'S RELATION TO THE SOIL, WASTES WORKERS, AND ROBS LABORERS AND SOIL

In the sphere of agriculture, modern industry has a more revolutionary effect than elsewhere, for this reason, that it annihilates the peasant, that bulwark of the old society, and replaces him by the wage-labourer. Thus the desire for social changes, and the class antagonisms are brought to the same level in the country as in the towns. The irrational, old-fashioned methods of agriculture are replaced by scientific ones. Capitalist production completely tears asunder the old bond of union which held together agriculture and manufacture in their infancy. But at the same time it creates the material conditions for a higher synthesis in the future, viz., the union of agriculture and industry on the basis of the more perfected forms they have each acquired during their temporary separation. Capitalist production, by collecting the population in great centres, and causing an ever-increasing preponderance of town population, on the one hand concentrates the historical motive power of society; on the other hand, it disturbs the circulation of matter between man and the soil, i.e., prevents the return to the soil of its elements consumed by man in the form of food and clothing; it therefore violates the conditions necessary to lasting fertility of the soil. By this action it destroys at the same time the health of the town labourer and the intellectual life of the rural labourer. But while upsetting the naturally grown conditions for the maintenance of that circulation of matter, it imperiously calls for its restoration as a system, as a regulating law of social production, and under a form appropriate to the full development of the human race. In agriculture as in manufacture, the transformation of production under the sway of capital, means, at the same time, the martyrdom of the producer; the instrument of labour becomes the means of enslaving, exploiting, and impoverishing the labourer; the social combination and organization of labour-processes is turned into an organized mode of crushing out the workman's individual vitality, freedom, and independence. The dispersion of the rural labourers over larger areas breaks their power of resistance while concentration increases that of the town operatives. In modern agriculture, as in the urban industries, the increased productiveness and quantity of the labour set in motion are bought at the cost of laying waste and consuming by disease labour-power itself. Moreover, all progress in capitalistic agriculture is a progress in the art, not only of robbing the labourer, but of robbing the soil; all progress in increasing the fertility of the soil for a given time, is a progress towards ruining the lasting sources of that fertility. The more a country starts its development on the foundation of modern industry, like the United States, for example, the more rapid is this process of destruction. Capitalist production, therefore, develops technology, and the combination together of various processes into a social whole, only by sapping the original sources of all wealth—the soil and the labourer.

Karl Marx, *Capital*, 1:505–7.

CAPITALIST FAILURE TO UTILIZE THE WASTE PRODUCTS OF INDUSTRY, AGRICULTURE, AND HUMAN CONSUMPTION, AND WAYS OF UTILIZING WASTE

The capitalist mode of production extends the utilization of the excretions of production and consumption. By the former we mean the waste of industry and agriculture, and by the latter partly the excretions produced by the natural exchange of matter in the human body and partly the form of objects that remains after their consumption. In the chemical industry, for instance, excretions of production are such by-products as are wasted in production on a smaller scale; iron filings accumulating in the manufacture of machinery and returning into the production of iron as raw material, etc. Excretions of consumption are the natural waste matter discharged by the human body, remains of clothing in the form of rags, etc. Excretions of consumption are of the greatest importance for agriculture. So far as their utilization is concerned, there is an enourmous waste of them in the capitalist economy. In London, for instance, they find no better use for the excretion of four and a half million human beings than to contaminate the Thames with it at heavy expense.

Rising prices of raw materials naturally stimulate the utilization of waste products. . . .

Karl Marx, *Capital* 3:101–3.

IN ALTERING NATURE MAN PRODUCES UNFORESEEN AND HARMFUL CONSEQUENCES

Animals, as already indicated, change external nature by their activities just as man does, even if not to the same extent, and these changes made by them in their environment, as we have seen, in turn react upon and change their originators. For in nature nothing takes place in isolation. Everything affects every other thing and vice-versa, and it is mostly because this all-sided motion and interaction is forgotten that our natural scientists are prevented from clearly seeing the simplest things. We have seen how goats have prevented the regeneration of forests in Greece; on St. Helena, goats and pigs brought by the first navigators to arrive there succeeded in exterminating almost completely the old vegetation of the island, and so prepared the ground for the spreading of plants brought by later sailors and colonists. But if animals exert a lasting effect on their environment, it happens unintentionally, and as far as the animals themselves are concerned, it is an accident. The further men become removed from animals, however, the more their effect on nature assumes the character of premeditated, planned action directed toward definite ends known in advance. The animal destroys the vegetation of a locality without realizing what it is doing. Man destroys it in order to sow field crops on the soil thus released, or to plant trees or vines which he knows will yield many times the amount sown. He transfers useful plants and domestic animals from one country to another and thus changes the flora and fauna of whole continents. More than this. Through artificial breeding, both plants and animals are so changed by the hand of man that they become unrecognizable. The wild plants from which our grain varieties originated are still being sought in vain. The question of the wild animal from which our dogs are descended, the dogs themselves being so different from one another, or our equally numerous breeds of horses, is still under dispute. . . .

In short, the animal merely *uses* external nature, and brings about changes in it simply by his presence; man by his changes makes it serve his ends, *masters* it. This is the final, essential distinction between man and other animals, and once again it is labour that brings about this distinction.

Let us not, however, flatter ourselves overmuch on account of our human conquests over nature. For each such conquest takes its revenge on us. Each of them, it is true, has in the first place the consequences on which we counted, but in the second and third places it has quite different, unforeseen effects which only too often cancel out the first. The people who, in Mesopotamia, Greece, Asia Minor, and elsewhere, destroyed the forests to obtain cultivable land, never dreamed that they were laying the basis for the present devastated condition of those countries, by removing along with the forests the collecting centres and reservoirs of moisture. When, on the southern slopes of the mountains, the Italians of the Alps used up the fir forests so carefully cherished on the northern slopes, they had no inkling that by doing so they were cutting at the roots of the dairy industry in their region; they had still less inkling that they were thereby depriving their mountain springs of water for the greater part of the year, making it possible for these to pour still more furious flood torrents on the plains during the rainy season. Those who spread the potato in Europe were not aware that with these farinaceous tubers they were at the same time spreading the disease of scrofula. Thus at every step we are reminded that we by no means rule over nature like a conqueror over a foreign people, like someone standing outside nature—but that we, with flesh, blood, and brain, belong to nature, and exist in its midst, and that all our mastery of it consists in the fact that we have the advantage over all other creatures of being able to know and correctly apply its laws.

And, in fact, with every day that passes we are learning to understand these laws more correctly, and getting to know both the more immediate and the more remote consequences of our interference with the traditional course of nature. . . .

Friedrich Engels, Dialectics of Nature, 239–46.

CAPITALISM WASTES AND EXHAUSTS THE SOIL

Here, in small-scale agriculture, the price of land, a form and result of private landownership, appears as a barrier to production itself. In large-scale agriculture, and large estates operating on a capitalist basis, ownership likewise acts as a barrier, because it limits the tenant farmer in his productive investment of capital which in the final analysis benefits not him, but the landlord. In both forms, exploitation and squandering of the vitality of the soil (apart from making exploitation dependent upon the accidental and unequal circumstances of individual produce rather than the attained level of social development) takes the place of conscious rational cultivation of the soil as eternal communal property, an inalienable condition for the existence and reproduction of a chain of successive generations of the human race.

Karl Marx, Capital 3:812.

DEFORESTATION UNDER CAPITALISM

The long production time (which comprises a relatively small period of working time) and the great length of the periods of turnover entailed make forestry an industry of little attraction to private and therefore capitalist enterprise, the latter being essentially private even if the associated capitalist takes the place of the individual capitalist. The development of culture and of industry in general has ever evinced itself in such energetic destruction of forests that everything done by it conversely for their preservation and restoration appears infinitesimal.

Karl Marx, Capital 2:244.

CAPITALISM RUINS THE WORKER'S HEALTH AND THE SOIL'S FERTILITY

But in its blind unrestrainable passion, its were-wolf hunger for surplus-labour, capital oversteps not only the moral, but even the merely physical maximum bounds of the working-day. It usurps the time for growth, development, and healthy maintenance of the body. It steals the time required for the consumption of fresh air and sunlight. It higgles over a meal-time, incorporating it where possible with the process of production itself, so that food is given to the labourer as to a mere means of production, as coal is supplied to the boiler, grease and oil to the machinery. It reduces the sound sleep needed for the restoration, reparation, refreshment of the bodily pow-ers to just so many hours of torpor as the revival of an organism, absolutely exhausted, renders essential. It is not the normal maintenance of the labour-power which is to determine the limits of the working-day; it is the great-est possible daily expenditure of labour-power, no matter how diseased, compulsory, and painful it may be, which is to determine the limits of the labourers' period of repose. Capital cares nothing for the length of life of labour-power. All that concerns it is simply and solely the maximum of labour-power, that can be rendered fluent in a working-day. It attains this end by shortening the extent of the labourer's life, as a greedy farmer snatches increased produce from the soil by robbing it of its fertility.

Karl Marx, *Capital* 1:264–65.

Notes

1. The quotations from the works of Marx and Engels are from the following:
   Friedrich Engels, *Dialectics of Nature* (New York: International Publishers, 1954).
   ———, *Herr Eugen Dühring's Revolution in Science (Anti-Dühring)*, ed. C. P. Dutt and trans. Emile Burns (New York: International Publishers, 1939, 1966).
   Karl Marx, *Capital: A Critique of Political Economy*, vol. 1, trans. Samuel Moore and Edward Aveling and ed. Friedrich Engels; vol. 2, trans. Ernest Untermann and ed. Friedrich Engels; vol. 3, trans. Ernest Untermann and ed. Friedrich Engels (New York: International Publishers, 1967).
   ———, *The Economic and Philosophic Manuscripts of 1844*, trans. Martin Milligan and ed. Dirk J. Struik (New York: International Publishers, 1964).
   ———, *Grundrisse: Foundations of the Critique of Political Economy* (rough draft), trans. Martin Nicolaus (New York: Random House, 1973).
   Karl Marx and Friedrich Engels, *The German Ideology*, pts. 1 and 3, trans. and ed. R. Pascal (New York: International Publishers, 1947).
   ———, *The Holy Family* (Los Angeles: Progress Publishers, 1956).
2. Science, generally speaking, costs the capitalist nothing, a fact that by no means hinders him from exploiting it. The science of others is as much annexed by capital as the labour of others. Capitalistic appropriation and personal appropriation, whether of science or of material wealth, are, however, totally different things. Dr. Ure him-self deplores the gross ignorance of mechanical science existing among his dear machinery-exploiting manufacturers, and Liebig can a tale unfold about the astounding ignorance of chemistry displayed by English chemical manufacturers.

The Concept of Enlightenment

MAX HORKHEIMER AND THEODOR ADORNO
Translated by John Cumming

IN THE MOST GENERAL sense of progressive thought, the Enlightenment has always aimed at liberating men from fear and establishing their
sovereignty. Yet the fully enlightened earth radiates disaster triumphant.
The program of the Enlightenment was the disenchantment of the world,
the dissolution of myths, and the substitution of knowledge for fancy.
Bacon, the "father of experimental philosophy,"[1] had defined its motives.
He looked down on the masters of tradition, the "great reputed authors"
who first

> believe that others know that which they know not; and after them-
> selves know that which they know not. But indeed facility to believe,
> impatience to doubt, temerity to answer, glory to know, doubt to contra-
> dict, end to gain, sloth to search, seeking things in words, resting in part
> of nature; these and the like have been the things which have forbidden
> the happy match between the mind of man and the nature of things; and
> in place thereof have married it to vain notions and blind experiments:
> and what the posterity and issue of so honorable a match may be, it is
> not hard to consider. Printing, a gross invention; artillery, a thing that
> lay not far out of the way; the needle, a thing partly known before: what
> a change have these three things made in the world in these times; the
> one in state of learning, the other in the state of war, the third in the
> state of treasure, commodities, and navigation! And those, I say, were
> but stumbled upon and lighted upon by chance. Therefore, no doubt, the
> sovereignty of man lieth hid in knowledge; wherein many things are re-
> served, which kings with their treasure cannot buy, nor with their force
> command; their spials and intelligencers can give no news of them, their
> seamen and discoverers cannot sail where they grow: now we govern nature

From: *Dialectic of Enlightenment* (1944; reprint, New York: Herder and Herder, 1972), 3–28,
42, excerpts.

in opinions, but we are thrall unto her in necessity: but if we would be led by her in invention, we should command her by action.[2]

Despite his lack of mathematics, Bacon's view was appropriate to the scientific attitude that prevailed after him. The concordance between the mind of man and the nature of things that he had in mind is patriarchal: the human mind, which overcomes superstition, is to hold sway over a disenchanted nature. Knowledge, which is power, knows no obstacles: neither in the enslavement of men nor in compliance with the world's rulers. As with all the ends of bourgeois economy in the factory and on the battlefield, origin is no bar to the dictates of the entrepreneurs: kings, no less directly than businessmen, control technology; it is as democratic as the economic system with which it is bound up. Technology is the essence of this knowledge. It does not work by concepts and images, by the fortunate insight, but refers to method, the exploitation of others' work, and capital. The "many things" which, according to Bacon, "are reserved," are themselves no more than instrumental: the radio as a sublimated printing press, the dive bomber as a more effective form of artillery, radio control as a more reliable compass. What men want to learn from nature is how to use it in order wholly to dominate it and other men. That is the only aim. Ruthlessly, in despite of itself, the Enlightenment has extinguished any trace of its own self-consciousness. The only kind of thinking that is sufficiently hard to shatter myths is ultimately self-destructive. In face of the present triumph of the factual mentality, even Bacon's nominalist credo would be suspected of a metaphysical bias and come under the same verdict of vanity that he pronounced on scholastic philosophy. Power and knowledge are synonymous.[3] . . .

The disenchantment of the world is the extirpation of animism. Xenophanes derides the multitude of deities because they are but replicas of the men who produced them, together with all that is contingent and evil in mankind; and the most recent school of logic denounces—for the impressions they bear—the words of language, holding them to be false coins better replaced by neutral counters. The world becomes chaos, and synthesis salvation. There is said to be no difference between the totemic animal, the dreams of the ghost-seer, and the absolute Idea. On the road to modern science, men renounce any claim to meaning. They substitute formula for concept, rule and probability for cause and motive. Cause was only the last philosophic concept which served as a yardstick for scientific criticism: so to speak because it alone among the old ideas still seemed to offer itself to scientific criticism, the latest secularization of the creative principle. . . .

The pre-Socratic cosmologies preserve the moment of transition. The moist, the indivisible, air, and fire, which they hold to be the primal matter of nature, are already rationalizations of the mythic mode of apprehension. Just as the images of generation from water and earth, which came from the Nile to the Greeks, became here hylozoistic principles, or elements, so all the equivocal multitude of mythical demons were intellectualized in the pure form of ontological essences. Finally, by means of the Platonic ideas, even the patriarchal gods of Olympus were absorbed in the philosophical logos. The Enlightenment, however, recognized the old powers in the Platonic and Aristotelian aspects of metaphysics, and opposed as superstition the claim that truth is predicable of universals. It asserted that in the authority of universal concepts, there was still discernible fear of the demonic spirits which men sought to portray in magic rituals, hoping thus to influence nature. From now on, matter would at last be mastered without any illusion of ruling or inherent powers, of hidden qualities. For the Enlightenment, whatever does not conform to the rule of computation and utility is suspect. So long as it can develop undisturbed by any outward repression, there is no holding it. In the process, it treats its own ideas of human rights exactly as it does the older universals. Every spiritual resistance it encounters serves merely to increase its strength.[4] Which means that enlightenment still recognizes itself even in myths. Whatever myths the resistance may appeal to, by virtue of the very fact that they become arguments in the process of opposition, they acknowledge the principle of dissolvent rationality for which they reproach the Enlightenment. Enlightenment is totalitarian.

Enlightenment has always taken the basic principle of myth to be anthropomorphism, the projection onto nature of the subjective.[5] In this view, the supernatural, spirits and demons, are mirror images of men who allow themselves to be frightened by natural phenomena. Consequently, the many mythic figures can all be brought to a common denominator, and reduced to the human subject. . . .

Formal logic was the major school of unified science. It provided the Enlightenment thinkers with the schema of the calculability of the world. The mythologizing equation of Ideas with numbers in Plato's last writings expresses the longing of all demythologization: number became the canon of the Enlightenment. The same equations dominate bourgeois justice and commodity exchange. . . . To the Enlightenment, that which does not reduce to numbers, and ultimately to the one, becomes illusion; modern positivism writes it off as literature. Unity is the slogan from Parmenides to [Bertrand] Russell. The destruction of gods and qualities alike is insisted upon. . . . In place of the local spirits and demons there appeared heaven and its hierarchy; in place of the invocations of the magician and the tribe, the distinct gradation of sacrifice and the labor of the unfree mediated through the word of command. The Olympic deities are no longer directly identical with elements, but signify them. In Homer, Zeus represents the sky and the weather, Apollo controls the sun, and Helios and Eos are already shifting to an allegorical function. The gods are distinguished from material elements as their quintessential concepts. From now on, being divides into the logos (which with the progress of philosophy contracts to the monad, to a mere point of reference), and into the mass of all things and creatures without. This single distinction between existence proper and reality engulfs all others. Without regard to distinctions, the world becomes subject to man. In this the Jewish creation narrative and the religion of Olympia are at one: "... and let them have dominion over the fish of the sea, and over the fowl of the air, and over the cattle, and over all the earth, and over every creeping thing that creepeth upon the earth."[6] "O Zeus, Father Zeus, yours is the dominion of the heavens, and you oversee the works of man, both wicked and just, and even the wantonness of the beasts; and righteousness is your concern."[7] ...

Myth turns into enlightenment, and nature into mere objectivity. Men pay for the increase of their power with alienation from that over which they exercise their power. Enlightenment behaves toward things as a dictator toward men. He knows them in so far as he can manipulate them. The man of science knows things in so far as he can make them. In this way their potentiality is turned to his own ends. In the metamorphosis the nature of things, as a substratum of domination, is revealed as always the same. This identity constitutes the unity of nature. It is a presupposition of the magical invocation as little as the unity of the subject. The shaman's rites were directed to the wind, the rain, the serpent without, or the demon in the sick man, but not to materials or specimens. Magic was not ordered by one, identical spirit: it changed like the cultic masks which were supposed to accord with the various spirits. Magic is utterly untrue, yet in it domination is not yet negated by transforming itself into the pure truth and acting as the very ground of the world that has become subject to it. The magician imitates demons; in order to frighten them or to appease them, he behaves frighteningly or makes gestures of appeasement. Even though his task is impersonation, he never conceives of himself as does the civilized man for whom the unpretentious preserves of the happy hunting-grounds become the unified cosmos, the inclusive concept for all possibilities of plunder. The magician never interprets himself as the image of the invisible power; yet this is the very image in which man attains to the identity of self that cannot disappear through identification with another, but takes possession of itself once and for all as an impenetrable mask. It is the identity of the spirit and its correlate, the unity of nature, to which the multiplicity of qualities falls victim. Disqualified nature becomes the chaotic matter of mere classification, and the all-powerful set becomes mere possession—abstract identity. ...

Abstraction, the tool of enlightenment, treats its objects as did fate, the notion of which it rejects: it liquidates them. Under the leveling domination of abstraction (which makes everything in nature repeatable), and of industry (for which abstraction ordains repetition), the freedom themselves finally came to form that "herd" which Hegel has declared to be the result of the Enlightenment.[8]

The distance between subject and object, a presupposition of abstraction, is grounded in the distance from the thing itself, which the master achieved through the mastered. The lyrics of Homer and the hymns of the Rig-Veda date from the time of territorial dominion and the secure locations in which a dominant warlike race established themselves over the mass of vanquished natives.[9] . . . A proprietor like Odysseus "manages from a distance a numerous, carefully gradated staff of cowherds, shepherds, swineherds and servants. In the evening, when he has seen from his castle that the countryside is illumined by a thousand fires, he can compose himself for sleep with a quiet mind: he knows that his upright servants are keeping watch lest wild animals approach, and to chase thieves from the preserves which they are there to protect."[10] The universality of ideas as developed by discursive logic, domination in the conceptual sphere, is raised up on the basis of actual domination. The dissolution of the magical heritage, of the old diffuse ideas, by conceptual unity, expresses the hierarchical constitution of life determined by those who are free. The individuality that learned order and subordination in the subjection of the world, soon wholly equated truth with the regulative thought without whose fixed distinctions universal truth cannot exist. Together with mimetic magic, it tabooed the knowledge which really concerned the object. Its hatred was extended to the image of the vanquished former age and its imaginary happiness. The chthonic gods of the original inhabitants are banished to the hell to which, according to the sun and light religion of Indra and Zeus, the earth is transformed. . . .

For science the word is a sign: as sound, image, and word proper it is distributed among the different arts, and is not permitted to reconstitute itself by their addition, by synesthesia, or in the composition of the *Gesamtkunstwerk*. As a system of signs, language is required to resign itself to calculation in order to know nature, and must discard the claim to be like her. As image, it is required to resign itself to mirror-imagery in order to be nature entire, and must discard the claim to know her. With the progress of enlightenment, only authentic works of art were able to avoid the mere imitation of that which already is. The practicable antithesis of art and science, which tears them apart as separate areas of culture in order to make them both manageable as areas of culture ultimately allows them, by dint of their own tendencies, to blend with one another even as exact contraries. In its neopositivist version, science becomes aestheticism, a system of detached signs devoid of any intention that would transcend the system: it becomes the game which mathematicians have for long proudly asserted is their concern. . . .

For enlightenment is as totalitarian as any system. Its untruth does not consist in what its romantic enemies have always reproached it for: analytical method, return to elements, dissolution through reflective thought; but instead in the fact that for enlightenment the process is always decided from the start. When in mathematical procedure the unknown becomes the unknown quantity of an equation, this marks it as the well-known even before any value is inserted. Nature, before and after the quantum theory, is that which is to be comprehended mathematically; even what cannot be made to agree, indissolubility and irrationality, is converted by means of mathematical theorems. In the anticipatory identification of the wholly conceived and mathematized world with truth, enlightenment intends to secure itself against the return of the mythic. It confounds thought and mathematics. In this way the latter is, so to speak, released and made into an absolute instance. "An infinite world, in this case a world of idealities, is conceived as one whose objects do not accede singly, imperfectly, and as if by chance to our cognition, but are attained by a rational, systematically unified method—in a process of infinite progression—so that each object is ultimately apparent according to its full inherent being ... In the Galilean mathematization of the world, however, *this selfness* is idealized under the guidance of the new mathematics: in modern terms, it becomes itself a mathematical multiplicity."[11] Thinking objectifies itself to become an automatic, self-activating process; an impersonation of the machine that it produces itself so that ultimately the machine can replace it. . . .

Mathematical procedure became, so to speak, the ritual of thinking. In spite of the axiomatic self-restriction, it establishes itself as necessary and objective: it turns thought into a thing, an instrument—which is its own term for it. . . .

In the enlightened world, mythology has entered into the profane. In its blank purity, the reality which has been cleansed of demons and their conceptual descendants assumes the numinous character which the ancient world attributed to demons. Under the title of brute facts, the social injustice from which they proceed is now as assuredly sacred a preserve as the medicine man was sacrosanct by reason of the protection of his gods. It is not merely that domination is paid for by the alienation of men from the objects dominated: with the objectification of spirit, the very relations of men— even those of the individual to himself—were bewitched. The individual is reduced to the nodal point of the conventional responses and modes of operation expected of him. Animism spiritualized the object; whereas industrialism objectifies the spirits of men. Automatically, the economic apparatus, even before total planning, equips commodities with the values which decide human behavior. Since, with the end of free exchange, commodities lost all their economic qualities except for fetishism, the latter has extended its arthritic influence over all aspects of social life. Through the countless agencies of mass production and its culture the conventionalized modes of behavior are impressed on the individual as the only natural, respectable, and rational ones. . . .

While bourgeois economy multiplied power through the mediation of the market, it also multiplied its objects and powers to such an extent that for their administration not just the kings, not even the middle classes are no longer necessary, but all men. They learn from the power of the things to dispense at last with power. Enlightenment is realized and reaches its term when the nearest practical ends reveal themselves as the most distant goal now attained, and the lands of which "their spials and intelligencers can give no news," that is, those of the nature despised by dominant science, are recognized as the lands of origin. Today, when Bacon's utopian vision that we should "command nature by action"—that is, in practice—has been realized on a tellurian scale, the nature of the thralldom that he ascribed to unsubjected nature is clear. It was domination itself. And knowledge, in which Bacon was certain the "sovereignty of man lieth hid," can now become the dissolution of domination. But in the face of such a possibility, and in the service of the present age, enlightenment becomes wholesale deception of the masses.

Notes

1. François de Voltaire, "Lettre XII," Œvres complètes, Lettres Philosophiques (Paris: Garnier, 1879), vol. 22: 118; François de Voltaire, "Letter Twelve, On Chancellor Bacon," Philosophical Letters (New York: Bobbs-Merrill, 1961), 46–51; see p. 48.
2. Francis Bacon, "In Praise of Human Knowledge" (Miscellaneous Tracts upon Human Knowledge), in The Works of Francis Bacon, ed. Basil Montagu (London, 1825), vol. 1:254ff.
3. Cf. Francis Bacon, Novum Organum, Works 14:31.
4. Cf. G. W. F. Hegel, Phänomenologie des Geistes (The Phenomenology of Spirit), in Werke (Frankfurt am Main: Suhrkampf, 1969–71), 2:410ff.
5. Xenophanes, Montaigne, Hume, Feuerbach, and Salomon Reinach are at one here. See, for Reinach, Orpheus, trans. F. Simmons (London, 1909), 9ff.
6. Genesis 1:26 (Authorized Version).
7. Archilochos, Carmina: The Fragments of Archilochus, trans. Guy Davenport (Berkeley: University of California Press, 1964), fragment 87; quoted by Deussen, Allgemeine Geschichte der Philosophie (Leipzig, 1911), vol. 2, pt. 1, p. 18.
8. Hegel, Phänomenologie des Geistes, 424.
9. Cf. W. Kirfel, Geschichte Indiens, in Propyläenweltgeschichte, 3:261ff; and G. Glotz, Histoire Grècque, vol. 1 of Histoire ancienne (Paris, 1938), 137ff.
10. Glotz, Histoire Grècque 1:140.
11. Edmund Husserl, "Die Krisis europäischen Wissenschaften und die transzendentale Phänomenologie," in Philosophia 1 (1936): 95ff.

Ecology and Revolution

HERBERT MARCUSE

COMING FROM THE UNITED STATES, I am a little uneasy discussing the ecological movement, which has already by and large been co-opted [there]. Among militant groups in the United States, and particularly among young people, the primary commitment is to fight, with all the means (severely limited means) at their disposal, against the war crimes being committed against the Vietnamese people. The student movement—which had been proclaimed to be dead or dying, cynical and apathetic—is being reborn all over the country. This is not an organized opposition at all, but rather a spontaneous movement which organizes itself as best it can, provisionally, on the local level. But the revolt against the war in Indochina is the only oppositional movement the establishment is unable to co-opt because neocolonial war is an integral part of that global counterrevolution which is the most advanced form of monopoly capitalism.

So, why be concerned about ecology? Because the violation of the earth is a vital aspect of the counterrevolution. The genocidal war against people is also "ecocide" insofar as it attacks the sources and resources of life itself. It is no longer enough to do away with people living now; life must also be denied to those who aren't even born yet by burning and poisoning the earth, defoliating the forests, blowing up the dikes. This bloody insanity will not alter the ultimate course of the war, but it is a very clear expression of where contemporary capitalism is at: the cruel waste of productive resources in the imperialist homeland goes hand in hand with the cruel waste of destructive forces and consumption of commodities of death manufactured by the war industry.

In a very specific sense, the genocide and ecocide in Indochina are the capitalist response to the attempt at revolutionary ecological liberation: the bombs are meant to prevent the people of North Vietnam from undertaking

From: "Ecology and Revolution," *Liberation*, 16 (September 1972): 10–12.

the economic and social rehabilitation of the land. But in a broader sense, monopoly capitalism is waging a war against nature—human nature as well as external nature. For the demands of ever more intense exploitation come into conflict with nature itself, since nature is the source and locus of the life-instincts which struggle against the instincts of aggression and destruction. And the demands of exploitation progressively reduce and exhaust resources: the more capitalist productivity increases, the more destructive it becomes. This is one sign of the internal contradictions of capitalism.

One of the essential functions of civilization has been to change the nature of man and his natural surroundings in order to "civilize" him—that is, to make him the subject-object of the market society, subjugating the pleasure principle to the reality principle and transforming man into a tool of ever more alienated labor. This brutal and painful transformation has crept up on external nature very gradually. Certainly, nature has always been an aspect (for a long time the only one) of labor. But it was also a dimension *beyond* labor, a symbol of beauty, of tranquility, of a nonrepressive order. Thanks to these values, nature was the very negation of the market society, with its values of profit and utility.

However, the natural world is a historical, a social world. Nature may be a negation of aggressive and violent society, but its pacification is the work of man (and woman), the fruit of his/her productivity. But the structure of capitalist productivity is inherently expansionist: more and more, it reduces the last remaining natural space outside the world of labor and of organized and manipulated leisure.

The process by which nature is subjected to the violence of exploitation and pollution is first of all an economic one (an aspect of the mode of production), but it is a political process as well. The power of capital is extended over the space for release and escape represented by nature. This is the totalitarian tendency of monopoly capitalism: in nature, the individual must find only a repetition of his own society; a dangerous dimension of escape and contestation must be closed off.

At the present stage of development, the absolute contradiction between social wealth and its destructive use is beginning to penetrate people's consciousnesses, even in the manipulated and indoctrinated conscious and unconscious levels of their minds. There is a feeling, a recognition, that it is no longer necessary to exist as an instrument of alienated work and leisure. There is a feeling and a recognition that well-being no longer depends on a perpetual increase in production. The revolt of youth (students, workers, women), undertaken in the name of the values of freedom and happiness, is an attack on all the values which govern the capitalist system. And this revolt is oriented toward the pursuit of a radically different natural and technical environment; this perspective has become the basis for subversive experiments such as the attempts by American "communes" to establish non-alienated relations between the sexes, between generations, between man and nature—attempts to sustain the consciousness of refusal and of renovation.

In this highly political sense, the ecological movement is attacking the "living space" of capitalism, the expansion of the realm of profit, of waste production. However, the fight against pollution is easily co-opted. Today, there is hardly an ad which doesn't exhort you to "save the environment," to put an end to pollution and poisoning. Numerous commissions are created to control the guilty parties. To be sure, the ecological movement may serve very well to spruce up the environment, to make it pleasanter, less ugly, healthier and hence, more tolerable. Obviously, this is a sort of co-optation, but it also a progressive element because, in the course of this co-optation, a certain number of needs and aspirations are beginning to be expressed within the very heart of capitalism and a change is taking place in people's behavior, experience, and attitudes towards their work. Economic and technical demands are transcended in a movement of revolt which challenges the very mode of production and model of consumption.

Increasingly, the ecological struggle comes into conflict with the laws which govern the capitalist system: the law of increased accumulation of capital, of the creation of sufficient surplus value, of profit, of the necessity of perpetuating alienated labor and exploitation. Michel Bosquet put it very well: the ecological logic is purely and simply the negation of capitalist logic; the earth can't be saved within the framework of capitalism, the Third World can't be developed according to the model of capitalism.

In the last analysis, the struggle for an expansion of the world of beauty, nonviolence and serenity is a political struggle. The emphasis on these values, on the restoration of the earth as a human environment, is not just a romantic, aesthetic, poetic idea which is a matter of concern only to the privileged; today, it is a question of survival. People must learn for themselves that it is essential to change the model of production and consumption, to abandon the industry of war, waste and gadgets, replacing it with the production of those goods and services which are necessary to a life of reduced labor, of creative labor, of enjoyment.

As always, the goal is well-being, but a well-being defined not by ever-increasing consumption at the price of ever-intensified labor, but by the achievement of a life liberated from the fear, wage slavery, violence, stench and infernal noise of our capitalist industrial world. The issue is not to beautify the ugliness, to conceal the poverty, to deodorize the stench, to deck the prisons, banks and factories with flowers; the issue is not the purification of the existing society but its replacement.

Pollution and poisoning are mental as well as physical phenomena, subjective as well as objective phenomena. The struggle for an environment ensuring a happier life could reinforce, in individuals themselves, the instinctual roots of their own liberation. When people are no longer capable of distinguishing between beauty and ugliness, between serenity and cacophony, they no longer understand the essential quality of freedom, of happiness. Insofar as it has become the territory of capital rather than of man, nature serves to strengthen human servitude. These conditions are rooted in the basic institutions of the established system, for which nature is primarily an object of exploitation for profit.

This is the insurmountable internal limitation of any capitalist ecology. Authentic ecology flows into a militant struggle for a socialist politics which must attack the system at its roots, both in the process of production and in the mutilated consciousnesses of individuals.

The Domination of Nature

WILLIAM LEISS

Technology reveals the active relation of man to nature, the immediate process of production of his life, and thereby also his social life-relationships and the cultural representations that arise out of them.
— Marx, *Capital*

1. INTRODUCTION

TECHNOLOGY HAS BEEN DESCRIBED as the concrete link between the mastery of nature through scientific knowledge and the enlarged disposition over the resources of the natural environment which supposedly constitutes mastery of nature in the everyday world. Normally the rubric "conquest of nature" is applied to modern science and technology together, simply on account of their manifest interdependence in the research laboratory and industry. When they are considered in isolation, as two related aspects of human activity among many others, the fact of their necessary connection must indeed be recognized if their progress in modern times is to be understood. But it does not follow automatically that they function as a unity with respect to the mastery of nature, since they are not identical with the latter: mastery of nature develops also in response to other aspects in the social dynamic, for example the process whereby new human needs are formed, and therefore its meaning with respect to technology may be quite different than it is in the case of science.

Nature per se is not the object of mastery; . . . instead various senses of mastery are appropriate to various perspectives on nature. If this proposition is correct, then the converse is likewise true, namely, that mastery of nature is not a project of science per se but, rather, a broader social task.[1] In this larger context, technology plays a far different role than does science,

From: "Technology and Domination," in *The Domination of Nature* (New York: George Braziller, 1972), 145–55, 161–65.

for it has a much more direct relationship to the realm of human wants and thus to the social conflicts which arise out of them. This is what Marx meant in referring to the "immediate" process of production in which technology figures so prominently—the direct connection between men's technical capacities and their ability to satisfy their desires, which is a constant feature of human history and is not bound to any specific form of scientific knowledge. On the other hand, science, like similar advanced cultural formations (religion, art, philosophy, and so forth), is indirectly related to the struggle for existence: in technical language, these are all *mediated* by reflective thought to a far greater extent. Of course this by no means implies that they lack a social content altogether, but only that it is present in highly abstract form and that by virtue of their rational impulse they transcend to some extent the specific historical circumstances which gave them birth. Therefore scientific rationality and technological rationality are not the same and cannot be regarded as the complementary bases of something called the domination of nature.

The character of technological rationality . . . must . . . be explored. Two considerations are especially relevant to the discussion. In the first place, the immediate connection of technology with practical life activity determines a priori the kind of mastery over nature that is achieved through technological development: caught in the web of social conflict, technology constitutes one of the means by which mastery of nature is linked to mastery over man. Secondly, the employment of technological rationality in the extreme forms of social conflict in the twentieth century—in weapons of mass destruction, techniques for the control of human behavior, and so forth—precipitates a crisis of rationality itself; the existence of this crisis necessitates a critique of reason that attempts to discover (and thus to aid in overcoming) the tendencies uniting reason with irrationalism and terror. These two themes have been brilliantly presented in the work of the contemporary philosopher Max Horkheimer.

The attempt to understand the significance of the domination of nature is a problem with which Horkheimer has been concerned during his entire career: one can find scattered references to it in books and essays of his spanning a period of forty years. In my estimation his analysis, although quite unsystematic, contains greater insight into the full range of the problem under discussion here than does any other single contribution, although one can of course find many affinities between his work and that of others. . . . The fundamental question which he poses is similar to the one raised by Husserl (although from a very different philosophical point of view) in *The Crisis of European Sciences*: What is the concept of rationality that underlies modern social progress? He also shares with Husserl the conviction that the social dilemmas related to scientific and technological progress have reached a critical point in the twentieth century. Horkheimer describes this situation as one in which "the antagonism of reason and nature is in an acute and catastrophic phase."[2]

2. THE CRITIQUE OF REASON

Horkheimer follows Nietzsche's pathbreaking thought and argues that the domination of nature or the expansion of human power in the world is a universal characteristic of human reason rather than a distinctive mark of the modern period:

> If one were to speak of a disease affecting reason, this disease should be understood not as having stricken reason at some historical moment, but as being inseparable from the nature of reason in civilization as we have known it so far. The disease of reason is that reason was born from man's urge to dominate nature. . . . One might say that the collective madness that ranges today, from the concentration camps to the seemingly most harmless mass-culture reactions, was already present in germ in primitive objectivization, in the first man's calculating contemplation of the world as a prey.[3]

Again like Nietzsche he finds the first clear expression of this will to power in the rationalist conceptions of ancient Greek philosophy, conceptions which determined the predominant course of subsequent Western philosophy. The concept (*Begriff*) itself, and especially the concept of the "thing," serves as a tool through which the chaotic, disorganized data of sensations and perceptions can be organized into coherent structures and thus into forms of experience that can give rise to exact knowledge. The deductive form of logic which emerges in Greek philosophy reinforces this original tendency and raises it to a much higher level. The deductive form of thought "mirrors hierarchy and compulsion" and first clearly reveals the social character of the structure of knowledge: "The generality of thought, as discursive logic develops it—domination in the realm of concepts—arises on the basis of domination in reality." The categories of logic suggest the power of the universal over the particular and in this respect they testify to the "thoroughgoing unity of society and domination," that is, to the ubiquity of the subjection of the individual to the whole in human society.[4]

Horkheimer differs from Nietzsche in attempting to distinguish two basic types of reason. Although in themselves all structures of logic and knowledge reflect a common origin in the will to domination, there is one type of reason in which this condition is transcended and another in which it is not: the former he calls objective reason, the latter, subjective reason.[5] The first conceives of human reason as a part of the rationality of the world and regards the highest expression of that reason (truth) as an ontological category, that is, it views truth as the grasping of the essence of things. Objective reason is represented in the philosophies of Plato and Aristotle, the Scholastics, and German idealism. It includes the specific rationality of man (subjective reason) by which man defines himself and his goals, but not exclusively, for it is oriented toward the whole of the realm of beings; it strives to be, as Horkheimer remarks, the voice of all that is mute in nature. On the other hand, subjective reason exclusively seeks mastery over things and does not attempt to consider what extrahuman things may be in and for themselves. It does not ask whether ends are intrinsically rational, but only how means may be fashioned to achieve whatever ends may be selected; in effect it defines the rational as that which is serviceable for human interests. Subjective reason attains its most fully developed form in positivism.

The two concepts do not exist for Horkheimer as static historical constants. Objective reason both undergoes a process of self-dissolution and also succumbs to the attack of subjective reason, and this fate in a sense represents necessary historical progress. The conceptual framework and hierarchy established by objective reason is too static, condemning men (as the subjects of historical change) to virtual imprisonment in an order which valiantly tries to maintain its traditional foundations intact. By the seventeenth century, for example, the combination of Aristotelianism and Christian dogma in late medieval philosophy had become intellectually sterile, a system in which the repetition of established formulas was substituted for original thought. The declining period of great philosophical systems is characterized by the increasingly effective onslaught of skepticism, such as that which struck Greek metaphysics in Hellenistic times and scholastic philosophy in the sixteenth century. The skeptics represent movements of "enlightenment"—a recurring pattern of which the eighteenth-century French Enlightenment is the most famous example—wherein thinkers undermine concepts and dogmas that were once vital but have subsequently ossified, often in the process turning into ideological masks for the material interests of certain social groups.

But in its later stages the movement of enlightenment reveals its own internal contradictions, represented by Horkheimer and Adorno in their famous notion of the "dialectic of enlightenment."[6] What marks the general program of enlightenment as a unitary phenomenon despite its various historical guises is the "demythologization of the world." In its earliest period it combatted the traditional religious mythologies (for example, in Greek civilization); in the modern West it takes the form of a struggle against mystification in religion and philosophy, and in its most advanced stages— as in positivism—it carries this campaign to the heart of conceptual thought itself, finally upholding the position that only propositions conforming to one particular notion of "verifiable knowledge" have any meaning at all. The sustained effort of demythologizing in modern times ends by stripping the world of all inherent purpose. Nature, for example, appears to scientific thought only as a collection of bodies in eternally lawful motion, and the social reflection of this scientific vision is the idea of a set of natural laws of economic behavior which blindly follow their established course and which exhibit no inherent rationality. The consequence of this view is to set the relationship of man and the world inescapably in the context of domination: man must either meekly submit to these natural laws (physical and economic) or attempt to master them; for since they possess no purpose, or at least none that he can understand, there is no possibility of reconciling his objectives with those of the natural order.

The dialectical nature of enlightenment becomes clear only at this advanced level. All other purposes having been driven out of the world, only one value remains as evident and fundamental: self-preservation. This is sought through mastery of the world to assure the self-preservation of the species and, within the species, through mastery of the economic process to assure the self-preservation of the individual. Yet the puzzling fact remains that adequate security (as the goal of self-preservation) is never attained, either for the species or the individual, and sometimes seems to be actually diminishing for both. Thus the struggle for mastery tends to perpetuate itself endlessly and to become an end in itself.

In the course of enlightenment the predominant function of reason is to serve as an instrument in the struggle for mastery. Reason becomes above all the tool by which man seeks to find in nature adequate resources for self-preservation. It separates itself from the nature given in sense perception and finds a secure point in the thinking self (the *ego cogito*), on the basis of which it tries to discover the means for subjecting nature to its requirements. In the new natural philosophy of the seventeenth century, this procedure is adopted as the theme of science, and as the principal mode of behavior through which the mastery of nature is pursued, science assumes an increasingly influential role in society. For Horkheimer the attributes of the modern scientific conception of nature which predispose it for the purposes of mastery are, in part: the principle of the uniformity of nature, the inherent technological applicability of its findings, the reduction of nature to pure "stuff" or abstract matter through the elimination of qualities as essential features of natural phenomena, and especially the primacy of mathematics in the representation of natural processes.[7]

Horkheimer does not present this picture of science in isolation, but rather tries to understand what complementary conditions are necessary in order for that science to become, as it has, a historical reality of great dimensions. He contends that the "mastery of inner nature" is a logical correlate of the mastery of external nature; in other words, the domination of the world that is to be carried out by subjective reason presupposes a condition under which man's reason is already master in its own house, that is, in the domain of human nature. The prototype of this connection can be found in Cartesian philosophy, where the ego appears as dominating internal nature (the passions) in order to prevent the emotions from interfering with the judgments that form the basis of scientific knowledge. The culmination of the development of the transcendental subjectivity inaugurated by Descartes is to be found in Fichte, in whose early works "the relationship between the ego and nature is one of tyranny," and for whom the "entire universe becomes a tool of the ego, although the ego has no substance or meaning except in its own bound-less activity.[8]

In the social context of competition and cooperation the abstract possibilities for an increase in the domination of nature are transformed into actual technological progress. But in the ongoing struggle for existence the desired goal (security) continues to elude the individual's grasp, and the technical mastery of nature expands as if by virtue of its own independent necessity, with the result that what was once clearly seen as a means gradually becomes an end in itself:

> As the end result of the process, we have on the one hand the self, the abstract ego emptied of all substance except its attempt to transform everything in heaven and on earth into means for its preservation, and on the other hand an empty nature degraded to mere material, mere stuff to be dominated, without any other purpose than that of this very domination.[9]

On the empirical level the mastery of inner nature appears as the modern form of individual self-denial and instinctual renunciation required by the social process of production. For the minority this is the voluntary, calculating self-denial of the entrepreneur; for the majority, it is the involuntary renunciation enforced by the struggle for the necessities of life.

The crucial question is: What is the historical dynamic that spurs on the mastery of internal and external nature in the modern period? Two factors shape the answer. One is that the domination of nature is conceived in terms of an intensive exploitation of nature's resources, and the other is that a level of control over the natural environment which would be sufficient (given a peaceful social order) to assure the material well-being of men has already been attained. But external nature continues to be viewed primarily as an object of potentially increased mastery, despite the fact that the level of mastery has risen dramatically. The instinctual renunciation—the persistent mastery and denial of internal nature—which is required to support the project for the mastery of external nature (through the continuation of the traditional work-process for the sake of the seemingly endless productive applications of technological innovations) appears as more and more irrational in view of the already attained possibilities for the satisfaction of needs.

Horkheimer answers the question posed in the preceding paragraph as follows: "The warfare among men in war and in peace is the key to the insatiability of the species and to its ensuing practical attitudes, as well as to the categories and methods of scientific intelligence in which nature appears increasingly under the aspect of its most effective exploitation."[10] The persistent struggle for existence, which manifests itself as social conflict both within particular societies and also among societies on a global scale, is the motor which drives the mastery of nature (internal and external) to ever greater heights and which precludes the setting of any a priori limit on this objective in its present form. Under these pressures the power of the whole society over the individual steadily mounts and is exercised through techniques uncovered in the course of the increasing mastery of nature. Externally, this means the ability to control, alter, and destroy larger and larger segments of the natural environment. Internally, terroristic and nonterroristic measures for manipulating consciousness and for internalizing heteronomous needs (where the individual exercises little or no independent reflective judgment) extend the sway of society over the inner life of the person. In both respects the possibilities and the actuality of domination over men have been magnified enormously.

As a result of its internal contradictions, the objectives which are embodied in the attempted domination of nature are thwarted by the enterprise itself. For the mastery of nature has been and remains a social task, not the appurtenance of an abstract scientific methodology or the happy (or unhappy) coincidence of scientific discovery and technological application. As such its dynamic is located in the specific societal process in which those objectives have been pursued, and the overriding feature of that context is bitter social conflict. Even in those nations where the fruits of technical progress are most evident.

despite all improvements and despite fantastic riches there rules at the same time the brutal struggle for existence, oppression, and fear. That is the hidden basis of the decay of civilization, namely that men cannot utilize their power over nature for the rational organization of the earth but rather must yield themselves to blind individual and national egoism under the compulsion of circumstances and of inescapable manipulation.[11]

The more actively is the pursuit of the domination of nature undertaken, the more passive is the individual rendered; the greater the attained power over nature, the weaker the individual vis-à-vis the overwhelming presence of society. . . .

3. The Revolt of Nature

The growing domination of men through the development of new techniques for mastering the natural environment and for controlling human nature does not go unresisted. Horkheimer analyzes the reaction to it under the heading of the "revolt of nature," a brilliant and original conception that has never received the attention it deserves.[12] The revolt of nature means the rebellion of human nature which takes place in the form of violent outbreaks of persistently repressed instinctual demands. As such it is of course not at all unique to modern history, but is rather a recurrent feature of human civilization. What is new in the twentieth century, on the other hand, is the fact that the potential scope of destructiveness which it entails is so much greater.

There are many reasons for this. The most fundamental is the fact that—at least in the industrially advanced countries—the traditional grounds for the repression of instinctual demands have been vitiated but yet continue to operate: the denial of gratification, the requirements of the work process, and the struggle for existence persist almost unchanged despite the feasibility of gradually mitigating those conditions by means of a rational organization of the available productive forces. The reality principle which has prevailed throughout history has lost most of its rational basis but not its force, and this introduces an element of irrationality into the very core of human activity: "Since the subjugation of nature, in and outside of man, goes on without a meaningful motive, nature is not really transcended or reconciled but merely repressed."[13] The denial of instinctual gratification—the subjugation of internal nature—is enforced in the interests of civilization; release from this harsh regime of the reality principle was to be found in the subjugation of external nature, which would permit the fuller satisfaction of instinctual demands while preserving the order of civilization. The persistence of social conflict thwarts that objective, however, and prompts a search for new means of repressing the sources of conflict in human nature. Security is sought in the power over external nature and over other men, a power that seems possible on the basis of the remarkable accomplishments of scientific and technological rationality, but the need for security, arising always afresh out of the irrational structure of social relations, is never appeased. The dialectic of rationality and irrationality feeds the periodic outbursts of destructive passions with ever more potent fuels.

Secondly, the revolt of human nature, directed against the structure of domination and its rationality, is proportional in intensity to that of the prevailing domination itself. Greater pressures produce correspondingly more violent explosions; the magnified level of domination in modern society, achieved in respect to both external and internal nature (as we have seen, both make their effects felt in everyday social life despite their differing immediate objects), is also a measure of the heightened potential of the revolt of nature. Thirdly, in recent times this revolt itself has been manipulated and encouraged by ruling social forces as an element in the struggle for socio-political mastery. Horkheimer refers here to fascism, which cultivates the latent irrationalities in modern society as material to be managed by rational techniques (propaganda, rallies, and so forth) in the service of political objectives.

The idea of the revolt of nature suggests that there may be an internal limit within the process of enlarging domination that was outlined above. Certainly, at every level of technological development the irrationalities present in the structure of social relations have prevented the realization of the full benefits that might have been derived from the instruments (including human labor) available for the exploitation of nature's resources. The misuse, waste, and destruction of these resources at every stage is at least partially responsible for the continued search for new technological capabilities, as if the possession of more refined techniques could somehow compensate for the misapplication of the existing ones. And because of the lasting institutional frameworks through which particular groups control the behavior of others, the new techniques are utilized sooner or later in the service of domination.

Yet it does not seem possible that this process can continue indefinitely, for at the higher levels the gap between the rational organization of labor and instrumentalities on the one hand, and the irrational uses to which that organization is put on the other, widens to the point where the objectives themselves are called into question. The problem is not only that the degree of waste and misuse of resources has increased enormously, but also that the implements of destruction now threaten the biological future of the species as a whole. This is the point beyond which the nexus of rational techniques and irrational applications ceases to have any justification at all; it represents the internal limit in the exercise of domination over internal and external nature, to exceed which entails that the intentions are inevitably frustrated by the chosen means.

The purpose of mastery over nature is the security of life—and its enhancement—alike for individuals and the species. But the means presently available for pursuing these objectives encompass such potential destructiveness that their full employment in the struggle for existence would leave in ruins all the advantages so far gained at the price of so much suffering. In the intensified social conflicts of the contemporary period, and especially in the phenomenon of fascism, Horkheimer sees this dialectic at work, and this is what he has tried to describe in the notion of the revolt of nature. The use of the most advanced rational techniques of domination over external and internal nature to prevent the emergence of the free social institutions envisaged in the utopian tradition represents for him the blind, irrational outbreak of human nature against a process of domination that has become self-destructive.

In the interval since Horkheimer first presented this notion a related aspect of the problem has been recognized: in a different sense the concept of the revolt of nature may be applied in relation to ecological damage in the natural environment. There is also an inherent limit in the irrational exploitation of external nature itself, for under present conditions the natural functioning of various biological ecosystems is threatened. It is possible that permanent and irreversible damage to some parts of the major planetary ecosystems may have already occurred; the consequences of this are not yet clear.[14] If it is the case that the natural environment cannot tolerate the present level of irrational technological applications without suffering breakdowns in the mechanisms that govern its cycles of self-renewal, then we would be justified in speaking of a revolt of external nature which accompanies the rebellion of human nature.

The dialectic of reason and unreason in our time is epitomized in the social dynamic which sustains the scientific and technological progress through which the resources of nature are ever more artfully exploited. This triumph of human rationality derives its impetus from the uncontrolled interaction of processes that are rooted in irrational social behavior: the wasteful consumption of the advanced capitalist societies, the fearful military contest between capitalist and socialist blocs, the struggles within and among socialist societies concerning the correct road to the future, and the increasing pressure on Third-World nations and their populations to yield fully to economic development and ideological commitment. In the passions that prompt such behavior are forged the ineluctable chains which bind together technology and political domination at present.

Notes

1. Max Horkheimer, "Soziologie und Philosophie," in *Sociologica II: Reden und Vorträge* by Max Horkheimer and Theodor Adorno (Frankfurt: Europäische Verlagsanstalt, 1962).
2. Max Horkheimer, *Eclipse of Reason* (New York: Columbia University Press, 1947), 177.
3. Ibid., 176.
4. Max Horkheimer and Theodor Adorno, *Dialektik der Aufklärung* (Frankfurt: S. Fischer, 1969), 20, 27–29.
5. Horkheimer, *Eclipse of Reason*, chap. 1; and Max Horkheimer, "Zum Begriff der Vernunft," in Horkheimer and Adorno, *Sociologica II*.
6. Horkheimer and Adorno, *Dialektik der Aufklärung*, 9ff; a brief outline is given in Horkheimer, *Eclipse of Reason*.
7. Horkheimer and Adorno, *Dialektik der Aufklärung*, 31ff, 189–91.
8. Horkheimer, *Eclipse of Reason*, 108. In one of his essays, Marcuse cited the following passage from Fichte's *Staatslehre* (1813): "The real station, the honor and worth of the human being, and quite particularly of man in his morally natural existence, consists without doubt in his capacity as original progenitor to produce out of himself new men, new commanders of nature: beyond his earthly existence and for all eternity to establish new masters of nature." Quoted in "On Hedonism," in *Negations: Essays in Critical Theory* (Boston: Beacon Press, 1968), 186.
9. Horkheimer, *Eclipse of Reason*, 97.
10. Ibid., 109
11. Max Horkheimer, "Zum Begriff des Menschen," in *Zur Kritik der instrumentellen Vernunft*, ed. A. Schmidt (Frankfurt: S. Fischer, 1967), 198.
12. See generally Horkheimer, *Eclipse of Reason*, chap. 3, esp. 109ff.
13. Ibid., 94.
14. See UNESCO, Intergovernmental Conference of Experts on the Scientific Basis for Rational Use and Conservation of the Resources of the Biosphere. "Final Report." UNESCO Document SC/MD/9 (1969).

The Failed Promise of Critical Theory

ROBYN ECKERSLEY

THE CRITICAL THEORY DEVELOPED by the members of the Frankfurt Institute of Social Research ("the Frankfurt school") has revised the humanist Marxist heritage in ways that directly address the wider emancipatory concerns of ecocentric theorists [i.e., the valuing (for their own sake) not just of individual living organisms, but also of ecological entities at different levels of aggregation].[1] In particular, Critical Theorists have laid down a direct challenge to the Marxist idea that "true freedom" lies beyond socially necessary labor. They have argued that the more we try to "master necessity" through the increasing application of instrumental reason to all spheres of life, the *less* free we will become.

Critical Theory represents an important break with orthodox Marxism— a break that was undertaken in order to understand, among other things, why Marx's original emancipatory promise had not been fulfilled. Like many other strands of Western Marxism, Critical Theory turned away from the scientism and historical materialism of orthodox Marxism. In the case of the Frankfurt school, however, it was not through a critique of political economy but rather through a critique of culture, scientism, and instrumental reason that Marxist debates were entered. One of the enduring contributions of the first generation of Frankfurt school theorists (notably, Max Horkheimer, Theodor Adorno, and Herbert Marcuse) was to show that there are different levels and dimensions of domination and exploitation *beyond* the economic sphere and that the former are no less important than the latter. The most radical theoretical innovation concerning this broader understanding of domination came from the early Frankfurt school theorists' critical examination

From: *Environmentalism and Political Theory* (Albany: State University of New York Press, 1992), 97–106.

of the relationship between humanity and nature. This resulted in a fundamental challenge to the orthodox Marxist view concerning the progressive march of history, which had emphasized the liberatory potential of the increasing mastery of nature through the development of the productive forces. Far from welcoming these developments as marking the "ascent of man from the kingdom of necessity to the kingdom of freedom" (to borrow Engels's phrase), Horkheimer, Adorno, and Marcuse saw them in essentially negative terms, as giving rise to the domination of both "outer" and "inner" nature.[2]

These early Frankfurt school theorists regarded the rationalization process set in train by the Enlightenment as a "negative dialectics." This was reflected, on the one hand, in the apprehension and conversion of nonhuman nature into resources for production or objects of scientific inquiry (including animal experimentation) and, on the other hand, in the repression of humanity's joyful and spontaneous instincts brought about through a repressive social division of labor and a repressive division of the human psyche. Hence their quest for a human "reconciliation" with nature. Instrumental or "purposive" rationality—that branch of human reason that is concerned with determining the most efficient means of realizing pregiven goals and which accordingly apprehends only the instrumental (i.e., use) value of phenomena—should not, they argued, become the exemplar of rationality for society. Human happiness would not come about simply by improving our techniques of social administration, by treating society and nature as subject to blind, immutable laws that could be manipulated by a technocratic elite.

The early Frankfurt school's critique of instrumental rationality has been carried forward and extensively revised by Jürgen Habermas, who has sought to show, among other things, how political decision making has been increasingly reduced to pragmatic instrumentality, which serves the capitalist and bureaucratic system while "colonizing the life-world."[3] According to Habermas, this "scientization of politics" has resulted in the lay public ceding ever greater areas of system-steering decision making to technocratic elites.

All of these themes have a significant bearing on the green critique of industrialism, modern technology, and bureaucracy, and the green commitment to grass-roots democracy. Yet Critical Theory has not had a major direct influence in shaping the theory and practice of the green movement in the 1980s, whether in West Germany or elsewhere.[4] The ideas of Marcuse and Habermas did have a significant impact on the thinking of the New Left in the 1960s and early 1970s and ... the general "participatory" theme that characterized that era has remained an enduring thread in the emancipatory stream of ecopolitical thought. Yet this legacy is largely an indirect one. Of course, there are some emancipatory theorists who have drawn upon Habermas's social and political theory in articulating and explaining some aspects of the green critique of advanced industrial society.[5] However, this can be contrasted with the much greater general influence of post-Marxist green theorists such as Murray Bookchin, Theodore Roszak, and Rudolf Bahro and non-Marxist green theories such as bioregionalism, deep/transpersonal ecology, and ecofeminism—a comparison that further underscores the distance green theory has had to travel away from the basic corpus of Marxism and neo-Marxism in order to find a comfortable "theoretical home."

It is important to understand why Critical Theory has not had a greater direct impact on green political theory and practice given that two of its central problems—the triumph of instrumental reason and the domination of nature—might have served as useful theoretical starting points for the green critique of industrial society. This possibility was indeed a likely one when it is remembered that both the Frankfurt school and green theorists acknowledge the dwindling revolutionary potential of the working class (owing to its integration into the capitalist order); both are critical of totalitarianism, instrumental rationality, mass culture, and consumerism; and both have strong German connections. Why did these two currents of thought not come together?

There are many possible explanations as to why Critical Theory has not been more influential. One might note, for example, the early Frankfurt school's pessimistic outlook (particularly that of Adorno and Horkheimer), its ambivalence toward nature romanticism (acquired in part from its critical inquiry into Nazism), its rarefied language, its distance from the imperfect world of day-to-day political struggles (Marcuse being an important exception here), and its increasing preoccupation with theory rather than praxis (despite its original project of uniting the two). Yet a more fundamental explanation lies in the direction in which Critical Theory has developed since the 1960s in the hands of Jürgen Habermas, who has, by and large, remained preoccupied with and allied to the fortunes of democratic socialism (represented by the Social Democratic party in West Germany) rather than the fledgling green movement and its parliamentary representatives.[6] Of course, the green movement has not escaped Habermas's attention. However, he has tended to approach the movement more as an indicator of the motivational and legitimacy problems in advanced capitalist societies rather than as the historic bearer of emancipatory ideas (this is to be contrasted with Marcuse, who embraced the activities of new social movements).[7] Habermas has analyzed the emergence of new social movements and green concerns as a grass-roots "resistance to tendencies to colonize the life-world."[8] With the exception of the women's movement (which Habermas does consider to be emancipatory), these new social movements (e.g., ecology, antinuclear) are seen as essentially *defensive* in character.[9] While acknowledging the ecological and bureaucratic problems identified by these movements, Habermas regards their proposals to develop counterinstitutions and "liberated areas" from *within the life-world* as essentially unrealistic. What is required, he has argued, are "technical and economic solutions that must, in turn, be planned globally and implemented by administrative means."[10] Yet as Anthony Giddens has pointedly observed, if the pathologies of advanced industrialism are the result of the triumph of purposive rationality, how can the life-world be defended against the encroachments of bureaucratic and economic steering mechanisms without transforming those very mechanisms?[11] In defending the revolutionary potential of new social movements, Murray Bookchin has accused Habermas of intellectualizing new social movements, "to a point where they are simply incoherent, indeed, atavistic."[12] According to Bookchin, Habermas has no sense of the potentiality of new social movements.

Yet Habermas's general aloofness from the green movement (most notably, its radical ecocentric stream) goes much deeper than this. It may be traced to Habermas's theoretical break with the "negative dialectics" of the early Frankfurt school theorists and with their utopian goal of a "reconciliation with nature." Habermas has argued that such a utopian goal is neither necessary nor desirable for human emancipation. Instead, he has welcomed the rationalization process set in train by the Enlightenment as a *positive* rather than negative development. This chapter will be primarily concerned to locate this theoretical break and outline the broad contours of the subsequent development of Habermas's social and political theory in order to identify what I take to be the major theoretical stumbling blocks in Habermas's oeuvre. This will help to explain, on the one hand, why Habermas regards the radical ecology movement as defensive and "neo-romantic" and, on the other hand, why ecocentric theorists would regard many of Habermas's theoretical categories as unnecessarily rigid and anthropocentric.

In contrast, a central theme of the early Frankfurt school theorists, namely, the hope for a reconciliation of the negative dialectics of enlightenment that would liberate both human and nonhuman nature, speaks directly to ecocentric concerns. While Adorno and Horkheimer were pessimistic as to the prospect of such a reconciliation ever occurring, Marcuse remained hopeful of the possibility that a "new science" might be developed, based on a more expressive and empathic relationship to the nonhuman world. This stands in stark contrast to Habermas's position—that science and technology can know nature only in instrumental terms since that is the only way in which it can be effective in terms of securing our survival as a species. Unlike Habermas, Marcuse believed that a qualitatively different society might produce a qualitatively different science and technology. Ultimately, however, Marcuse's notion of a "new science" remained vague and undeveloped and, in any event, was finally overshadowed—indeed contradicted—by his overriding concern for the emancipation of the human senses and the freeing up of the instinctual drives of the individual. . . . This required nothing short of the total abolition of necessary labor and the rational mastery of nature, a feat that could be achieved *only* by advanced technology and widespread automation.

THE LEGACY OF HORKHEIMER, ADORNO, AND MARCUSE

The contributions of Horkheimer and Adorno in the 1940s, and Marcuse in the 1950s and 1960s, contain a number a theoretical insights that foreshadowed the ecological critique of industrial society that was to develop from the late 1960s.[13] Indeed, these insights might have provided a useful starting point for ecocentric theorists by providing a potential theoretical linkage between the domination of the human and nonhuman worlds. By drawing back from the preoccupation with class conflict as the "motor of history" and examining instead the conflict between humans and the rest of nature, Horkheimer and Adorno developed a critique that sought to transcend the socialist preoccupation with questions concerning the control and distribution of the fruits of the industrial order. In short, they replaced the critique of political economy with a critique of technological civilization. As Martin Jay has observed, they found a conflict whose origins predated capitalism and whose continuation (and probable intensification) appeared likely to survive the demise of capitalism.[14] Domination was recognized as increasingly assuming a range of noneconomic guises, including the subjugation of women and cruelty to animals—matters that had been overlooked by most orthodox Marxists.[15] The Frankfurt school also criticized Marxism for reifying nature as little more than raw material for exploitation, thereby foreshadowing aspects of the more recent ecocentric critique of Marxism. Horkheimer and Adorno argued that this stemmed from the uncritical way in which Marxism had inherited and perpetuated the paradoxes of the Enlightenment tradition—their central target. In this respect, Marxism was regarded as no different from liberal capitalism.

Horkheimer and Adorno's contribution was essentially conducted in the form of a philosophical critique of reason. Their goal was to rescue reason in such a way as to bring instrumental reason under the control of what they referred to as "objective" or "critical" reason. By the latter, Adorno and Horkheimer meant that synthetic faculty of mind that engages in critical reflection and goes beyond mere appearances to a deeper reality in order to reconcile the contradictions between reality and appearance. This was to be contrasted with "instrumental reason," that one-sided faculty of mind that structures the phenomenal world in a commonsensical, functional way and is concerned with efficient and effective adaptation, with means, not ends. The Frankfurt school theorists sought to defend reason from attacks on both sides, that is, from those who reacted against the rigidity of abstract rationalism (e.g., the romanticists) and from those who asserted the epistemological supremacy of the methods of the natural sciences (i.e., the positivists). The task of Critical Theory was to foster a mutual critique and reconciliation of these two forms of reason. In particular, reason was hailed by Marcuse as an essential "critical tribunal" that was the core of any progressive social theory; it lay at the root of Critical Theory's utopian impulse.[16]

According to Horkheimer and Adorno, the Age of Enlightenment had ushered in the progressive replacement of tradition, myth, and superstition with reason, but it did so at a price. The high ideals of that period had become grossly distorted as a result of the ascendancy of instrumental reason over critical reason, a process that Max Weber decried as simultaneously leading to the rationalization *and* disenchantment of the world. The result was an inflated sense of human self-importance and a quest to dominate nature. Horkheimer and Adorno argued that this overemphasis on human self-importance and sovereignty led, paradoxically, to a loss of freedom. This arose, they maintained, because the instrumental manipulation of nature that flowed from the anthropocentric view that humans were the measure of all things and the masters of nature inevitably gave rise to the objectification and manipulation of humans:

> Men pay for the increase in their power with alienation from that over which they exercise their power. Enlightenment behaves toward things as a dictator toward men. He knows them in so far as he can manipulate them. The man of science knows things in so far as he can make them. In this way their potentiality is turned to his own ends. In the metamorphosis the nature of things, as the substratum of domination, is revealed as always the same. This identity constitutes the unity of nature.[17]

The first generation of Critical Theorists also argued that the "rational" domination of outer nature necessitated a similar domination of inner nature by means of the repression and renunciation of the instinctual, aesthetic, and expressive aspects of our being. Indeed, this was seen to give rise to the paradox that lay at the heart of the growth of reason. The attempt to create a free society of autonomous individuals via the domination of outer nature was self-vitiating because this very process also distorted the subjective conditions necessary for the realization of that freedom.[18] The more we seek material expansion in our quest for freedom from traditional and natural constraints, the more we become distorted psychologically as we deny those aspects of our own nature that are incompatible with instrumental reason. As Alford has observed, Horkheimer and Adorno condemned "not merely science but the Western intellectual tradition that understands reason as effective adaptation."[19] Whereas Weber had described the process of rationalization as resulting in the disenchantment of the world, Horkheimer and Adorno described it as resulting in the "revenge of nature." This was reflected in the gradual undermining of our biological support system and, more significantly, in a new kind of repression of the human psyche. Such "psychic repression" was offered as an explanation for the modern individual's blind susceptibility, during times of social and economic crisis, to follow those demagogues (Hitler being the prime example) who offer the alienated individual a sense of meaning and belonging. From a Critical Theory perspective, then, just as the totalitarianism of Nazism was premised on the will to *engineer* social problems out of existence, the bureaucratic state and corporate capitalism may be seen as seeking to *engineer* ecological problems out of existence.

Adorno, Horkheimer, and Marcuse longed for "the resurrection of nature"—a new kind of mediation between society and the natural world. Whitebook has described this resurrection as referring to "the transformation of our relation to and knowledge of nature such that nature would once again be taken as purposeful, meaningful or as possessing value."[20] This did not mean a nostalgic regress into primitive animism or pre-Enlightenment mythologies that sacrificed critical reason—the phenomenon of Nazism demonstrated the dangers of such a simplistic solution. Rather, their utopia required the *integrated* recapture of the past. This involved remembering rather than obliterating the experiences and ways of being of earlier human cultures and realizing that the modern rationalization process and the increasing differentiation of knowledge (particularly the factual, the normative, and the expressive) has been both a learning *and* unlearning process. What was needed, Adorno, Horkheimer, and Marcuse believed, was a new harmonization of our rational faculties and our sensuous nature.[21]

Yet Adorno and Horkheimer recognized that their utopia was very much against the grain of history. Unlike Marx, they stressed the *radical discontinuity* between the march of history and the liberated society they would like to see. As we saw, this sprang from the lack of a revolutionary subject that would be able to usher in the reconciliation of humanity with inner and outer nature. After all, how could there be a revolutionary subject when the individual in mass society had undergone such psychological distortion and was no longer autonomous? Accordingly, they were unable to develop a revolutionary praxis to further their somewhat vague utopian dream. However, they insisted that the utopian impulse that fueled that dream, although never fully realizable, must be maintained as providing an essential source of critical distance that guarded against any passive surrender to the status quo.

Although Marcuse explored the same negative dialectics as Adorno and Horkheimer, he reached a more optimistic conclusion concerning the likelihood of a revolutionary praxis developing. In particular, he saw the counterculture and student movements of the 1960s and early 1970s as developing a more expressive relationship to nature that was cooperative, aesthetic—even erotic. Here, he suggested, were the seeds of a new movement that could expose the ideological functions of instrumental rationality and mount a far-reaching challenge to the "false" needs generated by modern consumer society that had dulled the individual's capacity for critical reflection.[22] Marcuse saw aesthetic needs as subversive force because they enable things to be seen and appreciated *in their own right*.[23] Indeed, he argued that the emancipation of the senses and the release of instinctual needs was a prerequisite to the liberation of nature (both internal and external).

In the case of the former, this meant the liberation of our primary impulses and aesthetic senses. In the case of the latter, it meant the overcoming of our incessant struggle with our environment and the recovery of the "life-enhancing forces in nature, the sensuous aesthetic qualities which are foreign to a life wasted in unending competitive performance."[24]

Marcuse also advanced the provocative argument that this kind of "sensuous perception" might form the epistemological basis of a new science that would overcome the one-dimensionality of instrumental reason that he believed underpinned modern science. Under a new science, Marcuse envisaged that knowledge might become a source of pleasure rather than the means of extending human control. The natural world would be perceived and responded to in an open, more passive, and more receptive way and be guided by the object of study (rather than by human purposes). Such a new science might also reveal previously undisclosed aspects of nature that could inspire and guide human conduct.[25] This was to be contrasted with modern "Galilean" science, which Marcuse saw as "the 'methodology' of a pre-given historical reality within whose universe it moves"; it reflects an interest in experiencing, comprehending, and shaping "the world in terms of calculable, predictable relationships among exactly identifiable units. In this project, universal quantifiability is a prerequisite for the *domination* of nature."[26]

Habermas has taken issue with Marcuse, claiming that it is logically impossible to imagine that a new science could be developed that would overcome the manipulative and domineering attitude toward nature characteristic of modern science.[27] There are certainly passages in Marcuse's *One-Dimensional Man* that suggest that it is the scientific method itself that has ultimately led to the domination of humans and that therefore a change in the very method of scientific inquiry is necessary to usher in a liberated society.[28] Against Habermas's interpretation, however, William Leiss has argued that these are isolated, inconsistent passages that run contrary to the main line of Marcuse's argument, which is that the problem is not with science or instrumental rationality per se but "with the repressive social institutions which exploit the achievements of that rationality to preserve unjust relationships."[29]

Yet these inconsistencies in Marcuse's discussion of the relationship between science and liberation do not appear to be resolvable either way. Indeed, it is possible to discern a third position that lies somewhere between Habermas's and Leiss's interpretations (although it is closer to Leiss's): that the fault lies neither with science nor instrumental rationality per se nor repressive social institutions per se but rather with the instrumental and anthropocentric character of the modern worldview. In *One-Dimensional Man*, Marcuse was concerned to highlight the inextricable interrelationship between science and society. He conceded that pure as distinct from applied science "does not project particular practical goals nor particular forms of domination," but it does proceed in a certain universe of discourse and cannot transcend that discourse.[30] According to Marcuse, scientific rationality was in itself, in its very abstractness and purity, operational in as much as it developed under an instrumental horizon. . . . This interpretation would tie the scientific project (method and theory), *prior* to all application and utilization, to a specific societal project, and would see the tie precisely in the inner form of scientific rationality, i.e., the functional character of its concepts.[31]

It is clear that Marcuse regarded the scientific method as being dependent on a preestablished universe of ends, in which and *for* which it has developed.[32] It follows, as he points out in *Counterrevolution and Revolt*, that:

> A free society may well have a very different a priori and a very different object; the development of the scientific concepts may be grounded in an experience of nature as a totality of life to be protected and "cultivated," and technology would apply this science to the reconstruction of the environment of life.[33]

Marcuse's point is a very general one: that a new or liberatory science can only be inaugurated by a liberatory society. It would be a "new" science because it would serve a new preestablished universe of ends, including a qualitatively new relationship between humans and the rest of nature. This third interpretation is much closer to Leiss's interpretation than Habermas's since it argues that we must reorder social relations before we reorder science if we wish to "resurrect" nature. Only then would we be able to cultivate a liberatory rather than a repressive mastery of nature.

Yet it is important to clarify what Marcuse meant by a "liberatory mastery of nature." As Alford has convincingly shown, Marcuse's new science appears as mere rhetoric when judged against the overall thrust of his writings.[34] As we saw in the previous chapter, Marcuse's principal Marxist reference point was the *Paris Manuscripts*, which Marcuse saw as providing the philosophical grounding for the realization of the emancipation of the senses and the reconciliation of nature. Moreover, his particular Marx/Freud synthesis was concerned to overcome repressive dominance, that is, the repression of the pleasure principle (the gratification of the instincts) by the reality principle (the need to transform and modify nature in order to survive, which is reflected in the work ethic and the growth of instrumental reason). Marcuse saw the reality principle as being culturally specific to an economy of scarcity. In capitalist society, the forces of production had developed to the point where scarcity (which gave rise to the "reality principle") need no longer be a permanent feature of human civilization. That is, the technical and productive apparatus was seen to be capable of meeting basic necessities with minimum toil so that there was no longer any basis for the repression of the instincts via the dominance of the work ethic. The continuance of this ethic must be seen as "surplus repression," which Marcuse maintained was secured, inter alia, by the manipulation of "false" consumer needs.[35] Marcuse ultimately wished to reap the full benefits promised by mainstream science, namely, a world where humans would be spared the drudgery of labor and be free to experience "eros and peace."

However, the necessary quid pro quo for the reassertion of the pleasure principle over the reality principle was that the nonhuman world would continue to be sacrificed in the name of human liberation. Marcuse shared Marx's notion of two mutually exclusive realms of freedom and necessity and, like Marx, he believed that "true freedom" lay beyond the realm of labor. Accordingly, total automation, made possible by scientific and technological progress, was essential on the ground that necessary labor was regarded as inherently unfree and burdensome. It demanded that humans subordinate their desires and expressive instincts to the requirements of the "objective situation" (i.e., economic laws, the market, and the need to make a livelihood. . . . Socialist stewardship under humanist eco-Marxism would usher in a "reconciliation with nature" of a kind that would see to the total domestication or humanization of the nonhuman world. As Malinovich has observed, "For Marcuse the concept of the 'development of human potentiality for its own sake' became *the* ultimate socialist value."[36] In Marcuse's own words, the emancipation of the human senses under a humanistic socialism would enable

> "the human appropriation of nature," i.e., through the transformation of nature into an environment (medium) for the human being as "species being"; free to develop the specifically human faculties: the creative, aesthetic faculties.[37]

Despite his intriguing discussion of the notion of a new, nondomineering science, then, Marcuse's major preoccupation with human self-expression, gratification, and the free play of the senses ultimately overshadowed his concern for the liberation of nonhuman nature. Any nonanthropocentric gloss that Marcuse may have placed on Marx's *Paris Manuscripts* must be read down in this context. Nonetheless, Marcuse's "ecocentric moments" (i.e., his discussion of a qualitatively different science and society that approach the nonhuman world as a partner rather than as an object of manipulation) serve as a useful foil to Habermas's more limited conceptualization of the scientific project.

Notes

1. The Frankfurt school was founded in 1923 as an independently endowed institute for the exploration of social phenomena. For a historical overview, see Martin Jay, *The Dialectical Imagination: A History of the Frankfurt School and the Institute of Social Research 1923–1970* (Boston: Little, Brown, 1973). [For a definition and discussion of ecocentrism, see Robyn Eckersley, *Environmentalism and Political Theory* (Albany: State University of New York Press, 1992), 47 and chap. 3, "Ecocentrism Explained and Defended."]

2. Friedrich Engels, "Socialism: Utopian and Scientific," in *The Marx-Engels Reader*, ed. Robert C. Tucker (New York: Norton, 1972), 638.

3. By "life-world" Habermas means "the taken-for-granted universe of daily social activity." Anthony Giddens, "Reason Without Revolution? Habermas's *Theories des kommunikativen Handelns*," in *Habermas and Modernity*, ed. Richard J. Bernstein (Cambridge, England: Polity, 1985), 101.

4. See, for example, Werner Hülsberg, *The German Greens: A Social and Political Profile* (London: Verso, 1988), 8–9 and John Ely, "Marxism and Green Politics in West Germany," *Thesis Eleven*, 1, no. 13 (1986): 27 and n. 11. It should be noted, however, that the themes of the *early* Frankfurt school theorists (Adorno, Horkheimer, and Marcuse) have had an important influence on the writings of Murray Bookchin, who has been an influential figure in the Green movement in North America. Bookchin was to *invert* the early Frankfurt school's thesis concerning the domination of human and nonhuman nature (see chap. 7).

5. For example, William Leiss, *The Domination of Nature* (Boston: Beacon, 1974); Timothy W. Luke and Stephen K. White, "Critical Theory, the Informational Revolution, and an Ecological Path to Modernity," in *Critical Theory and Public Life*, ed. John Forester (Cambridge: MIT Press, 1985), 22–53; and John Dryzek, *Rational Ecology: Environment and Political Economy* (Oxford: Blackwell, 1987).

6. See, for example, Peter Dews, ed., *Habermas: Autonomy and Solidarity* (London: Verso, 1986), 210.

7. Marcuse saw the ecology and feminist movements in particular as the most promising political movements, and he foreshadowed many of the insights of ecofeminism. For example, in *Counterrevolution and Revolt* (London: Allen Lane, 1972), he argued for the elevation of the "female principle," describing the women's movement as a radical force that was undermining the sphere of aggressive needs, the performance principle, and the social institutions by which these are fostered (75).

8. Jürgen Habermas, "New Social Movements," *Telos* 49 (1981): 35. This article is extracted from the final chapter of Jürgen Habermas, *The Theory of Communicative Action*, vol. 2, *Life-world and System: A Critique of Functionalist Reason*, trans. Thomas McCarthy (Boston: Beacon Press, 1987).

9. Habermas, "New Social Movements," 34.

10. Ibid., 35.

11. See Giddens, "Reason without Revolution?" 121.

12. Murray Bookchin, "Finding the Subject: Notes on Whitebook and Habermas Ltd.," *Telos* 52 (1982): 83.

13. Theodor Adorno and Max Horkheimer, *Dialectic of Enlightenment*, trans. John Cumming (London: Verso, 1979). This work was written during the Second World War and first published in 1944. See also Herbert Marcuse, *Eros and Civilization: A Philosophical Inquiry into Freud* (London: Routledge and Kegan Paul, 1956) and *One-Dimensional Man* (London: Routledge and Kegan Paul, 1964; London: Abacus, 1972).

14. Jay, *Dialectical Imagination*, 256.

15. Ibid., 257 (see Adorno and Horkheimer, *Dialectic of Enlightenment*, 84 and 245–55). Friedrich Engels's discussion of the subjugation of women in *The Origin of the Family, Private Property and the State* (London: Lawrence and Wishart, 1940) is, of course, an important exception.

16. See Martin Jay, "The Frankfurt School and the Genesis of Critical Theory," in *The Unknown Dimension: European Marxism Since Lenin*, ed. Dick Howard and Karl E. Klare (New York: Basic, 1972), 240–41.

17. Adorno and Horkheimer, *Dialectic of Enlightenment*, 9.

18. This theme has also been pursued by Eric Fromm in his *Escape from Freedom* (New York: Holt, Rinehart, and Winston, 1969).
19. C. Fred Alford, *Science and the Revenge of Nature: Marcuse and Habermas* (Gainesville: University Presses of Florida, 1985), 16.
20. Joel Whitebook, "The Problem of Nature in Habermas," *Telos* 40 (1979): 55.
21. Albrecht Wellmer, "Reason, Utopia, and the Dialectic of Enlightenment," *Praxis International* 3 (1983): 91.
22. Marcuse, *One-Dimensional Man.*
23. Marcuse, *Counterrevolution and Revolt,* 74.
24. Ibid., 60.
25. Ibid. Marcuse argued that instead of seeing nature as mere utility, "the emancipated senses, in conjunction with a natural science proceeding on their basis, would guide the 'human appropriation' of nature" (ibid).
26. Marcuse, *One-Dimensional Man,* 133–34.
27. Jürgen Habermas, *Toward a Rational Society: Student Protest, Science, and Politics,* trans. Jeremy J. Shapiro (London: Heinemann Educational Books, 1971), 85–87.
28. For example, Marcuse has stated: "The principles of modern science were *a priori* structured in such a way that they could serve as conceptual instruments for a universe of self-propelling, productive control; theoretical operationalism came to correspond to practical operationalism. The scientific method [which] led to the ever-more-effective domination of nature thus came to provide the pure concepts as well as the instrumentalities for the ever-more-effective domination of man by man *through* the domination of nature" (*One-Dimensional Man,* 130). And later: "The point which I am trying to make is that science, *by virtue of its own method* and concepts, has projected and promoted a universe in which the domination of nature has remained linked to the domination of man—a link which tends to be fatal to the universe as a whole" (ibid., 136).
29. William Leiss, "Technological Rationality: Marcuse and His Critics," *Philosophy of the Social Sciences* 2 (1972): 34–35. This essay also appears as an appendix to Leiss, *Domination of Nature,* 199–212.
30. Marcuse, *One-Dimensional Man,* 129.
31. Ibid., 129 and 131.
32. Ibid., 137.
33. Marcuse, *Counterrevolution and Revolt,* 61.
34. Alford, *Science and the Revenge of Nature,* 49–68.
35. Herbert Marcuse, *Eros and Civilization,* esp. 35, 37, and 87–88.
36. Myriam Miedzian Malinovich, "On Herbert Marcuse and the Concept of Psychological Freedom," *Social Research* 49 (1982): 164.
37. Marcuse, *Counterrevolution and Revolt,* 64.

PART II

Environmental Economics and Politics

Economics and the
Environment

ALAN MILLER

THE PRESENT ENVIRONMENTAL CRISIS has two distinct geograph-
ical focuses. The primary characteristics of the ecological disruptions in the
Northern Hemisphere are the pollution problems arising from rapid and
often uncontrolled industrial development. Disruptions in the Southern
Hemispheric countries, while sometimes including pollution of the land,
air, and water, tend to be focused on the issues of economic underdevelopment
and its corollary—a consequent lack of consumer demand, food and hunger
problems, and rapid population growth.[1] This chapter indicates how these
problems are interconnected and then suggests a factor of common origin
for both sets of ecological disturbance.

First, imagine a map of the world at the end of World War II. Northern
countries (U.S.S.R., Europe, North America, Japan) are in white; southern
continents (Africa, Asia, and Latin America) are in gray. Most of the nations
in this shaded area were traditionally characterized by relationships of
dependency (for example, colonies or dependent economies) with the primary
industrial powers. Were we then to take four additional transparency maps,
which similarly indicated by darker coloration the extent of poverty, hunger,
population-growth rates, and economic stagnation, the overlay of coloration
would almost identically match the map of the colonial empires. This basic
object lesson in political geography highlights the fact that there is a causal
relationship between the particular forms of environmental disturbance in
the First (overdeveloped) World and in the Third (underdeveloped) World.
The causal link in the equation appears to have been the operation of the
world market or free enterprise system.

From: A Planet to Choose: Value Studies in Political Ecology (New York: Pilgrim Press,
1978), 23–35, 49.

The capitalist mode of production provides the key to understanding the nature of the existing global environmental crisis. It is the mode of production that sets the context within which a nation's social structure and its external relationships with other countries develop. Economic relations thus establish the limits on the possible ways in which certain elements of a society can develop. The mode of production also fosters conditions that further the development of certain social outcomes rather than others.

> Karl Polanyi argues that capitalist institutions (he prefers the term "market economy") organize production by linking all of the elements of the production process—including man himself—together by means of markets. The work process, rather than being determined on the basis of tradition, force, or collective decision, gets determined according to profitability or market criteria. . . .
> Under capitalism the economic system begins to dominate the rest of society rather than being submerged in it. *Homo economicus* (economic man) faces the market alone, deprived of the counteracting supportive tradition of community and kinship ties. The possibility of individual starvation amid group plenty replaced the previous assurance of an individual's "just" share of the community's collective material resources.[2]

It is in the dependency relationships that exist between the rich nations and the poor nations (and similarly between the rich and the poor classes within each developed and underdeveloped nation) that we can find the root of most of today's environmental problems.

From the dawn of civilization human beings have been engaged in the production and distribution of essential goods and services. It has been the manner of this production—by the individual or the guild, for private profit or public good—that has determined the relative success or failure of a particular group or nation. More than any other single factor these material conditions of production have determined the patterns of justice or injustice within societies and have similarly been the principal regulators of environmental quality.

One of the basic keys to power within the human productive enterprise has been the ability of an economic unit to use production to develop an economic surplus. The development of such a surplus—a return from production greater than the cost of the labor and the materials involved—has been the primary determinant of a nation's ability to promote the growth of its internal productive enterprise, to provide work for its people, and to make advances in its external power relative to other nations.

Economic evolution, however, has always been a seesaw operation, with one group or nation prospering at the cost of a restriction in the economic development of another. Parallel and equal economic growth between nations—and between groups within nations—has been rare.

Thus, it is important to analyze the basic forms of surplus accumulation and to note their impact on environmental quality. Historically, five primary forms of accumulation have been used by economic systems and/ or nation states to develop the surplus required for internal industrial development.

SURPLUS ACCUMULATION

SURPLUS VALUE FROM LABOR: Economic enterprises are always more or less efficient. Under capitalist economic orders, production systems are inefficient when the goods produced have a market value less than or equal to the cost of the labor and the means of production involved.[3] In either of these situations no surplus is obtained, and thus no capital is accumulated for further investment in plant and equipment.[4] These conditions are typical of the overproduction phase of the classic capitalist economic cycle. Typically, a "crisis" is produced—a recession or depression where internal demand is inadequate to consume the goods and the services.

Thus, to accumulate capital it is necessary for the value of the goods produced to be higher than the cost of the labor, the materials, and the equipment invested in production. When a worker is paid one hundred dollars to produce two hundred dollars worth of goods and the cost of materials and equipment is fifty dollars, then capital is theoretically accumulated in the amount of fifty dollars for the individual or the group owning or controlling the means of production.

Economic systems have been notoriously rapacious in exploiting this surplus value potential of labor. Lowering wages and/or increasing productivity have been traditional means of accelerating accumulation. Child labor, the sixteen-hour working day and the piecework wage of capitalist nations during the Industrial Revolution in Europe, and the "primitive socialist accumulation" of the Soviet Union from 1921 through the end of World War II are all classic illustrations of the use of labor to produce extra value.[5]

Most socialist economists make a strong case for some form of a "labor theory of value," that is, that ultimately *all* value derives from the productivity of human labor. Many other factors may intervene in the process of accumulating surplus, but when everything goes back to its source, labor stands (according to this theory) as the bedrock of capital formation. At any rate, all social and economic systems have understood that labor is, at least, a primary means of surplus accumulation.

PRIMITIVE ACCUMULATION: Economic systems backed by a strong military arm, at certain stages of their economic development, have made maximum use of primitive accumulation—the outright theft or exploitation of the wealth of another country. During the fifteenth and sixteenth centuries, for example, the Portuguese and the Spaniards, in order to develop "free" capital for the development of their own and other European economies, devised a simple form of acquiring surplus. They sent in soldiers, settlers, and priests to steal the gold and the silver from the mines of Latin America for transshipment to their native country. The Dutch followed much the same practice in exploiting the riches of Southeast Asia.

> In the middle ages a small bag of pepper was worth more than a man's life, but gold and silver were the keys used by the Renaissance to open the doors of paradise in heaven and of capitalist mercantilism on earth. The epic of the Spaniards and Portuguese in America combined propagation of the Christian faith with usurpation and plunder of native wealth.[6]

Historically, for those with the power to enforce primitive accumulation, it was easier to extract the wealth of others than to go through the routine but laborious process of the more traditional forms of capital acquisition. Over the years thousands of tons of gold and of silver were seized by European nations to fuel their own internal development. The impact of this form of accumulation on people and on the environment was great. Karl Marx stated:

> The discovery of gold and silver in America, the extirpation, enslavement and entombment in mines of the aboriginal population, the beginning of the conquest and looting of the East Indies, the turning of Africa into a warren for the commercial hunting of black skins, signalized the rosy dawn of the era of capitalist production. These idyllic proceedings are the chief momenta of primitive accumulation.[7]

ACCUMULATION THROUGH THE DEVELOPMENT OF EXTERNAL COLONIES: Hard on the heels of primitive accumulation was the development of overseas colonies by the wealthy maritime nations. Spain and Portugal in Africa and in Latin America, Britain in North and South America and in South Asia, France and Germany in Africa and in the Pacific, the Dutch in Southeast Asia—all had a single intention: the exploitation of the people and the natural resources of the poorer countries in order to develop surplus capital for domestic advancements.

In its first phase colonial surplus accumulation usually took two forms. Indigenous labor was used to provide cheap raw material for shipment to the manufacturing countries. Finished goods, ordinarily high-priced luxury items for consumption by the native bourgeoisie, were returned to the colony for purchase. From the fifteenth to the twentieth centuries every other continent came under the political economic hegemony of European nations. It was not until the end of World War II, in 1945, and the emergence of a new power balance in the world that colonial peoples were finally able to begin the process of throwing off the chains of oppression.

For many the victory was short lived. To replace traditional colonial rule, effected by arms and religion, came a new era of neocolonialism. The wealth exploited earlier from these emerging Third-World nations became the new means of their continued subjugation. Economic investments and foreign-aid programs became as effective as guns and Bibles in maintaining the political and economic superiority of the developed countries. For every dollar invested after World War II in one of the new Third-World nations by private foreign concerns, two to three dollars in earnings were extracted and returned to the home country.[8]

The old colonial powers of imperial Europe were soon of secondary importance, however, to the new economic colossus of the United States. Although more rational and certainly more sophisticated, economic colonialism was as heavy a burden to the poor as its more primitive forerunner. The chains of dependence, now reforged by the merger of American and European capital, continued virtually as strong and as intact as ever. Indeed, the gap in wealth between the developed and the developing countries has grown faster since World War II than during any previous era of formal colonial repression. Environmental destruction has accelerated along with it.

ACCUMULATION THROUGH THE DEVELOPMENT OF INTERNAL COLONIES: Slavery had its origins in the early needs of dominant groups and nations to accumulate surplus wealth. It is a variation of primitive accumulation based on the acquisition of surplus value from labor. Slave camps and wage-rate factories have certain common historical roots.

To own slaves was a simple form of rapid capital accumulation, since the total productivity of the slave or indentured servant—minus purchase price and cost of maintenance—accrued to the owner. When brought into the home country, slave populations became internal colonies, that is, groupings whose exploited skills and labor directly benefited the internal economy of the nation. Africans who were brought into the United States served essentially this purpose.

During the twentieth century the process of internal colonization became more sophisticated in the United States. With the abolition of formal slavery, blacks, Asians, and, later, Hispanic Americans became a reserve work force to be used by American industry to provide cheap labor and to maintain pressure on other immigrant workers to keep wage demands low. Even now, blacks and Hispanic Americans provide this basic function in the society: assisting the capital accumulation process of American industry. Thus, persons of color in the United States often legitimately consider themselves to be members of the Third World. The prime distinction is that they have been colonized internally rather than in the classic external fashion of earlier centuries.[9]

ACCUMULATION THROUGH TRADE SURPLUS: The nation that sells more abroad than it purchases ends up with a balance of payments surplus—a trade balance. Traditionally, trade balances have been the province of the developed nations, which, through colonial enterprises and manufacturing and marketing skills, have been able to buy raw materials at minimum prices and sell manufactured goods at market prices. Such surpluses are then reinvested in the internal economy (sometimes in the political or economic colony) to gain even more rapid capital formation.

Although trade surpluses are still a reality primarily within the developed world, certain new producer cartels (most notably the Organization of

Petroleum Exporting Countries—OPEC) have begun to show substantial trade surpluses.

The evolution of such trade balances within the less-developed nations has only occurred through a unilateral increase in the price of the basic export commodity and not through any pattern of overall economic growth. Only when such surpluses are reinvested to stimulate internal demand can a nation begin the more important process of fueling social and economic development plans.

Since the less developed countries are, as a rule, only primary product exporters, it is unusual for such a grouping to have any major impact on the international commodity markets. Thus far, petroleum is the exception proving the basic rule. However, as the rich nations become more and more dependent on the developing nations for minerals and for other raw materials to enable them to continue their rapid economic growth programs, substantial pressure to disrupt the economic hegemony of the industrial powers may be brought to bear by poorer countries.

UNEQUAL, PERIPHERAL, ECONOMIC DEVELOPMENT[10]

Although many developing nations are seriously studying the economic models of Cuba and of the People's Republic of China for insights that may be helpful in their own programs and for clues to the containment of industrialization's pollution, the Third World continues to pattern most of its development theory on the First (industrialized) rather than the Second (socialist) World. The reason for this is clear: the Third World has traditionally played a key role in the functioning of the world capitalist market economy.

With brutal efficiency the Industrial Revolution integrated most of the poorer countries into the global market system of capitalism. As noted earlier, this process of economic integration carried a high price for these dependent countries. The primary cost of surplus accumulation was borne by Asians, by Africans, and by Latin Americans. And although the market system today provides most of the foreign exchange earnings of the developing nations, it is important to understand how this process has led to economic dependency (including its inevitable consequence—industrial stagnation) and to environmental destruction.

Those nations able to develop an economic surplus underwent significant industrial (and oftentimes political) revolution. They became the *center* economies of the world, with the capacity to exchange manufactured goods for inexpensive raw materials produced by cheap labor. It was the *peripheral* economies, supplying labor, primary commodities, and markets, that fueled the rapid growth of the developed countries. It was this growing relationship of dependency that made it virtually impossible for real economic growth to develop on the periphery.

The key strategic role of capital accumulation in this process is apparent. Since the completion of the first industrial revolutions in Europe (c. 1825–50) and in America (c. 1850–1900), no nation outside the socialist bloc has been able to conclude its own unassisted industrial revolution.[11] The basic dependency linkages established during the periods of mercantilist and early capitalist development have been maintained to the present with almost no exceptions. With the Industrial Revolution in the West, the massive accumulation of economic surplus, and the internal industrial development of Europe and of North America, the economic spheres of the center and the periphery became concretized.

The peripheral connection (that obtaining for most of the Third World) is a relationship between the sector of primary export goods and of luxury import goods. In all peripheral economies basic national income and foreign exchange earnings come from the export of low-cost primary goods (rubber, jute, coffee, tobacco, sugar, minerals, other raw materials), with the surplus expended on the importation of "hardware" (often military equipment) and high-priced luxury goods for the small national ruling class.

There is little domestic production of goods and services needed for the people because of the stagnation of internal development. While the economy has produced a surplus, most of it has either been extracted by the related center economy or has been used up by the tiny national bourgeoisie. Wages must be kept low so that the primary export goods will have low prices and ready markets. Low wages mean that the internal demand for goods (except for the wealthy class) is stifled. Without internal demand the economy stagnates even more. Macroenvironmental problems—hunger, rapid population growth, misuse of land and resources—inevitably follow. Within the center economies, however, a vast array of mass consumer goods are produced for domestic consumption and for export abroad. In addition, there is a primary emphasis on the manufacture of capital goods (the machines that make the machines). Thus, jobs are provided, new economic surplus is developed, and a relatively constant (although uneven—due to the recurrent cycles of overproduction and underconsumption) internal demand for goods and for services is stimulated.

For the center economies labor is a cost and a benefit. Wages must be paid for labor, but it is these wages that generate internal demand for the products that keep industry moving. In the peripheral economies labor is largely a cost and is of only slight benefit. Since there is little basis for the creation of internal demand (few jobs, low wages, minimal industrial plant), labor can never really stimulate the overall economy. The small amount of surplus that is retained within the peripheral economy goes to support the ruling elite and its projects. In some Third-World countries the top 5 percent of the population consumes up to 40 percent of the national income, while the top 20 percent take 60 percent—sometimes more—of the total.[12] The masses become increasingly marginalized—under- and unemployed, more and more impoverished, and of only slight benefit to the economy apart from their ability to provide cheap labor.

This is therefore the framework for the *essential* theory of *unequal exchange*. The products exported by the periphery are important to the extent that . . . the return to labor will be less than what it is at the center. And it can be less to the extent that it will, by every means—economic and non-economic, be made subject to this new function, i.e., providing cheap labor to the export sector. . . . Once society is subjected to this new function—becoming in this sense dependent—it loses its traditional character since it is not the function of real, traditional societies (i.e., pre-capitalist) to supply cheap labor for capitalism.[13]

Thus, the vicious cycle of dependency continues for Third-World economies under the integrated world market system. The implications for environmental value reformulation are immense. As mentioned earlier, the great ecological problems in the less-developed world are hunger and starvation, rapid population growth, poverty, and dependency. If these problems are, in fact, largely the result of traditional colonialism and of unequal, peripheral development, then some reconstitution of the global market economy will be required to solve the environmental dilemma.

Virtually no possibility for a resolution of these problems exists under the present systems. Economic independence requires internal demand stimulation and a form of industrial development. Central to this process would be the initial erection of high tariff barriers to keep out manufactured goods from the industrial nations, thus stimulating local industry.

The mythologies of free trade inevitably work to the detriment of the nation seeking internal development. During much of the eighteenth and the nineteenth centuries in Latin America (the period of British economic hegemony), the absence of import duties—a basic rule of free trade—meant that the mills of Manchester provided the clothing for much of the South American population. As a consequence, native industry stagnated and internal demand was stifled. Free trade normally means an open door for the products of the already industrialized nations and unemployment for those dislocated from work at home. A first step in the American Revolution, in 1776, was the erection of high import duties in order to keep British and other foreign goods out and thus stimulate the American industrial sector.

Whether or not such a course of action could be peacefully effected today is doubtful. Developing nations must interrupt the historic patterns of dependency and peripherality. However, the military might of the industrial powers has traditionally been felt by developing nations seeking independence. In the last decade examples abound of the U.S. attempt to maintain economic hegemony over nations seeking an end to dependency relationships (Guatemala, Vietnam, Cuba, the Dominican Republic, Chile).

Also, the cost of servicing foreign debt (the debt-service ratio) has become an almost intolerable burden for those countries that are trying to acquire a greater degree of freedom. It has been estimated that by 2000 the cost of interest on foreign debt may equal the total export earnings of some Third-World countries. Hence, internal structural change within the economies of the peripheral nations seems to be the sine qua non of development. . . .

What is certain is that neither the pollution problems of the overdeveloped world nor the macroenvironmental issues facing the economically peripheral nations can be resolved apart from massive structural change in the primary global economic systems.

Notes

1. This general description is not intended to minimize the fact that there is also widespread hunger and poverty in the northern industrial nations and increasing industrial pollution in the Southern Hemisphere, due in large part to the export of the most polluting industries from the developed to the less-developed world.
2. Michael Reich, Richard Edwards, and Thomas Weisskopf, eds., *The Capitalist System* (Englewood Cliffs, N.J.: Prentice-Hall, 1972), 92.
3. "Means of production" means simply the land, raw materials, machines, factories, and other goods used in production.
4. To a large degree the same factors apply in socialist production. One essential difference is whether the surplus ends up in private hands or in the public treasury. Also, under socialism, central planning allows deficit surplus operations, for example, public services, to be carried by more value productive sectors. In capitalist economies this is done more fitfully by government intervention to soak up some of the surplus through taxation programs in order to cover needed public services.
5. See the economic works of Eugeny A. Preobrazhensky on the theory of primitive socialist accumulation. Although he is considered by many to be a Trotskyist, his work provided the theoretical framework of Stalinist surplus accumulation from c. 1925–54. Note especially *The New Economics*, trans. Brian Pearce (New York: Oxford University Press, 1965).
6. Eduardo Galeano, *Open Veins of Latin America: Five Centuries of the Pillage of a Continent* (New York: Monthly Review Press, 1973), 24.
7. Karl Marx, *Capital* (New York: Vintage, 1977), pt. 8, chap. 31, 914–30.
8. Thomas Weisskopf, "U.S. Foreign Private Investment: An Empirical Survey" *Monthly Review*, June 1970, passim.
9. The U.S. Department of Labor has estimated black unemployment in the last three years at varying rates between 15 and 20 percent. This figures does not include those persons who have given up the possibility of ever finding employment. In 1975, 48.6 percent of black teenagers were reported to be unemployed.
10. The theory of unequal, peripheral, economic development has been developed and described in the following works. My analysis is based on these viewpoints: Samir Amin, "Accumulation and Development: A Theoretical Model," *Review of African Political Economy*, vol. 1 (Aug.–Nov. 1974); and Khieu Samphan in Samir Amin, *Accumulation on a World Scale*, trans. Brian Pearce (New York: Monthly Review Press, 1974). I am also indebted to Alain DeJanvry, professor of agricultural economics at the University of California, Berkeley, for insights into the theory of unequal development.
11. Following the Meiji restoration in the nineteenth century, Japan did effect a form of an industrial revolution. This buildup in the economy and industry,

however, was made possible only by the cooperation of the major industrial nations. Since Japan was not a resource-rich nation, it was not to the benefit of the industrial nations to keep it underdeveloped.

12. Weisskopf, "U.S. Foreign Private Investment."
13. Amin, "Accumulation and Development."

Poverty and Population

BARRY COMMONER

THE WORLD POPULATION PROBLEM is a bewildering mixture of the simple and the complex, the clear and the confused.

What is relatively simple and clear is that the population of the world is getting larger, and that this process cannot go on indefinitely because there are, after all, limits to the resources, such as food, that are needed to sustain human life. Like all living things, people have an inherent tendency to multiply geometrically—that is, the more people there are the more people they tend to produce. In contrast, the supply of food rises more slowly, for unlike people it does not increase in proportion to the existing rate of food production. This is, of course, the familiar Malthusian relationship and leads to the conclusion that the population is certain eventually to outgrow the food supply (and other needed resources), leading to famine and mass death unless some other countervailing force intervenes to limit population growth. One can argue about the details, but taken as a general summary of the population problem, the foregoing statement is one which no environmentalist can successfully dispute.

When we turn from merely stating the problem to analyzing and attempting to solve it, the issue becomes much more complex. The simple statement that there is a limit to the growth of the human population, imposed on it by the inherent limits of the earth's resources, is a useful but abstract idea. In order to reduce it to the level of reality in which the problem must be solved, what is required is that we find the *cause* of the discrepancy between population growth and the available resources. Current views on this question are neither simple nor unanimous.

One view is that the cause of the population problem is uncontrolled fertility, the countervailing force—the death rate—having been weakened

---

From: "How Poverty Breeds Overpopulation (and Not the Other Way Around)," *Ramparts* (1974): 21–25, 58–59, excerpts; revised 1990.

by medical advances. According to this view, given the freedom to do so people will inevitably produce children faster than the goods needed to support them. It follows, then, that the birthrate must be deliberately reduced to the point of "zero population growth."

The methods that have been proposed to achieve this kind of direct reduction in birthrate vary considerably. Among the ones advanced in the past are: (1) providing people with effective contraception and access to abortion facilities and with education about the value of using them (i.e., family planning); (2) enforcing legal means to prevent couples from producing more than some standard number of children ("coercion"); (3) withholding of food from the people of starving developing countries which, having failed to limit their birthrate sufficiently, are deemed to be too far gone or too unworthy to be saved (the so-called "lifeboat ethic")....

The author of the "lifeboat ethic" is Garrett Hardin, who stated ... that

> so long as nations multiply at different rates, survival requires that we adopt the ethic of the lifeboat. A lifeboat can hold only so many people. There are more than two billion wretched people in the world—ten times as many as in the United States. It is literally beyond our ability to save them all..: Both international granaries and lax immigration policies must be rejected if we are to save something for our grandchildren.[1]

But there is another view of population which is much more complex. It is based on the evidence, amassed by demographers, that the birthrate is not only affected by biological factors, such as fertility and contraception, but by equally powerful *social* factors.

Demographers have delineated a complex network of interactions among these social factors. This shows that population growth is not the consequence of a simple arithmetic relationship between birthrate and death rate. Instead, there are circular relationships in which, as in an ecological cycle, every step is connected to several others.

Thus, while a reduced death rate does, of course, increase the rate of population growth, it can also have the opposite effect—since families usually respond to a reduced rate of infant mortality by opting for fewer children. This negative feedback modulates the effect of a decreased death rate on population size. Similarly, although a rising population increases the demand on resources and thereby worsens the population problem, it also stimulates economic activity. This, in turn, improves educational levels. As a result the average age at marriage tends to increase, culminating in a reduced birthrate, which mitigates the pressure on resources.

In these processes, there is a powerful social force which, paradoxically, both reduces the death rate (and thereby stimulates population growth) and also leads people voluntarily to restrain the production of children (and thereby reduces population growth). That force, simply stated, is the quality of life—a high standard of living, a sense of well-being and of security in the future. When and how the two opposite effects of this force are felt differs with the stages in a country's economic development. In a premodern society, such as England before the Industrial Revolution or India before the advent of the English, both death rates and birthrates were high. But they were in balance and population size was stable. Then, as agricultural and industrial production began to increase and living conditions improved, the death rate began to fall. With the birthrate remaining high the population rapidly increased in size. However, later, as living standards continued to improve, the decline in death rate persisted but the birthrate began to decline as well, reducing the rate of population growth.[2]

For example, at around 1800, Sweden had a high birthrate (about 33/1,000), but since the death rate was equally high, the population was in balance. Then as agriculture and, later, industrial production advanced, the death rate dropped until, by the mid-nineteenth century, it stood at about 20/1,000. Since the birthrate remained constant during that period of time, there was a large excess of births over deaths and the population increased rapidly. Then, however, the birthrate began to drop, gradually narrowing the gap until in the mid-twentieth century it reached about 14/1,000, when the death rate was about 10/1,000.[3] Thus, under the influence of a constantly rising standard of living the population moved, with time, from a position of balance *at a high death rate* to a new position of near-balance *at a low death rate*. But in between the population increased considerably.

This process, the *demographic transition*, is clearly characteristic of all Western countries. In most of them, the birthrate does not begin to fall appreciably until the death rate is reduced below about 20/1,000. However, then the drop in birthrate is rapid.

In the [developed] countries, the death rate began to decline in the mid-eighteenth century, reaching an average of 30/1,000 in 1850, 24/1,000 in 1900, 16/1,000 in 1950, and 9/1,000 in 1985. In contrast, the birthrate remained constant at about 40/1,000 until 1850, then dropping rapidly, reaching 32/1,000 in 1900, 23/1,000 in 1950, and 14/1,000 in 1985. As a result, populations grew considerably, especially in the nineteenth century, then slowed to the present net rate of growth of 0.4 percent per year.

In developing countries, the average death rate was more or less constant, at about 38/1,000 until 1850, then declining to 33/1,000 in 1900, 23/1,000 in 1950, and 10/1,000 in 1985. The average birthrate, on the other hand, remained at a constant high level, 43/1,000, until about 1925; it has since declined at an increasing rate, reaching 37/1,000 in 1950, and 30/1,000 in 1985. . . . Thus, in developing countries the progressively rapid drop in birthrate will accelerate progress toward populations that, like those of developed countries, are approximately in balance.[4] . . .

A drop in infant mortality from about 80/1,000 to 25/1,000 is a critical turning point which can lead to a very rapid decline in birthrate in response to reduced infant mortality. The latter, in turn, is always very responsive to improved living conditions, especially with respect to nutrition. Consequently, there is a kind of critical standard of living which, if achieved, can lead to a rapid reduction in birthrate and an approach to a balanced population.

Thus, in human societies, there is a built-in control on population size: If the standard of living, which initiates the rise in population, *continues* to increase, the population eventually begins to level off. This self-regulating process begins with a population in balance, but at a high death rate and low standard of living. It then progresses toward a population which is larger, but once more in balance, at a low death rate and a high standard of living.

The chief reason for the rapid rise in population in developing countries is that this basic condition has not been met. The explanation is a fact about developing countries which is often forgotten—that they were recently, and in the economic sense often still remain, colonies of more developed countries. In the colonial period, Western nations introduced improved living conditions (roads, communications, engineering, agricultural and medical services) as part of their campaign to increase the labor force needed to exploit the colony's natural resources. This increase in living standards initiated the first phase of the demographic transition.

But most of the resultant wealth did not remain in the colony. As a result, the second (or population-balancing) phase of the demographic transition could not take place. Instead, the wealth produced in the colony was largely diverted to the advanced nation—where it helped *that* country achieve for itself the second phase of the demographic transition. Thus colonialism involves a kind of demographic parasitism: The second population-balancing phase of the demographic transition in the advanced country is fed by the suppression of that same phase in the colony.

It has long been known that the accelerating curve of wealth and power of Western Europe, and later of the United States and Japan, has been heavily based on exploitation of resources taken from the less powerful nations: colonies, whether governed legally, or—as in the case of the U.S. control of certain Latin American countries—by extralegal and economic means. The result has been a grossly inequitable rate of development among the nations of the world. As the wealth of the exploited nations was diverted to the more powerful ones, their power, and with it their capacity to exploit, increased. The gap between the wealth of nations grew, as the rich were fed by the poor.

What is evident from the above considerations is that this process of international exploitation has had another very powerful but unanticipated effect: rapid growth of the population in the former colonies. An analysis by the demographer Nathan Keyfitz leads him to conclude that the growth of industrial capitalism in the Western nations in the period 1800–1950 resulted in the development of a one-billion excess in the world population, largely in the tropics. Thus the present world population crisis—the rapid growth of population in developing countries (the former colonies)—is the result not so much of policies promulgated by these countries but of a policy, colonial exploitation, forced on them by developed countries.[5]

Given this background, what can be said about the various alternative methods of achieving a balanced world population? In India, there has been an interesting, if partially inadvertent, comparative test of two of the possible approaches: family planning programs and efforts (also on a family basis) to elevate the living standard. The results of this test show that while the family planning effort itself failed to reduce the birthrate, improved living standards succeeded.

In 1954, a Harvard team undertook the first major field study of birth control in India. The population of a number of test villages was provided with contraceptives and suitable educational programs; birthrates, death rates, and health status in this population were compared with the comparable values in an equivalent population in control villages. The study covered the six-year period 1954–60.

A follow-up in 1969 showed that the study was a failure. Although in the test population the crude birthrate dropped from 40/1,000 in 1957 to 35/1,000 in 1968, a similar reduction also occurred in the control population. The birth-control effort had no measurable effect on birthrate.

We now know *why* the study failed, thanks to a remarkable book by Mahmood Mamdani (*The Myth of Population Control*). He investigated in detail the impact of the study on one of the test villages, Manupur. What Mamdani discovered is a total confirmation of the view that population control in a country like India depends on the economically motivated desire to limit fertility. Talking with the Manupur villagers he discovered why, despite the study's statistics regarding ready "acceptance" of the offered contraceptives, the birthrate was not affected:

> One such "acceptance" case was Asa Singh, a sometime land laborer who is now a watchman at the village high school. I questioned him as to whether he used the tablets or not: "Certainly I did. You can read it in their books—From 1957 to 1960, I never failed." Asa Singh, however, had a son who had been born sometime in late 1958 or 1959. At our third meeting I pointed this out to him . . . Finally he looked at me and responded, "Babuu someday you'll understand. It is sometimes better to lie. It stops you from hurting people, does no harm, and might even help them." The next day Asa Singh took me to a friend's house . . . and I saw small rectangular boxes and bottles, one piled on top of the other, all arranged as a tiny sculpture in a corner of the room. This man had made a sculpture of birth-control devices. Asa Singh said: "Most of us threw the tablets away. But my brother here, he makes use of everything."[6]

Such stories have been reported before and are often taken to indicate how much "ignorance" has to be overcome before birth control can be effective in countries like India. But Mamdani takes us much further into the problem, by finding out why the villagers preferred not to use the contraceptives. In one interview after another he discovered a simple, decisive factor that in order to advance their economic condition, to take advantage of the opportunities newly created by the development of independent India, *children were essential.* Mamdani makes this very explicit:

> To begin with, most families have either little or no savings, and they can earn too little to be able to finance the education of *any* children, even through high school. Another source of income must be found, and the only solution is, as one tailor told me, "to have enough children so that there are at least three or four sons in the family." Then each son can finish high school by spending part of the afternoon working... After high school, one son is sent on to college while the others work to save and pay the necessary fees... Once his education is completed, he will use his increased earnings to put his brother through college. He will not marry until the second brother has finished his college education and can carry the burden of educating the third brother... What is of interest is that, as the Khanna Study pointed out, it was the rise in the age of marriage—from 17.5 years in 1956 to 20 in 1969—and not the birth-control program that was responsible for the decrease in the birthrate in the village from 40 per 1,000 in 1957 to 35 per 1,000 in 1968. While the birth control program was a failure, the net result of the technological and social change in Manupur was to bring down the birthrate.[7]

Here, then, in the simple realities of the village of Manupur, are the principles of the demographic transition at work. There *is* a way to control the rapid growth of populations in developing countries. It is to help them develop—and more rapidly achieve the level of welfare that everywhere in the world is the real motivation for a balanced population.

Against this success, the proponents of the "lifeboat ethic" would argue that it is too slow, and they would take steps to *force* developing nations to reduce their birthrate even though the incentive for reduced fertility—the standard of living and its most meaningful index, infant mortality—is still far inferior to the levels which have motivated the demographic transition in the Western countries. And where, in their view, it is too late to save a poor, overpopulated country the proponents of this so-called ethic would withdraw support (in the manner of the hopelessly wounded in military "triage") and allow it to perish.

This argument is based (at least in the realm of logic) on the view, to quote Hardin, that "it is literally beyond our ability to save them all." Hardin's assertion, if not the resulting "ethic," reflects a commonly held view that there is simply insufficient food and other resources in the world to support the present world population at the standard of living required to motivate the demographic transition. It is commonly pointed out, for example, that the United States consumes about one-third of the world's resources to support only 6 percent of the world's population, the inference being that there are simply not enough resources in the world to permit the rest of the world to achieve the standard of living and low birthrate characteristic of the United States.

The fault in this reasoning is readily apparent if one examines the actual relationship between the birthrates and living standards of different countries. The only available comparative measure of standard of living is GNP per capita. Neglecting for a moment the faults inherent in GNP as a measure of the quality of life, a plot of birthrate against GNP per capita is very revealing. The poorest countries (GNP per capita less than five hundred dollars per year) have the highest birthrates, [32–55] per 1,000 population per year. When GNP per capita per year [reaches] four to five thousand dollars, the birthrate drops sharply, reaching about [15–19]/1,000. . . . What this means is that in order to bring the birthrates of the poor countries down to the low levels characteristic of the rich ones, the poor countries do not need to become as affluent (at least as measured . . . by GNP per capita) as the United States [which had a GNP per capita of $15,541 in 1984]. Achieving a per capita GNP only, let us say, one-third of that of the United States . . ., these countries could, according to the above relationship, reach birthrates almost as low as that of the European and North American countries. . . .[8]

In a sense the demographic transition is a means of translating the availability of a decent level of resources, especially food, into a voluntary reduction in birthrate. It is a striking fact that the efficiency with which such resources can be converted into a reduced birthrate is much higher in the developing countries than in the advanced ones. . . . The per capita cost of bringing the standard of living of poor countries with rapidly growing populations to the level which—based on the behavior of peoples all over the world—would motivate voluntary reduction of fertility is very small, compared to the per capita wealth of developed countries. . . .

My own purely personal conclusion is, like all of these, not scientific but political, that the world population crisis, which is the ultimate outcome of the exploitation of poor nations by rich ones, ought to be remedied by returning to the poor countries enough of the wealth taken from them to give their peoples both the reason and the resources voluntarily to limit their own fertility.

In sum, I believe that if the root cause of the world population crisis is poverty, then to end it we must abolish poverty. And if the cause of poverty is the grossly unequal distribution of world's wealth, then to end poverty, and with it the population crisis, we must redistribute that wealth, among nations and within them.

Notes

1. Garrett Hardin, "Lifeboat Ethics," *Bioscience* 24 (1974): 561.
2. E. A. Wrigley, *Population and History* (New York: McGraw-Hill, 1969).
3. D. J. Bogue, *Principles of Demography* (New York: Wiley 1969), 59.
4. Barry Commoner, *Making Peace with the Planet* (New York: Pantheon, 1990), 158–59. Data from Nathan Keyfitz, "The Growing Human Population," *Scientific American* 261, no. 3 (Sept. 1989): 118–26.
5. Nathan Keyfitz, "National Populations and the Technological Watershed," *Journal of Social Issues* 23 (1967): 62–78.
6. Mahmood Mamdani, *The Myth of Population Control* (New York: Monthly Review Press, 1972), 32.
7. Mamdani, *Myth of Population Control*, 115–16, 117–18.
8. R. L. Sivard, *World Military and Social Expenditures 1987–1988*, 12th ed. (Washington, D.C.: World Priorities, 1987); World Population Prospects: Estimates and Projections as Assessed in 1984, United Nations, New York, 1986.

Steady-State Economics

HERMAN DALY

A STEADY-STATE ECONOMY is a necessary and desirable future state of affairs and ... its attainment requires quite major changes in values, as well as radical, but nonrevolutionary, institutional reforms. Once we have replaced the basic premise of "more is better" with the much sounder axiom that "enough is best," the social and technical problems of moving to a steady state become solvable, perhaps even trivial. But *unless* the underlying growth paradigm and its supporting values are altered, all the technical prowess and manipulative cleverness in the world will not solve our problems and, in fact, will make them worse.

The recognition that there are problems of political economy that have no technical solution but do have a moral solution goes very much against the grain of modern economic theory. Yet economics began as a branch of moral philosophy, and the ethical content was at least as important as the analytic content up through the writings of Alfred Marshall.[1] From then on, the structure of economic theory became more and more top-heavy with analysis. Layer upon layer of abstruse mathematical models were erected higher and higher above the shallow concrete foundation of fact. The behavior of a peasant selling a cow was analyzed in terms of the calculus of variations and Lagrangian multipliers. From the angelic perspective of hyperplanes cavorting in n-space, economists overlooked some critical biophysical and moral facts. The biophysical facts have asserted themselves in the form of increasing ecological scarcity: depletion, pollution, and ecological disruption. The moral facts are asserting themselves in the form of increasing existential scarcity: anomie, injustice, stress, alienation, apathy, and crime. . . .

In the face of these now undeniable facts, modern economic thought

From: *Steady-State Economics* (San Francisco: W. H. Freeman, 1977), 2–18, excerpts.

cuts its losses in two ways: (1) It argues that the newly revealed dimension of ecological scarcity simply requires more clever technology and more growth, albeit growth of a slightly different kind. (2) It argues that existential scarcity (resulting from a shortage of whatever does in fact make people whole, well, and happy) is simply not real. This point has been well discussed by Walter Weisskopf.[2] Whatever the public chooses is assumed to be in the public interest, and there is no distinction between what people of the present age of advertising *think* will make them whole and happy and what would *in fact* make them so.

It is not easy (beyond the level of basic necessities) to make factual statements about what is good for people, but it is rash to assume that no such statements are possible—that all of ethics can be reduced to the level of personal tastes and that the community is nothing but an aggregate of isolated individuals.

The attraction of these simple, and I believe quite erroneous, assumptions is that by emasculating the concepts of ecological and existential scarcity, the orthodox economic growth paradigm covers up the weaknesses in its factual foundations and can thus continue building its analytical tower of babel up to a theoretical bliss point.

Only by returning to its moral and biophysical foundations and shoring them up, will economic thinking be able to avoid a permanent commitment to misplaced concreteness and crackpot rigor. Scientistic pretention and blind aping of the mechanistic methods of physics, even after physics has abandoned the mechanistic philosophy, should be replaced by value-based thinking in the mode of classical political economy.[3] Separation of "is" from "ought" is an elementary rule of clear thinking. But this separation belongs within the mind of the individual thinker. It should never have become the basis for division of labor between people and professions, much less an excuse for "running to hide in thickets of Algebra, while abandoning the really tough questions to journalists and politicians".[4] Of all fields of study, economics is the last one that should seek to be "value-free," lest it deserve Oscar Wilde's remark that an economist is a man who knows the price of everything and the value of nothing.

Not all physical scientists have been flattered by the economists' emulation. For example, Norbert Wiener observed:

> The success of mathematical physics led the social scientists to be jealous of its power without quite understanding the intellectual attitudes that had contributed to this power. The use of mathematical formulae had accompanied the development of the natural sciences and become the mode in the social sciences. Just as primitive peoples adopt the Western modes of denationalized clothing and of parliamentarism out of a vague feeling that these magic rites and vestments will at once put them abreast of modern culture and technique, so the economists have developed the habit of dressing up their rather imprecise ideas in the language of the

infinitesimal calculus. . . . To assign what purports to be precise values to such essentially vague quantities is neither useful nor honest, and any pretense of applying precise formulae to these loosely defined quantities is a sham and a waste of time.[5]

The challenge is to develop a political economics that recognizes both ecological and existential scarcity and develops its propositions at a low to intermediate level of abstraction, understandable by the layman or average citizen, rather than dictated by a priesthood of "technically competent" obscurantists. If economic reality is actually so complex that it can only be described by complicated mathematical models that add epicycles to epicycles and externalities to externalities, then the reality should be simplified. Human institutions should not be allowed to grow beyond the human scale in size and complexity.[6] Otherwise, the economic machine becomes too heavy a burden on the shoulders of the citizen, who must continually grind and regrind himself to fit the imperatives of the overall system, and who becomes ever more vulnerable to the failure of other interdependent pieces that are beyond his control and even beyond awareness.[7] Lack of control by the individual over institutions and technologies that not only affect his life but determine his livelihood is hardly democratic and is, in fact, an excellent training in the acceptance of totalitarianism. . . .

We need to recognize . . . that a U.S.-style high-mass consumption, growth-dominated economy for a world of four billion people is impossible. Even more impossible is the prospect of an ever-growing standard of per capita consumption for an ever-growing world population. The minerals in concentrated deposits in the earth's crust, and the capacity of ecosystems to absorb large quantities or exotic qualities of waste materials and heat, set a limit on the number of person-years that can be lived in the "developed" state, as that term is understood today in the United States. How the limited number of person-years of "developed" living will be apportioned among nations, among social classes, and over generations will be the dominant economic and political issue for the future.[8]

The steady-state economy respects impossibilities and does not foolishly squander resources in vain efforts to overcome them. Our present institutions allow technology to be autonomous and force man to play the accommodating role. The steady-state economy seeks to change institutions in such a way that people become autonomous and technology is not abandoned, but is demoted to its proper accommodating role. Growth economics gave technology free rein. Steady-state economics channels technical progress in the socially benign directions of small scale, decentralization, increased durability of products, and increased long-run efficiency in the use of scarce resources. . . .

Probably the major disservice that experts provide in confronting the problems of mankind is dividing the problem in little pieces and parceling them out to specialists. Food problems belong to agriculture and energy problems to engineering or physics; employment and inflation belong to economics; adaptation belongs to psychologists and genetic engineers; and the "environment" is currently up for grabs by disciplinary imperialists. Although it is undeniable that each specialty has much of importance to say, it is very doubtful that the sum of all these specialized utterances will ever add up to a coherent solution, because the problems are not independent and sequential but highly interrelated and simultaneous. Someone has to look at the whole, even if it means foregoing full knowledge of all of the parts. Since "economics" as well as "ecology" come from the same Greek root (*oikos*), meaning "management of the household," and since man's household has extended to include not only nations but also the planet as a whole, economics is probably the discipline that has least justification for taking a narrow view. Let us take a minute to consider the economy, environmental quality, food, energy, and adaptation as interrelated subtopics within the framework of economics viewed as management of the household of man.

The economy, or household of mankind, consists of two things: the members of the family and their furniture and possessions, or, in purely physical terms, human bodies and physical commodities or artifacts. For the last century or more, the most salient characteristic of the human household has been its enormous quantitative growth. Population has grown at rates vastly in excess of any that have ever prevailed in the entire history of the species. This unprecedented population growth has been accompanied by, and in part made possible by, an even greater rate of increase in the production of artifacts. World population has grown at around 2 percent annually, doubling every thirty-five years, and world consumption has grown at about 4 percent annually, doubling every seventeen or eighteen years. But production and consumption are not the precise words, since man can neither produce nor destroy matter and energy but only transform them from one state to another. Man transforms raw materials into commodities and commodities into garbage. In the process of maintaining ever larger populations of both people and artifacts, the volume of raw materials transformed into commodities and ultimately into garbage has increased greatly. . . .

Furthermore, man cannot convert waste back into raw materials except by expending energy that inevitably degrades into waste heat, which cannot be recycled. Man can let nature recycle some wastes if he is not too impatient and refrains from overloading natural cycles. Recycling is a good idea, but it has limits provided by the second law of thermodynamics, which, in effect, says that energy cannot be recycled and that matter can only be recycled at something less than 100 percent.

Why has the human household grown so rapidly? Basically, because we made it grow. Since procreating is a more popular activity than dying, and is likely to remain so, we eagerly reduce death rates and only half-heartedly talk about reducing birthrates. . . .

Although many question whether further population growth is desirable, very few people question the desirability or possibility of further economic growth. Indeed, economic growth is the most universally accepted goal in the world. Capitalists, communists, fascists, and socialists all want economic growth and strive to maximize it. The system that grows fastest is considered best. The appeals of growth are that it is the basis of national power and that it is an alternative to sharing as a means of combating poverty. It offers the prospect of more for all with sacrifice by none—a prospect that is in conflict with the "impossibility theorem" discussed above. If we are serious about helping the poor, we shall have to face up to the moral issue of redistribution and stop sweeping it under the rug of aggregate growth.

What are the implications of this growth-dominated, imperialistic style of managing the human household for the specific issues of environmental quality, food, energy, and adaptation?

While the human household has been rapidly growing, the environment of which it is a part has steadfastly remained constant in its quantitative dimensions. Its size has not increased, nor have the natural rates of circulation of the basic biogeochemical cycles that man exploits. As more people transform more raw materials per person into commodities, we experience higher rates of depletion; as more people transform more commodities into waste, we experience higher rates of pollution. We devote more effort and resources to mining poorer mineral deposits and to cleaning up increased pollution, and we then count many of these extra expenses as an increase in GNP and congratulate ourselves on the extra growth! The problem with GNP is that it counts consumption of geological capital as current income. . . .[9]

While the growth-induced increases in depletion and pollution have adverse direct effects on the human household that are bad enough (e.g., lead and mercury poisoning, congestion, air and water pollution), they also have indirect effects that are likely to be worse. The indirect effects occur through interferences with natural ecosystems that inhibit their ability to perform the free life-support services that we take for granted.[10] For example, the most important service of all, photosynthesis, may be interfered with by changing the acidity of the soil that supports plant life, a change resulting from acid rains induced by air pollution caused by burning fossil fuels. In addition, the heat balance and temperature gradients of the earth can be changed by air pollution and by intensive local use of energy, with unpredictable effects on climate, rainfall, and agriculture. Deforestation results in the loss of water purification and flood and erosion control services formerly provided gratis by the forests, as well as the loss of wildlife habitats and of a potentially perennial source of timber. Ecologists have convincingly argued that the natural services provided by Louisiana marshlands (a spawning ground for much marine life of the Gulf of Mexico, a natural tertiary sewage treatment plant, a buffer zone for hurricane protection, and a recreation area) are probably much more valuable than the so-called development uses of new residential areas and shopping centers or even oil wells, at least beyond a limited number.[11]

Food, unlike coal or petroleum, is a renewable resource—a means of capturing the continual flow of solar energy. But the necessity to feed a large and growing population at an increasing level of per capita consumption in rich countries like the United States has made agriculture dependent on a continuous subsidy of nonrenewable fossil fuels, chemicals, and mineral fertilizers. . . . As Howard Odum says, industrial man no longer eats potatoes made from solar energy; he now eats potatoes made partly of oil.[12] As the fossil-fuel subsidy becomes scarcer and more expensive, agriculture will have to rely more on solar energy and human labor. It may be that (as is already happening in Brazil) more cropland will be devoted to sugarcane in order to make alcohol to mix with gasoline for fuel—just the reverse of the process of turning petroleum into food that was attracting attention a few years ago!

Growth of the human household within a finite physical environment is eventually bound to result in both a food crisis and an energy crisis and in increasingly severe problems of depletion and pollution. Within the context of overall growth, these problems are fundamentally insoluble, although technological stopgaps and palliatives are possible. Technological adaptation has been the dominant reaction, aided by the information and incentives provided by market prices. We need, however, to shift the emphasis toward ecological adaptation, that is, to accept natural limits to the size and dominion of the human household, to concentrate on moral growth and qualitative improvement rather than on the quantitative imperialist expansion of man's dominion. The human adaptation needed is primarily a change of heart, followed by a shift to an economy that does not depend so much on continuous growth. As Arnold Toynbee put it:

> More and more people are coming to realize that the growth of material wealth which the British industrial revolution set going, and which the modern British-made ideology has presented as being mankind's proper paramount objective, cannot in truth be the wave of the future. Nature is going to compel posterity to revert to a stable state on the material plane and to turn to the realm of the spirit for satisfying man's hunger for infinity.[13] . . .

WHAT IS A STEADY-STATE ECONOMY?

Economic analysis, or any analytic thought for that matter, must begin with what Joseph Schumpeter calls a "preanalytic vision" or what Thomas Kuhn calls a basic "paradigm."[14] Analytic thought carves up this vision into parts and shows the relationship among the parts. If the analytic knife is wielded skillfully, the pieces will be cut cleanly along natural seams rather than torn raggedly, and the relations among the parts will be simple and basic rather than contrived and complex. But prior to analytic thought there must be a basic vision of the shape and nature of the total reality to be analyzed and some feeling for where natural joints and seams lie, and for the way in which the whole to be analyzed fits into the totality of things. Our basic definitions arise out of this preanalytic vision, which limits the style and direction of our thinking.

The vision of the economy from which the steady-state concept arises is that of two physical populations—people and artifacts—existing as elements of a larger natural system. These physical populations have two important aspects. On the one hand, they yield services—artifacts (physical capital) serve human needs, and so do other human beings. The body of a skilled worker or doctor is a physical asset that yields services both to the immediate owner of the body and to others. On the other hand, these populations require maintenance and replacement. People continually get hungry, cold, and wet, and eventually they die. Artifacts wear out and must be replaced. These two populations may be thought of as a fund, like a lake, with an outflow necessitated by death and depreciation, which can be reduced but never eliminated. The outflow is offset by an inflow of births and production which may exceed, fall short of, or equal the outflow. Consequently the fund or lake may grow, decline, or remain constant.

From the physical nature of these populations, several things are apparent. Since, from the first law of thermodynamics, we know that matter-energy can be neither created nor destroyed, it is apparent that the fund of physical and human capital has some important relations with the rest of the world. The rest of the world is a source for its inputs of matter-energy and a sink for its outputs. Everything has to come from somewhere and go somewhere. "Somewhere" is in both cases the natural environment. The larger the lake, the larger must be the outflow, because death and depreciation cannot be reduced beyond some lower limit, and consequently the larger must be the offsetting inflow. If there were no death or depreciation, then our "lake of capital" (to use A. C. Pigou's phrase) would be a closed system rather than an open system and would be limited in its size only by the total amount of water, not by the conditions governing the flow of water through the total natural system. The second law of thermodynamics tells us that death and depreciation cannot be eliminated, so it is clear that our lake must remain an open system if it is to maintain a constant level. If inflow is less than outflow, the lake will eventually disappear; if inflow is greater than outflow, it will eventually contain all the water there is and will not be able to have any more inflow. But the outflow will continue and bring the lake down to some smaller equilibrium size which can be maintained by the natural hydrologic flows.

However, the lake analogy fails in several important aspects. First, the fund of water in the lake is homogeneous, whereas the fund of people and artifacts is highly varied and complex. Second, the water entering the lake is both quantitatively and qualitatively equal to the water flowing out, assuming an equilibrium lake. But while the equilibrium lake of people and artifacts is maintained by an inflow of matter-energy equal in *quantity* to the outflow, the two are very different in *quality*. The matter-energy going in is useful raw material, while that coming out is useless waste. The flowthrough, or throughput of matter-energy that maintains the fund of artifacts and people, is entropic in nature. Low-entropy inputs are imported and high-entropy outputs are exported. The high-entropy output cannot be directly used again as an input for the same reason that organisms cannot eat their own excrement. Although it would appear that the real lake's outflow is qualitatively the same as the inflow, this is not strictly true. The outflow water could not return to the inflow stream without being pumped or without being evaporated and lifted again by the hydrologic cycle powered by the sun. So even the throughput of water that maintains a lake is an entropic flow, although this is obscured by the fact that the water looks the same going in as it does coming out. The matter-energy throughput that maintains the fund of people and artifacts does not even look the same. Anyone can tell the difference between equal quantities of raw materials and waste.

In sum, the vision is that of a physical open system, a fund of service-yielding assets maintained by a throughput that begins with depletion of nature's sources of useful low entropy and ends with the pollution of nature's sinks with high-entropy waste. There are two physical magnitudes, a *stock* of capital (people and artifacts) and a *flow* of throughput. There is one psychic magnitude of service or want satisfaction that is rendered by the stocks and is, of course, their reason to be. Whatever value we attribute to the satisfaction of our wants and needs is imputed to the stocks that satisfy those needs and, in turn, is imputed to the throughput that maintains the stocks.

The important role of the laws of thermodynamics in this vision will be developed later, but for now it is enough to recognize that the entropy law is the basic physical coordinate of scarcity. Were it not for the entropy law, nothing would ever wear out; we could burn the same gallon of gasoline over and over, and our economic system could be closed with respect to the rest of the natural world.

From this general vision we must now distill a precise definition of a steady-state economy. What is it precisely that is not growing, or held in a steady state? Two basic physical magnitudes are to be held constant: the population of human bodies and the population of artifacts (stock of physical wealth). Since artifacts are, in a very real sense, extensions of the human body, the steady-state economy may be thought of as a logical continuation of the demographer's notion of a stationary population to include not only human bodies but also their multifarious physical extensions. What is held constant is capital stock in the broadest physical sense of the term, including capital goods, the total inventory of consumer goods, and the population of human bodies.

Of equal importance is what is *not* held constant. The culture, genetic inheritance, knowledge, goodness, ethical codes, and so forth embodied in human beings are not held constant. Likewise, the embodied technology, the design, and the product mix of the aggregate total stock of artifacts are not held constant. Nor is the current distribution of artifacts among the population taken as constant. Not only is quality free to evolve, but its development is positively encouraged in certain directions. If we use "growth" to mean quantitative change, and "development" to refer to qualitative change, then we may say that a steady-state economy develops but does not grow, just as the planet earth, of which the human economy is a subsystem, develops but does not grow.[15]

The maintenance of constant physical populations of people and artifacts requires births to offset inevitable deaths and new production to offset inevitable physical depreciation. Births should be equal to deaths at low rather than high levels so that life expectancy is long rather than short. Similarly, new production of artifacts should equal depreciation at low levels so that the durability or "longevity" of artifacts is high. New production implies increasing depletion of resources. Depreciation implies the creation of physical waste, which, when returned to the environment, becomes pollution. Depletion and pollution are costs, and naturally they should be minimized for any given level of stocks to be maintained.

Thus we may succinctly define a *steady-state economy* (hereafter abbreviated SSE) as *an economy with constant stocks of people and artifacts, maintained at some desired, sufficient levels by low rates of maintenance "throughput,"* that is, by the lowest feasible flows of matter and energy from the first stage of production (depletion of low-entropy materials from the environment) to the last stage of consumption (pollution of the environment with high-entropy wastes and exotic materials). It should be continually remembered that the SSE is a *physical* concept. If something is nonphysical, then perhaps it can grow forever. If something can grow forever, then certainly it is nonphysical.

How does this physical concept of growth relate to economic growth? As currently measured by real GNP, which is a value *index* of a *physical* flow, economic growth is strictly tied to physical quantities. Even services are always measured as the use of some*thing* or some*body* for a period of time, and these things and persons require physical maintenance; more of them require more physical maintenance. In calculating real GNP, efforts are made to correct for changes in price levels, in relative prices, and in product mix, so as to measure only real change in physical quantities produced. However, the SSE is defined in terms of constant *stocks* (a quantity measured at a point in time, like an inventory), not *flows* (a quantity measured over an interval in time, like annual sales). GNP is a flow and is logically irrelevant to the definition of an SSE. Nevertheless, to the considerable extent that GNP reflects throughput, then a policy of maximizing GNP growth would imply maximizing a cost. The steady-state perspective seeks to maintain a desired level of stocks with a minimum throughput, and if minimizing the throughput implies a reduction in GNP, that is totally acceptable. The steady-state paradigm assumes some *sufficient* level of stocks, an assumption that is absent from the growth paradigm. . . .

Although the idea of a SSE may seem strange to us who have always lived in a growth economy, neither the concept nor the reality is at all novel. John Stuart Mill discussed the notion with compelling clarity over a century ago. And it is instructive to remember that mankind has, for over 99 percent of its tenure on earth, existed in conditions closely approximating a SSE. Only in the last two hundred years has growth been sufficiently rapid to be felt within the span of a single lifetime, and only in the last forty years has it assumed top priority and become truly explosive. In the long run, stability is the norm and growth the aberration. It could not be otherwise.

Notes

1. For example, in the first textbook of political economy, Malthus's *Principles of Political Economy*, we find the following statement: "It has been said, and perhaps with truth that the conclusions of Political Economy partake more of the stricter sciences than those of most of the other branches of human knowledge. . . . There are indeed in Political Economy great general principles . . . [but] we shall be compelled to acknowledge that the science of Political Economy bears a nearer resemblance to the science of morals and politics than to that of mathematics" (T. R. Malthus, *Principles of Political Economy* [London: Kelley, 1820], 1). [Alfred Marshall, *Official Papers* (London: Macmillan, 1926).]
2. Walter A. Weisskopf, *Alienation and Economics* (New York: Dutton, 1971).
3. Nicholas Georgescu-Roegen, *The Entropy Law and the Economic Process* (Cambridge: Harvard University Press, 1971).
4. Joan Robinson, *Economic Philosophy* (London: C. A. Watts, 1962).
5. Norbert Wiener, *God and Golem, Inc.* (Cambridge: MIT Press, 1964), 89.
6. E. F. Schumacher, *Small Is Beautiful: Economics as if People Mattered* (New York: Harper and Row, 1973).
7. Roberto Vacca, *The Coming Dark Age*, trans. by J. S. Whale (Garden City, N.Y.: Doubleday, 1974).
8. Nathan Keyfitz, "Population Theory and Doctrine: A Historical Survey," in *Readings in Population*, ed. William Petersen (New York: Macmillan, 1972).
9. Schumacher, *Small Is Beautiful*.
10. Herman E. Daly, "On Economics as a Life Science," *Journal of Political Economy* (May/June 1968).
11. James Gosselink, Eugene Odum, and R. M. Pope, "The Value of the Tidal Marsh," Work Paper no. 3, Gainesville: Urban and Regional Development Center, University of Florida, 1973.
12. Howard T. Odum, *Environment, Power, and Society* (New York: Wiley, 1971).
13. Arnold Toynbee, *Observer*, 11 June 1972.
14. Joseph Schumpter, *History of Economic Analysis* (New York: Oxford, 1954); Thomas Kuhn, *The Structure of Scientific Revolutions* (Chicago: University of Chicago Press, 1962).

15. The capital stock is an aggregate of unlike things, and to speak of it as constant in the aggregate, yet variable in composition, implies some coefficients of equivalence among the various unlike things. This problem haunts standard economics as well. However, as will be seen later, we do not really need an operational measure of the aggregate stock. We can control throughput and let the stock grow to whatever maximum size can be supported by the limited throughput. Control over aggregate throughput will result from controls (depletion quotas) on particular resources. If, thanks to technological progress, it becomes possible to support a larger stock with the same throughput, that is all to the good and should be allowed to happen. Eventually diminishing returns requires that the mix of artifacts will shift more toward producer's capital and away from consumer's capital, and a given gross throughput will contain an ever smaller net amount of usable matter-energy.

Human Needs and Social Change

LEN DOYAL AND IAN GOUGH

BASIC INDIVIDUAL AND SOCIETAL NEEDS

THERE CAN BE NO question that, on the face of it, relativists who believe in the essential subjectivity of human need have a case. People do have strong feelings about what they need and these feelings do vary enormously between cultures and over time. But if any distinction is coherently to be drawn between wants and needs, a nonsubjective demarcation criterion must be found. . . .

Analyses of need are integrally related to ideas about what is necessary for persons to be capable of any successful action—irrespective of the aim of the action or the perceptions or culture of the actor.[1] "Action" here is more than "behavior"—it involves purpose, and understanding an action requires knowledge of the person's reason for doing it. It is this which distinguishes human beings from animals.

While it may be true that all human goals are specific to particular cultures, in order to achieve any of these goals people have to act. It follows that there are certain preconditions for such actions to be undertaken—people must have the mental ability to deliberate and to choose, and the physical capacity to follow through on their decisions. To put this another way, in order to act successfully, people need physically to survive and need enough sense of their own identity or autonomy to initiate actions on the basis of their deliberations. *Survival* and *autonomy* are therefore basic needs: they are both conceptual and empirical preconditions for the achievement of other goals.

From: Paul Ekins, ed., *The Living Economy: A New Economics in the Making* (New York: Routledge, 1986), 69–72, 78–79.

107

Really, it is health rather than survival—both physical and mental health—which is the most basic human need and the one which it is in the interests of individuals to satisfy before any others. As far as physical health is concerned, it must be recognized that there is a point beyond which the capacity for successful action is so reduced that the actor will be regarded by others as abnormal—however the abnormality may be culturally regarded. This notion of a base is also applicable to mental ill-health, although here the deciding factor is the incapacity to perform some tasks culturally regarded as normal without the excuse of a physical handicap, old age, or poor physical health. Since the capacity for successful action will be proportional to the satisfaction of health needs of both kinds, they can be ranked. Thus in all cultures, people will be regarded as more or less "healthy"—whatever the culture under consideration and however "health" is conceptualized.

The second set of basic needs which must be met for actions to be successful relates to individual identity or autonomy—the private and public sense of "self." In order for actions to be identified as such, they must be initiated by individual people. "Initiate" is a crucial word here, because a person who initiates an action is presumed to do so in a fundamentally different way from that in which a machine (e.g., a robot) does. The machine is understood with reference to its mechanism. But people consist of more than the relationships between their bodily components. They acquire their autonomy as individuals by being able to formulate aims and beliefs and to put them into practice. Hence, all things being equal, we hold individuals practically and morally responsible for their actions but we do not do the same for machines.

For the need for autonomy to be met in practice, teachers will be required who already possess their own autonomous identities and the physical and mental skills which go with them. People do not teach themselves to act, they have to learn;[2] and this is a further individual need which goes hand in hand with autonomy. Which skills are learned will vary from culture to culture, but they are not totally variable; there is a universal empirical base line defining what individuals must do to meet their needs for health and autonomy. For example, children have to learn to interact socially in minimally acceptable ways, irrespective of the specific cultural rules they follow in the process. Persistently lying to or punching one's fellows will never be a recipe for successful social interaction. Similarly, in all cultures, language skills are necessary as the medium through which people learn conceptually to order their world and to deliberate about what to do in it. In this sense, consciousness is essentially social, a byproduct of educational interaction with others. It is not, as the song goes, that "people who need people are the luckiest people in the world." Everyone needs people to be anyone.

Thus far in our consideration of individual needs, we have assumed that it was possible for health and autonomy through education to be achieved. However, since learning is a social process, it necessarily involves individuals interacting in social groups. This means that certain prerequisites must be met for any such groups or for a society as a whole to function with any degree of long-term success. Individuals therefore have basic *societal needs*— those social preconditions for the achievement of the individual needs we have just described. These can be analyzed in four categories.

The first concerns *material production*. In all cultures, it is necessary somehow to create the food, drink, clothing, and shelter required for a minimal level of health to be achieved. Such activities constitute the material base of the culture. Of course, they will vary between cultures, especially when these are radically different. Material production is essentially social. Humans are not genetically endowed with the physiology or mentality to enable them to survive, much less prosper, on their own. Group interaction in the form of a division of labor is thus required, with the character and complexity of the rules differing with specific cultures and types of production.

The second requirement for the survival of any culture is *successful repro— duction*—the process "which begins with ovulation and ends when the child is no longer dependent on others for necessities of survival."[3] This includes two separate elements: biological reproduction, and infant care and socialization. Historically, the mechanism for fulfilling both types of reproductive need has been some form of family structure, though with a wide variation of kinship patterns. To say this, is not, of course, to endorse any particular set of social relations. It is simply to underline the societal need for certain types of technical and educational practices, for example midwifery and the learning of basic principles of infant care.

The third requirement for the success of any social system concerns *communication*. People are not born with an understanding of theories and practices relevant to the modes of production and reproduction in their society. Nor will they necessarily be predisposed to accept the particular system of allocation and distribution which typifies it. As we have seen, such understanding and acceptance has to be learned. This will in turn be based on an already existing culture: a body of rules which partly defines the form of life of the society involved. Such rules enable social goals to be achieved because through their linguistic expression they constitute the medium by which aims and beliefs are individually and collectively acted upon. The resulting communication obviously takes many forms, including the sharing of practical skills and the legitimation of modes of exchange and distribution.

Finally, it is clear that the mere existence of sets of rules—of a culture in our sense—will not guarantee their perpetuation or implementation. There must be some system of *political authority*, ultimately backed up by sanctions, which will ensure that they will be taught, learned, and correctly followed. The exact character of such authority will vary enormously, depending on the size, complexity, and equity of different societies. Yet, however centralized or dispersed the authority may be, it must be effective in its own terms if the society in question is to survive, let alone flourish. One does not have to believe in Hobbes's "war of all against all" to accept this. It is a consequence of the importance which, as we have shown, attaches to rules as the social cement which ensures the possibility of social, and therefore of individual, survival.

In general then, these four societal needs—production, reproduction, cultural communication, and political authority—constitute the structural properties which any minimally successful mode of social organization must embody. The degree to which individual needs are met will depend in principle on such success, which will in turn depend on individuals who are healthy, autonomous, and educated enough to know what is expected of them, how to do it and the implications of not doing it. This interdependence between individual and societal needs should make it clear that we are not adopting the sort of abstract individualism which, for example, is so often attributed to utilitarian writers. Yet, it should be equally clear that we do not accept forms of functionalism which presuppose that individuals simply mirror the structural properties of their social environment. *The only criterion for evaluating social systems which we are advocating thus far is: How far do they enable individual basic needs to be met?* We are conceptualizing basic individual needs in a way that is independent of any particular social environment. . . .

OPTIMIZING BASIC NEED SATISFACTION AND HUMAN LIBERATION

. . . Ultimately human needs have to be defined as those levels of health and autonomy which are available to all people in all societies, given present and probable future levels of resources. Thus it is important to stress that the logic of the politics of human need is irrepressibly global in scope: no group of humans can have their basic needs excluded from consideration. Moreover, nothing can justify the lack of equity with which these resources are at present consumed. It follows that, other things being equal, global redistribution should have priority over national redistribution. Finally, ecological considerations dictate that the need-satisfaction of future individuals needs to be taken into account in the present allocation of resources.

This being said—and we recognize its utopian ring—three . . . points are in order. First, ecological strategies for conserving and protecting the environment from needless waste and pollution may have such immediate and dramatic consequences for basic individual needs that they will quickly become incorporated into debates about wider political issues. This is beginning to happen, for example in the struggles concerning the toxic consequences of many industrial processes. Perhaps nothing more dramatically illustrates the distinction between needs and wants, and the ways in which this distinction can be molded to serve arbitrary interests, than public demand for commodities which are known to be manufactured in ways which pollute the environment. Second, the conflicts of economic and other interests which plague attempts to protect the environment on a national level are exponentially compounded when one moves to the international arena. As a result most ecological struggles are at present restricted to those societies where the political process, at least in principle, gives them some chance of making an impact, and where a minimal level of basic need-satisfaction has already been achieved. Yet this situation runs the severe risk that the environmental needs of the developed countries could well be met at the expense of the basic needs of the rest of the world. The third point is that, if they are to be successful, international strategies for satisfying basic needs, such as those recently suggested by the ILO and the Brandt Report,[4] must conform to the same communicational and constitutional norms which we have argued are necessary for the success of national strategies. In particular, enormous constraints would have to be placed on the vested interests of the developed nations and of powerful groups within them. The political possibility of achieving such constraints and the circumstances under which it could be done is another open question. Indeed, at the moment, any effective answer seems almost inconceivable; but what is even more inconceivable is to stop trying to reach one—to relinquish the goal of human liberation on a global scale.

Many of the detailed implications for policy of these points still need to be elaborated. But it is clear that the Baconian vision of nature as an unending storehouse and unfillable cistern is over. The new reality is that awareness of the delicacy of the biosphere must now go hand in hand with democratically planned production for human need.

Notes

1. J. Galtung, "The Basic Needs Approach," in *Human Needs,* ed. K. Lederer (Cambridge, Mass.: Oelschlager, Gunn, and Hain, 1980); R. Plant, H. Lesser, and P. Taylor-Gooby, *Political Philosophy and Social Welfare* (London: Routledge and Kegan Paul, 1980).
2. When we refer to "teachers" or "education," we are using the terms in their broadest sense, not just referring to formal schooling.
3. M. O'Brien, *The Politics of Reproduction* (London: Routledge and Kegan Paul, 1981).
4. Cf. International Labour Organisation, "Conclusions of the 1976 World Employment Conference," in *Meeting Basic Needs: Strategies for Eradicating Mass Poverty and Unemployment* (Geneva: ILO, 1977) and the report of the Brandt Commission, *North South—A Programme for Survival* (London: Pan Books, 1980).

Creating a Green Future

BRIAN TOKAR

THE WORLD HAS CHANGED dramatically in the decade since the emergence of the international green movement. In Europe, some of the greens' most far-reaching demands have begun to be partially realized: the dissolution of military blocs; the withdrawal of nuclear weapons; the inclusion of ecological considerations in economic planning in some countries. But the future remains ominously uncertain. The economic unification of Europe may cement the foundations of peace, but also threatens to concentrate economic power in ever fewer hands and threaten the integrity of local cultures. The end of state censorship in the East revealed an ecological nightmare which will only worsen if Western corporations are allowed a free hand in reshaping Eastern Europe's economies. Global issues like the greenhouse effect, deforestation, and the thinning of the ozone layer affirm the need for new economic structures and new ways of thinking about human civilization's relationship to nature.

The green vision of a peaceful world moving beyond industrialism is now more compelling and more urgent than ever before. While many in Eastern Europe are rushing as fast as they can to import the structures and the values of the West, the breathtaking pace of change in the East has combined with growing awareness of the global ecological crisis to inspire a new blossoming of alternative currents. The 1989 revolutions in Eastern Europe offer a powerful example of the possibility of rapid and sweeping political change and the irresistible human impulse for freedom and liberation. The specter of impending ecological collapse adds a renewed sense of urgency to the call for fundamental change.

Across Europe and around the world, greens are in the forefront of evolving a vision of a different kind of society. Despite tremendous internal struggles

From: *The Green Alternative: Creating an Ecological Future*, rev. ed. (1987; San Pedro: R. & E. Miles, 1993), 141–48.

and sometimes debilitating compromises, greens remain the most articulate political opposition and the leading symbol of a different kind of politics. In many countries, they have tapped a deep-lying discontent with the present state of affairs that reaches across the conventional political spectrum. From Scandinavia to the shores of the Mediterranean, from the British Isles to the Danube and eastward, people are envisioning and creating the foundations for a green Europe, free of the pressures of either corporate or military domination.

From its European origins, green politics has spread throughout the world. Australia and New Zealand, each with their own growing ecological awareness and a heritage of political independence, have become centers of resistance to nuclear technology in all its forms. Australian greens have been at the forefront of efforts to protect rain forests and agricultural soils, combat ocean pollution and develop alternative economic policies. In Japan, the world's leading symbol of technological achievement, a green movement has developed, with ties to a blossoming cooperative movement and a strong awareness of the links between political and personal changes. Greens are appearing in the Third World, too—in Brazil, several long-time activists, including some prominent former guerrillas and labor activists, have formed a green party dedicated to preserving the rain forests and the rights of indigenous peoples, as well as nonviolence, racial and sexual equality, and political independence.

Can it happen here? Can a new ecological politics take hold in the United States, where the frontier myth of prosperity at the expense of nature reigns supreme in our history? The opinion shapers in the mass media tell us that the age of discontent that began in the 1960s is over, that the world is going "America's" way, that the era of complacency, materialism, and an acceptance of selfish, competitive values is here to stay. People are no longer interested in changing the world, we are told, just in getting ahead and making a comfortable life for themselves, whatever the consequences.

Here, too, a growing ecological awareness is beginning to change all that. Polls show three-quarters of the American public placing a higher priority upon the protection of the natural environment than almost any other concern. People have come to see that economic trends, military defense, and other such national obsessions are meaningless if we cannot breathe the air, drink the water, count upon the safety of our food, or trust the integrity of the countless other species with which we share this earth. People of all walks of life are participating in Earth Day celebrations, changing their buying habits, and pressing for changes in policies and institutions in the hope of bringing our society in line with a more respectful attitude toward the planet and our fellow inhabitants upon it.

Despite a growing unity around environmental concerns, however, our society remains deeply polarized. Behind a veil of prosperity and economic vitality, new economic pressures are afflicting both the urban and rural poor. The affluent rush to gentrify inner-city neighborhoods and buy new vacation homes in the countryside, while increasing numbers of long-time farm families and middle-class homeowners are facing foreclosure. Consumerism appears to be on the rise, while more Americans than ever before are going hungry and millions are abandoned to live on the streets.

While some people are seeking a gentler way of life and a healed relationship with the earth, many of their peers are finding newly fashionable rationalizations for joining the culture of exploitation and greed, sometimes with a veneer of an "enlightened" or even "environmental" form of consumerism. An ecological way of life is advertised as yet another commodity to be bought. People confronted with their own isolation in an increasingly insecure and atomized society have abandoned earlier dreams of community to pursue their own personal fortunes. This cult of selfishness and defense of the status quo that reached its apex during the 1980s flourishes right alongside the mistrust of powerful institutions that came to the fore during the 1960s.

It is a time of searching and a time of vision, however. A new spiritual yearning has reached segments of our society that seemed unaffected by the changes of the sixties and seventies. Many people are seeking a new ethical underpinning for their lives. Some seek the comforts of traditional religion, and our country's churches and synagogues are increasingly reflective of the moral tensions of our time. From the rise of liberation theologies and the nationwide sanctuary movement to protect Central American war refugees, to the politics of fear spread through the fundamentalist churches, religion and politics are no longer as easily separable as we used to think.

In this highly charged climate, conventional politics are increasingly suspect. Traditional liberalism is largely discredited, as the moral fervor it seemed to command in the early 1960s has given way to a managerial politics that seeks shallow, bureaucratic solutions to deep-lying social problems. People display a profound boredom and frustration with the methods and institutions of conventional politics. Conservatives claimed a moral upper hand through the Reagan years, but even a peek below the surface of modern conservatism reveals its subservience to corporate power and its profound disrespect for individual and community rights. A moral vacuum has developed, one which the ideologists and television preachers of the far Right sought to fill with their own manipulative calls to community and their overflowing confidence in the repressive values they call "American."

The ecological crisis and the shredding of the social safety net continue to remind us of the urgency of the situation. In the face of unprecedented threats to our species' survival, it is too late to expect a solution to emerge from anywhere on the conventional political spectrum. It has become necessary for peace-loving and ecologically minded people to articulate not just a new politics, but a new ethics and a new earth-centered moral sensibility that can reawaken the life-affirming impulses our society seeks to submerge.

A new ecological sensibility is emerging in a variety of ways, from many sectors of our society. The preceding pages have touched on just a few of these. Bioregionalists, ecofeminists, community activists, spiritual healers, earth-guided poets, and curious people of all walks of life are raising hope for the evolution of new attitudes and new ways of relating to the earth and to our fellow beings.

For all their talk about community, freedom and "traditional" values, the Right's political agenda seeks to limit individual rights and police people's private lives, while political constraints on corporations are increasingly relaxed. Economic activity, for them, should be free to expand without restriction, but the rights of people to live their own lives, and of communities to protect themselves from the system's abuses, are increasingly restricted from above.

The green movement represents an active political expression of these values. We have seen how greens are beginning to fit the pieces together, discovering how ecological principles can guide us on a path toward the necessary social transformation. But many questions remain. How can the various ecological currents evolve together to create a dynamic whole that is greater than the sum of its individual parts? How does a green outlook change the ways people work for social change? How can local autonomy be preserved and strengthened as green values are articulated and expressed on a regional and national scale? When is it appropriate for greens to participate in conventional politics? What is the relationship among personal, social, and political changes?

Greens in the United States are working today in many different spheres and using many different approaches. Some people focus their attention on specific issues of local and regional concern. They are working to curb the excesses of the corporate economy and to head off the social and ecological disruptions inherent in our present way of life. This current is basically oppositional in nature, embracing political methods such as community organizing, referendum campaigns, legal interventions, and direct nonviolent actions to protest or obstruct especially threatening policies and projects.

A second current is more reconstructive in its approach. It includes a wide variety of efforts to create living alternatives to our present ways—a wealth of experiments in cooperation and local democracy, both in the community and the work place. This includes developing alternative technologies, founding cooperatives of all kinds, and raising bioregional awareness. For a green movement to develop and grow in North America, there will need to be a merging of oppositional and reconstructive strategies that allows these two currents to support and strengthen each other.

Issue-oriented politics without an alternative vision can be politically limiting and personally frustrating. Many people are uncomfortable with the way things are, but are not motivated to act on their beliefs because they see no other way. Others might choose to work on a particular issue of concern, but are easily exhausted as each small victory reveals new complications.

One might work for many months to block a particularly devastating project or to achieve a particular reform in the system, only to find that new injustices crept in the back door while attention was focused upon one small piece of the problem. The ecological crisis cannot be simply controlled within the limits of the existing system. In fact, some greens believe that reformist efforts merely forestall the impending collapse of the industrial economies, a collapse which may need to occur before the real work of reconstruction can begin.

It can be equally limiting to work to create new institutions without actively seeking to understand and oppose the injustice of our present ways. Such efforts can be slowly bought off and accommodated into the service of the present system. One can point to food co-ops that have become more involved in elaborately marketing their goods than in providing an alternative to the existing food supply system. The alternative energy movement in New England became tied to the ecologically devastating vacation-home industry in the 1980s, as solar builders drifted toward affluent resort areas in their search for steady employment and the freedom to experiment. Should healthy food and solar-heated homes become the luxury goods of an affluent minority seeking to purchase an ecological "life-style"? How can a green sensibility guide us toward a better way?

The greens have borrowed a phrase (originally attributed to the ecologist René Dubos) that has become a slogan for the worldwide peace movement: "Think globally; act locally." Local ecological problems, local symbols of the military-industrial complex, and local attempts to create alternatives in housing, food distribution, and other basic needs all offer a focus for local activities that carry a global message. By working primarily on the local level, greens are demonstrating the power of people to really change things and creating the grass-roots basis for a real change in consciousness.

In local issue-oriented work, a green sensibility offers new opportunities to link together specific issues. For example, local weapons industries fuel the arms race, distort employment patterns in a region, and often produce the largest quantities of toxic wastes. Campaigns aimed toward converting such facilities to peaceful and ecologically sound uses can raise questions of community and worker control and the nature of economic and technological change. Large development projects often destroy prime farmland, wildlife habitats, and people's homes, while centralizing the control of a local economy in fewer hands. A green anti-development campaign can fully integrate these closely related concerns. By articulating the interconnections among issues, exploring the ecological dimensions of community struggles, and upholding a holistic social and ecological vision, a green approach can help transcend the limitations of traditional issue-oriented politics.

Traditionally, coalitions of political groups are very narrowly defined. Several organizations come together to deal with a very limited set of common concerns and the coalition's definition is confined to a lowest common denominator of what different groups can agree upon. Green politics suggest a different kind of coalition, in which people are united by a larger vision of an ecologically transformed society. The Ozark Community Congresses and other bioregional gatherings are examples of this new kind of coalition-building. In San Diego in the 1970s, a coalition of publicly supported educational and social service organizations created a visionary Community Congress to pool organizational resources and promote the greater empowerment of people in a community setting. The Earth Day direct action on Wall Street in 1990 brought together seventy different groups from around the country—antitoxics groups, peace and antinuclear organizations, students, gays and lesbians, bioregionalists, anarchists, surrealists, and many others—united around a green vision of a cooperative, ecological way of life.

In the environmental movement, a green outlook offers a new grass-roots focus and more democratic organizational models. As the large national environmental groups become increasingly focused on events and personalities in Washington, D.C., new kinds of groups have emerged to fill in the gaps. The national movement against hazardous waste and toxic chemical pollution is one example of a different kind of approach. Antitoxics activists have devoted their energies to helping people form local groups to confront local problems. National organizations gather and publish information, offer technical help, teach organizing skills, and engage in lobbying at the federal level. Their lobbying efforts, however, are sustained by hundreds of large and small citizens' groups organized and ready to take action to protect their communities. Through this work, the people most directly facing the hazards of an economy based on poison are beginning to raise fundamental questions about the nature of the economy and its institutions.

Another new model for environmental politics is offered by the Earth First! movement. Begun as a loose network of people committed to environmental direct action in the Southwest, Earth First! has grown to include over fifty affiliated groups all across the continent. Earth First! gained its notoriety through its advocacy of sabotage, or "monkeywrenching," to prevent the destruction of wilderness, but its focus has broadened to encompass a wide variety of approaches, from guerrilla theater in the National Parks to nonviolent blockades of logging and construction sites. They have developed detailed plans for the reclamation and expansion of wilderness areas threatened by development, carrying their plans to the appropriate officials with the backing of national letter-writing campaigns, as well as sit-ins, office takeovers and the direct obstruction of logging and road-building operations.

Earth First! groups are not tied to any formal organizational structure, but keep in close contact through the mails and through their newspaper, which has become a major national forum for the renewal of ecological activism and thought. They have also helped publicize action campaigns to save wilderness in other parts of the world. In recent years, people have put their bodies on the line to hold back bulldozers and chain saws in the Australian rain forest, on the last remaining wild stretch of the Danube River in Austria, and on Indian lands in Ontario and British Columbia.

A green perspective can also change the way people work to create alternatives to our present way of life. A green vision encourages strong ties between alternative efforts in different spheres, alliances between consumer and producer co-ops, and the adoption of more democratic forms of organization. Rather than simply providing goods and services that may or may not be available through conventional channels, green efforts to create working alternatives can help people to begin actively withdrawing from the system that oppresses us all. At a time when the dominant system is proving increasingly unable to meet people's basic needs, a green movement can help put the search for entirely new ways of living and working together—a search grounded in a traditional closeness to the land—back on the social agenda.

PART III

Deep, Social, and Socialist Ecology

Deep Ecology

ARNE NAESS

Ecologically responsible policies are concerned only in part with pollution and resource depletion. There are deeper concerns which touch upon principles of diversity, complexity, autonomy, decentralization, symbiosis, egalitarianism, and classlessness.

THE EMERGENCE OF ECOLOGISTS from their former relative obscurity marks a turning-point in our scientific communities. But their message is twisted and misused. A shallow, but presently rather powerful movement, and a deep, but less-influential movement, compete for our attention. I shall make an effort to characterize the two.

1. THE SHALLOW ECOLOGY MOVEMENT:

Fight against pollution and resource depletion. Central objective: the health and affluence of people in the developed countries.

2. THE DEEP ECOLOGY MOVEMENT:

1. Rejection of the man-in-environment image in favor of *the relational, total-field image*. Organisms as knots in the biospherical net or field of intrinsic relations. An intrinsic relation between two things A and B is such that the relation belongs to the definitions or basic constitutions of A and B, so that without the relation, A and B are no longer the same things. The total-field model dissolves not only the man-in-environment concept,

From: "The Shallow and the Deep, Long-Range Ecology Movement: A Summary," *Inquiry* 16 (1973): 95–100.

but every compact thing-in-milieu concept—except when talking at a superficial or preliminary level of communication.

2. *Biospherical egalitarianism*—in principle. The "in principle" clause is inserted because any realistic praxis necessitates some killing, exploitation, and suppression. The ecological field worker acquires a deep-seated respect, or even veneration, for ways and forms of life. He reaches an understanding from within, a kind of understanding that others reserve for fellow men and for a narrow section of ways and forms of life. To the ecological field worker, *the equal right to live and blossom* is an intuitively clear and obvious value axiom. Its restriction to humans is an anthropocentrism with detrimental effects upon the life quality of humans themselves. This quality depends in part upon the deep pleasure and satisfaction we receive from close partnership with other forms of life. The attempt to ignore our dependence and to establish a master-slave role has contributed to the alienation of man from himself.

Ecological egalitarianism implies the reinterpretation of the future-research variable, "level of crowding," so that *general* mammalian crowding and loss of life-equality is taken seriously, not only human crowding. (Research on the high requirements of free space of certain mammals has, incidentally, suggested that theorists of human urbanism have largely underestimated human life-space requirements. Behavioral crowding symptoms [neuroses, aggressiveness, loss of traditions . . .] are largely the same among mammals.)

3. *Principles of diversity and of symbiosis.* Diversity enhances the potentialities of survival, the chances of new modes of life, the richness of forms. And the so-called struggle of life, and survival of the fittest, should be interpreted in the sense of ability to coexist and cooperate in complex relationships, rather than ability to kill, exploit, and suppress. "Live and let live" is a more powerful ecological principle than "Either you or me."

The latter tends to reduce the multiplicity of kinds of forms of life, and also to create destruction within the communities of the same species. Ecologically inspired attitudes therefore favor diversity of human ways of life, of cultures, of occupations, of economies. They support the fight against economic and cultural, as much as military, invasion and domination, and they are opposed to the annihilation of seals and whales as much as to that of human tribes or cultures.

4. *Anticlass posture.* Diversity of human ways of life is in part due to (intended or unintended) exploitation and suppression on the part of certain groups. The exploiter lives differently from the exploited, but both are adversely affected in their potentialities of self-realization. The principle of diversity does not cover differences due merely to certain attitudes or behaviors forcibly blocked or restrained. The principles of ecological egalitarianism and of symbiosis support the same anticlass posture. The ecological attitude favors the extension of all three principles to any group conflicts, including those of today between developing and developed nations. The three principles

also favor extreme caution toward any overall plans for the future, except those consistent with wide and widening classless diversity.

5. Fight against *pollution and resource depletion*. In this fight ecologists have found powerful supporters, but sometimes to the detriment of their total stand. This happens when attention is focused on pollution and resource depletion rather than on the other points, or when projects are implemented which reduce pollution but increase evils of the other kinds. Thus, if prices of life necessities increase because of the installation of antipollution devices, class differences increase too. An ethics of responsibility implies that ecologists do not serve the shallow, but the deep ecological movement. That is, not only point (5), but all seven points must be considered together.

Ecologists are irreplaceable informants in any society, whatever their political color. If well organized, they have the power to reject jobs in which they submit themselves to institutions or to planners with limited ecological perspectives. As it is now, ecologists sometimes serve masters who deliberately ignore the wider perspectives.

6. *Complexity, not complication*. The theory of ecosystems contains an important distinction between what is complicated without any Gestalt or unifying principles—we may think of finding our way through a chaotic city—and what is complex. A multiplicity of more or less lawful, interacting factors may operate together to form a unity, a system. We make a shoe or use a map or integrate a variety of activities into a workaday pattern. Organisms, ways of life, and interactions in the biosphere in general, exhibit complexity of such an astoundingly high level as to color the general outlook of ecologists. Such complexity makes thinking in terms of vast systems inevitable. It also makes for a keen, steady perception of the profound *human ignorance* of biospherical relationships and therefore of the effect of disturbances.

Applied to humans, the complexity-not-complication principle favors division of labor, *not fragmentation of labor*. It favors integrated actions in which the whole person is active, not mere reactions. It favors complex economies, an integrated variety of means of living. (Combinations of industrial and agricultural activity, of intellectual and manual work, of specialized and nonspecialized occupations, of urban and non-urban activity, of work in city and recreation in nature with recreation in city and work in nature . . .)

It favors soft technique and "soft future-research," less prognosis, more clarification of possibilities. More sensitivity toward continuity and live traditions, and—most important—toward our state of ignorance.

The implementation of ecologically responsible policies requires in this century an exponential growth of technical skill and invention—but in new directions, directions which today are not consistently and liberally supported by the research policy organs of our nation-states.

7. *Local autonomy and decentralization*. The vulnerability of a form of life is roughly proportional to the weight of influences from afar, from outside the local region in which that form has obtained an ecological equilibrium.

This lends support to our efforts to strengthen local self-government and material and mental self-sufficiency. But these efforts presuppose an impetus toward decentralization. Pollution problems, including those of thermal pollution and recirculation of materials, also lead us in this direction, because increased local autonomy, if we are able to keep other factors constant, reduces energy consumption. (Compare an approximately self-sufficient locality with one requiring the importation of foodstuff, materials for house construction, fuel and skilled labor from other continents. The former may use only 5 percent of the energy used by the latter.) Local autonomy is strengthened by a reduction in the number of links in the hierarchical chains of decision. (For example a chain consisting of local board, municipal council, highest subnational decision maker, a statewide institution in a state federation, a federal national government institution, a coalition of nations, and of institutions, for example, EEC top levels, and a global institution can be reduced to one made up of local board, nationwide institution, and global institution.) Even if a decision follows majority rules at each step, many local interests may be dropped along the line, if it is too long.

Summing up, then, it should, first of all, be borne in mind that the norms and tendencies of the Deep Ecology movement are not derived from ecology by logic or induction. Ecological knowledge and the life-style of the ecological field worker have *suggested, inspired, and fortified* the perspectives of the Deep Ecology movement. Many of the formulations in the above seven-point survey are rather vague generalizations, only tenable if made more precise in certain directions. But all over the world the inspiration from ecology has shown remarkable convergencies. The survey does not pretend to be more than one of the possible condensed codifications of these convergencies.

Secondly, it should be fully appreciated that the significant tenets of the Deep Ecology movement are clearly and forcefully *normative*. They express a value priority system only in part based on results (or lack of results, cf. point [6]) of scientific research. Today, ecologists try to influence policy-making bodies largely through threats, through predictions concerning pollutants and resource depletion, knowing that policy makers accept at least certain minimum *norms* concerning health and just distribution. But it is clear that there is a vast number of people in all countries, and even a considerable number of people in power, who accept as valid the wider norms and values characteristic of the Deep Ecology movement. There are political potentials in this movement which should not be overlooked and which have little to do with pollution and resource depletion. In plotting possible futures, the norms should be freely used and elaborated.

Thirdly, in so far as ecology movements deserve our attention, they are *ecophilosophical* rather than ecological. Ecology is a *limited* science which makes *use* of scientific methods. Philosophy is the most general forum of debate on fundamentals, descriptive as well as prescriptive, and political philosophy is one of its subsections. By an *ecosophy* I mean a philosophy of ecological harmony or equilibrium. A philosophy as a kind of *sofia* wisdom, is openly normative, it contains *both* norms, rules, postulates, value priority announcements *and* hypotheses concerning the state of affairs in our universe. Wisdom is policy wisdom, prescription, not only scientific description and prediction.

The details of an ecosophy will show many variations due to significant differences concerning not only "facts" of pollution, resources, population, etc., but also value priorities. Today, however, the seven points listed provide one unified framework for ecosophical systems.

In general system theory, systems are mostly conceived in terms of causally or functionally interacting or interrelated items. An ecosophy, however, is more like a system of the kind constructed by Aristotle or Spinoza. It is expressed verbally as a set of sentences with a variety of functions, descriptive and prescriptive. The basic relation is that between subsets of premises and subsets of conclusions, that is, the relation of derivability. The relevant notions of derivability may be classed according to rigor, with logical and mathematical deductions topping the list, but also according to how much is implicitly taken for granted. An exposition of an ecosophy must necessarily be only moderately precise considering the vast scope of relevant ecological and normative (social, political, ethical) material. At the moment, ecosophy might profitably use models of systems, rough approximations of global systematizations. It is the global character, not preciseness in detail, which distinguishes an ecosophy. It articulates and integrates the efforts of an ideal ecological team, a team comprising not only scientists from an extreme variety of disciplines, but also students of politics and active policy makers.

Under the name of *ecologism*, various deviations from the deep movement have been championed—primarily with a one-sided stress on pollution and resource depletion, but also with a neglect of the great differences between under- and over-developed countries in favor of a vague global approach. The global approach is essential, but regional differences must largely determine policies in the coming years.

The Deep Ecology Movement

BILL DEVALL

THERE ARE TWO GREAT streams of environmentalism in the latter half of the twentieth century. One stream is reformist, attempting to control some of the worst of the air and water pollution and inefficient land-use practices in industrialized nations and to save a few of the remaining pieces of wildlands as "designated wilderness areas." The other stream supports many of the reformist goals but is revolutionary, seeking a new metaphysics, epistemology, cosmology, and environmental ethics of person/planet. This chapter is an intellectual archeology of the second of these streams of environmentalism, which I will call "Deep Ecology."

There are several other phrases that some writers are using for the perspective I am describing in this paper. Some call it "eco-philosophy" or "foundational ecology" or the "new natural philosophy." I use "Deep Ecology" as the shortest label. Although I am convinced that Deep Ecology is radically different from the perspective of the dominant social paradigm, I do not use the phrase "radical ecology" or "revolutionary ecology" because I think those labels have such a burden of emotive associations that many people would not hear what is being said about Deep Ecology because of their projection of other meanings of "revolution" onto the perspective of Deep Ecology.

I contend that both streams of environmentalism are reactions to the successes and excesses of the implementation of the dominant social paradigm. Although reformist environmentalism treats some of the symptoms of the environmental crisis and challenges some of the assumptions of the dominant social paradigm (such as growth of the economy at any cost), Deep Ecology questions the fundamental premises of the dominant social paradigm. In the future, as the limits of reform are reached and environmental problems become more serious, the reform environmental movement will have to come to terms with Deep Ecology.

From: *Natural Resources Journal* 20 (April 1980): 299–313.

The analysis in the present paper was inspired by Arne Naess's paper on "shallow and deep, long-range" environmentalism.[1] The methods used are patterned after John Rodman's seminal critique of the resources conservation and development movement in the United States.[2] The data are the writings of a diverse group of thinkers who have been developing a theory of deep ecology, especially during the last quarter of a century. Relatively few of these writings have appeared in popular journals or in books published by mainstream publishers. I have searched these writings for common threads or themes much as Max Weber searched the sermons of Protestant ministers for themes which reflected from and back to the intellectual and social crisis of the emerging Protestant ethic and the spirit of capitalism.[3] Several questions are addressed in this paper: What are the sources of Deep Ecology? How do the premises of Deep Ecology differ from those of the dominant social paradigm? What are the areas of disagreement between reformist environmentalism and Deep Ecology? What is the likely future role of the Deep Ecology movement?

THE DOMINANT PARADIGM

A paradigm is a shorthand description of the world view, the collection of values, beliefs, habits, and norms which form the frame of reference of a collectivity of people—those who share a nation, a religion, a social class. According to one writer, a *dominant* social paradigm is the mental image of social reality that guides expectations in a society.

The dominant paradigm in North America includes the belief that "economic growth," as measured by the Gross National Product, is a measure of progress, the belief that the primary goal of the governments of nation-states, after national defense, should be to create conditions that will increase production of commodities and satisfy material wants of citizens, and the belief that "technology can solve our problems." Nature, in this paradigm, is only a storehouse of resources which should be "developed" to satisfy ever-increasing numbers of humans and ever-increasing demands of humans. Science is wedded to technology, the development of *techniques* for control of natural processes (such as weather modification). Change ("planned obsolescence") is an end in itself. The new is valued over the old and the present over future generations. The goal of persons is personal satisfaction of wants and a higher standard of living as measured by possession of commodities (houses, autos, recreation vehicles, etc.).[4] Whatever its origin, this paradigm continues to be dominant, to be preached through publicity (i.e., advertising), and to be part of the world view of most citizens in North America.[5]

For some writers, the dominant social paradigm derives from Judeo-Christian origins.[6] For others, the excesses of air and water pollution, the demand for more and more centralization of political and economic power and the disregard for future generations, and the unwise use of natural resources derive from the ideology and structure of capitalism or from the Lockean view that property must be "improved" to make it valuable to the "owner" and to society.[7] For others, the dominant social paradigm derives from the "scientism" of the modern West (Europe and North America) as applied to the technique of domination.[8]

Following Thomas Kuhn's theory of the dominance of paradigms in modern science and the operation of scientists doing what he calls normal science within a paradigm, it can be argued that (1) those who subscribe to a given paradigm share a definition of what problems are and their priorities; (2) the general heuristics, or rules of the game, for approaching problems is widely agreed upon; (3) there is a definitive, underlying confidence among believers of the paradigm that solutions within the paradigm do exist; and (4) those who believe the assumptions of the paradigm may argue about the validity of data, but rarely are their debates about the definition of what the problem is or whether there are solutions or not. Proposed solutions to problems arising from following the assumptions of the paradigms are evaluated as "reasonable," "realistic," or "valid" in terms of the agreed upon "rules of the game." When the data is difficult to fit to the paradigm, frequently there is dissonance disavowal, an attempt to explain away the inconsistency.[9]

It is possible for a paradigm shift to occur when a group of persons finds in comparing its data with generally accepted theory that the conclusions become "weird" when compared with expectations. In terms of the shared views of the goals, rules, and perceptions of reality in a nation, a tribe, or a religious group, for example, a charismatic leader, a social movement, or a formation of social networks of persons exploring a new social paradigm may be at the vanguard of a paradigm shift.

Reformist environmentalism in this paper refers to several social movements which are related in that the goal of all of them is to change society for "better living" without attacking the premises of the dominant social paradigm. These reform movements each defined a problem—such as need for more open space—and voluntary organizations were formed to agitate for social changes. There has also been considerable coalition building between different voluntary organizations espousing reform environmentalism. Several reformist environmental movements, including at least the following, have been active during the last century: (1) the movement to establish urban parks, designated wilderness areas, and national parks;[10] (2) the movement to mitigate the health and public safety hazards created by the technology which was applied to create the so-called Industrial Revolution.[11] The Union of Concerned Scientists, for example, has brought to the attention of the general public some of the hazards to public health and safety of the use of nuclear power to generate electricity; (3) the movement to develop "proper" land-use planning. This includes the city beautiful movement of the late nineteenth century and the movement to zone and plan land use such as the currently controversial attempts to zone uses along the coastal zones;[12] (4) the resources conservation and development movement symbolized by the philosophy of multiple use of Gifford Pinchot and the U.S. Forest Service;[13] (5) the "back to the land" movement of the 1960s and 1970s and the "organic farming" ideology; (6) the concern with exponential growth of human population and formation of such groups as Zero Population Growth;[14] (7) the "humane" and "animal liberation" movement directed at changing the attitudes and behavior of humans toward some other aspects of animals;[15] and (8) the "limits to growth" movement which emphasizes we should control human population and move toward a "steady-state" or "conserver society" as rapidly as possible.[16]

SOURCES OF DEEP ECOLOGY

What I call Deep Ecology in this paper is premised on a gestalt of person-in-nature. The person is not above or outside of nature. The person is part of creation ongoing. The person cares for and about nature, shows reverence toward and respect for nonhuman nature, loves and lives with nonhuman nature, is a person in the "earth household" and "lets being be," lets nonhuman nature follow separate evolutionary destinies. Deep Ecology, unlike reform environmentalism, is not just a pragmatic, short-term social movement with a goal like stopping nuclear power or cleaning up the waterways. Deep Ecology first attempts to question and present alternatives to conventional ways of thinking in the modern West. Deep Ecology understands that some of the "solutions" of reform environmentalism are counterproductive. Deep Ecology seeks transformation of values and social organization.

The historian Lynn White, Jr., in his influential 1967 article, "The Historical Roots of Our Ecologic Crisis," provided one impetus for the current upwelling of interest in Deep Ecology by criticizing what he saw as the dominant Judeo-Christian view of man versus nature, or man at war with nature. But there are other writers, coming from diverse intellectual and spiritual disciplines, who have provided, in the cumulative impact of their work, a profound critique of the dominant social paradigm and the "single vision" of science in the modern (post-1500) West.[17]

One major stream of thought influencing the development of Deep Ecology has been the influx of Eastern spiritual traditions into the West which began in the 1950s with the writings of such people as Alan Watts[18] and Daisetz Suzuki.[19] Eastern traditions provided a radically different man/nature vision than that of the dominant social paradigm of the West. During the 1950s the so-called beat poets such as Alan Ginsberg seemed to be groping for a way through Eastern philosophy to cope with the violence, insanity, and alienation of people from people and people from nature they experienced in North America. Except for Gary Snyder, who developed into one of the most influential eco-philosophers of the 1970s, these beat poets, from the perspective of the 1970s, were naive in their understanding of Eastern philosophy, ecology, and the philosophical traditions of the West.

During the late 1960s and 1970s, however, philosophers, scientists, and social critics have begun to compare Eastern and Western philosophic traditions as they relate to science, technology, and man/nature relations. Fritjof Capra's *Tao of Physics*, for example, emphasizes the parallels between Eastern philosophies and the theories of twentieth-century physics.[20] Joseph Needham's massive work, *Science and Civilization in China*, brought to the consciousness of the West the incredibly high level of science, technology, and civilization achieved in the East for millennia and made available to Western readers an alternative approach to science and human values.[21] More recently, Needham has suggested that modern Westerners take the philosophies of the East as a spiritual and ethical basis for modern science.[22] Works by Huston Smith, among others, have also contributed to this resurgent interest in relating the environmental crisis to the values expressed in the dominant Western paradigm. Smith and others have looked to the Eastern philosophies for spiritual-religious guidance.[23]

Several social philosophers have written brilliant critiques of Western societies but have not presented a new metaphysical basis for their philosophy nor attempted to incorporate Eastern philosophy into their analyses. Jacques Ellul wrote on *technique* and the technological society.[24] Paul Goodman discussed the question "can there be a humane technology?"[25] Herbert Marcuse analyzed "one-dimensional man" as the prototypical "modern" urbanite.[26] The works of Theodore Roszak have also had considerable impact on those thinkers interested in understanding the malaise and contradictions of modern societies by examining the premises of the dominant social paradigm.[27]

A second stream of thought contributing to Deep Ecology has been the reevaluation of Native Americans (and other preliterate peoples) during the 1960s and 1970s. This is not a revival of the romantic view of Native Americans as "noble savages" but rather an attempt to evaluate traditional religions, philosophies, and social organizations of Native Americans in objective, comparative, analytic, and critical ways.

A number of questions have been asked. How did different tribes at different times cope with changes in their natural environment (such as prolonged drought) and with technological innovation? What were the "separate realities" of Native Americans and can modern Western man *understand* and *know*, in a phenomenological sense, these "separate realities"? The experiences of Carlos Castaneda, for example, indicate it may be very difficult for modern man to develop such understanding since this requires a major perceptual shift of man/nature. Robert Ornstein concludes, "Castaneda's experience demonstrates primarily that the Western-trained intellectual, even a 'seeker,' is by his culture almost completely unprepared to understand esoteric traditions."[28]

From the many sources on Native Americans which have become available during the 1970s, I quote a statement by Luther Standing Bear, an Oglala Sioux, from *Touch the Earth* to illustrate the contrast with the modern paradigm of the West:

> We did not think of the great open plains, the beautiful rolling hills, and winding streams with tangled growth, as "wild." Only to the white man was nature a "wilderness" and only to him was the land "infested" with "wild" animals and "savage" people. To us it was tame. Earth was bountiful and we were surrounded with the blessings of the Great Mystery. Not until the hairy man from the east came and with brutal frenzy heaped injustices upon us and the families we loved was it "wild" for us. When the very animals of the forest began fleeing from his approach, then it was that for us the "wild west" began.[29]

A third source of Deep Ecology is found in the "minority tradition" of Western religious and philosophical traditions. The philosopher George Sessions has claimed that

> in the civilized West, a tenuous thread can be drawn through the Presocratics, Theophrastus, Lucretius, St. Francis, Bruno and other neo-Platonic mystics, Spinoza, Thoreau, John Muir, Santayana, Robinson Jeffers, Aldo Leopold, Loren Eiseley, Gary Snyder, Paul Shepard, Arne Naess, and maybe that desert rat, Edward Abbey. This minority tradition, despite differences, could have provided the West with a healthy basis for a realistic portrayal of the balance and interconnectedness of three artificially separable components (God/Nature/Man) of an untimely seamless and inseparable Whole.[30]

Sessions, together with Arne Naess and Stuart Hampshire, has seen the philosopher Spinoza as providing a unique fusion of an integrated man/nature metaphysic with modern European science.[31] Spinoza's ethics is most naturally interpreted as implying biospheric egalitarianism, and science is endorsed by Spinoza as valuable primarily for contemplation of a pantheistic, sacred universe and for spiritual discipline and development. Spinoza stands out in a unique way in opposition to other seventeenth-century philosophers—for example, Bacon, Descartes, and Leibniz—who were at that time laying the foundations for the technocratic-industrial social paradigm and the fulfillment of the Christian imperative that man *must* dominate and control all nature. It has been claimed by several writers that the poet-philosopher Robinson Jeffers, who lived most of his life on the California coastline at Big Sur, was Spinoza's twentieth-century "evangelist" and that Jeffers gave Spinoza's philosophy an explicitly ecological interpretation.[32]

Among contemporary European philosophers, the two most influential have been Alfred North Whitehead and Martin Heidegger.[33] In particular, more American philosophers, both those with an interest in ecological consciousness and those interested in contemporary philosophers, are discussing

Heidegger's critique of Western philosophy and contemporary Western societies. Because Heidegger's approach to philosophy and language is so different from the language we are accustomed to in American academia, any summary of his ideas would distort the theory he is presenting. The reader is referred to the books and articles on Heidegger cited below.[34]

A fourth source of reference for the Deep Ecology movement has been the scientific discipline of *ecology*. For some ecology is a science of the "home," of the "relationships between," while for others ecology is a perspective. The difference is important, for ecology as a science is open for co-optation by the engineers, the "technological fixers" who want to "enhance," "manage," or "humanize" the biosphere. At the beginning of the "environmental decade" of the 1970s, two ecologists issued a warning against this approach:

> Even if we dispense with the idea that ecologists are some sort of environmental engineers and compare them to the pure physicists who provide scientific rules for engineers, do the tentative understandings we have outlined (in their article) provide a sound basis for action by those who would manage the environment? It is self-evident that they do not. . . . We submit that ecology as such probably cannot do what many people expect it to do; it cannot provide a set of "rules" of the kind needed to manage the environment.[35]

Donald Worster, at the conclusion of his scholarly and brilliant history of ecological thinking in the West, is of the same opinion.[36]

But ecologists do have an important task in the Deep Ecology movement. They can be *subversive* in their perspective. For human ecologist Paul Shepard, "the ideological status of ecology is that of a resistance movement" because its intellectual leaders such as Aldo Leopold challenge the major premises of the dominant social paradigm.[37] As Worster in his history of ecology points out:

> All science, though primarily concerned with the "Is," becomes implicated at some point in the "Ought." Conversely, spinning out moral visions without reference to the material world may ultimately be an empty enterprise; when . . . moral values depart too far from nature's ways. . . . Scientist and moralist might together explore a potential union of their concerns; might seek a set of empirical facts with ethical meaning, a set of moral truths. The ecological ethic of interdependence may be the outcome of just such a dialectical relation.[38]

A final source of inspiration for the deep, long-range ecology movement is those artists who have tried to maintain a sense of place in their work.[39] Some artists, standing against the tide of midcentury pop art, minimalist art, and conceptual art have shown remarkable clarity and objectivity in their perception of nature. This spiritual-mystical objectivism is found, for example, in the photographs of Ansel Adams.[40] For these artists, including

Morris Graves, who introduced concepts of Eastern thought (including Zen Buddhism) into his art, and Larry Gray, who reveals the eloquent light of revelation of nature in his skyscapes, men reaffirm their spiritual kinship with the eternity of God in nature through art.[41]

THEMES OF DEEP ECOLOGY

I indicated in preceding pages that many thinkers are questioning some of the premises of the dominant social paradigm of the modern societies. They are attempting to extend on an appropriate metaphysics, epistemology, and ethics for what I call an "ecological consciousness." Some of these writers are very supportive of reformist environmental social movements, but they feel reform while necessary is not sufficient. They suggest a new paradigm is required and a new utopian vision of "right livelihood" and the "good society." Utopia stimulates our thinking concerning alternatives to present society.[42] Some persons, such as Aldo Leopold, have suggested that we begin our thinking on utopia not with a statement of "human nature" or "needs of humans" but by trying to "think like a mountain." This profound extending, "thinking like a mountain," is part and parcel of the phenomenology of ecological consciousness.[43] Deep Ecology begins with Unity rather than dualism which has been the dominant theme of Western philosophy.[44]

Philosopher Henryk Skolimowski, who has written several papers on the options for the ecology movements, asserts:

> We are in a period of ferment and turmoil, in which we have to challenge the limits of the analytical and empiricist comprehension of the world as we must work out a new conceptual and philosophical framework in which the multitude of new social, ethical, ecological, epistemological, and ontological problems can be accommodated and fruitfully tackled. The need for a new philosophical framework is felt by nearly everybody. It would be lamentable if professional philosophers were among the last to recognize this.[45]

Numerous other writers on Deep Ecology, including William Ophuls, E. F. Schumacher, George Sessions, Theodore Roszak, Paul Shepard, Gary Snyder, and Arne Naess, have in one way or another called for a new social paradigm or a new environmental ethic. We must "think like a mountain," according to Aldo Leopold. And Roberick Nash says:

> Do rocks have rights? If the times comes when to any considerable group of us such a question is no longer ridiculous, we may be on the verge of a change of value structures that will make possible measures to cope with the growing ecologic crisis. One hopes there is enough time left.[46]

Any attempt to create artificially a "new ecological ethics" or a "new ontology of man's place in nature" out of the diverse strands of thought which make up the Deep Ecology movement is likely to be forced and futile. However, by explicating some of the major themes embodied in and presupposed by the intellectual movement I am calling Deep Ecology, some groundwork can be laid for further discussion and clarification.[47] Following the general outline of perennial philosophy, the order of the following statements summarizing Deep Ecology's basic principles are metaphysical-religious, psychological-epistemological, ethical, and social-economic-political. These concerns of Deep Ecology encompass most of reformist environmentalism's concerns but subsume them in its fundamental critique of the dominant paradigm.

According to Deep Ecology:

1. A new cosmic/ecological metaphysics which stresses the identity (I/ thou) of humans with nonhuman nature is a necessary condition for a viable approach to building an eco-philosophy. In Deep Ecology, the wholeness and integrity of person/planet together with the principle of what Arne Naess calls "biological equalitarianism" are the most important ideas. Man is an integral part of nature, not over or apart from nature. Man is a "plain citizen" of the biosphere, not its conqueror or manager. There should be a "democracy of all God's creatures" according to St. Francis; or as Spinoza said, man is a "temporary and dependent mode of the whole of God/Nature." Man flows with the system of nature rather than attempting to control all of the rest of nature. The hand of man lies lightly on the land. Man does not perfect nature, nor is man's primary duty to make nature more efficient.[48]

2. An objective approach to nature is required. This approach is found, for example, in Spinoza and in the works of Spinoza's twentieth-century disciple, Robinson Jeffers. Jeffers describes his orientation as a philosophy of "inhumanism" to draw a sharp and shocking contrast with the subjective anthropocentrism of the prevailing humanistic philosophy, art, and culture of the twentieth-century West.[49]

3. A new psychology is needed to integrate the metaphysics in the mind field of postindustrial society. A major paradigm shift results from psychological changes of perception. The new paradigm requires rejection of subject/object, man/nature dualisms and will require a pervasive awareness of total intermingling of the planet earth. Psychotherapy seen as adjustment to ego-oriented society is replaced by a new ideal of psychotherapy as spiritual development.[50] The new metaphysics and psychology leads logically to a posture of biospheric egalitarianism and liberation in the sense of autonomy, psychological/emotional freedom of the individual, spiritual development for *Homo sapiens*, and the right of other species to pursue their own evolutionary destinies.[51]

4. There is an objective basis for environmentalism, but objective science in the new paradigm is different from the narrow, analytic conception of the "scientific method" currently popular. Based on "ancient wisdom," science should be both objective and participatory without modern science's subject/

object dualism. The main *value* of science is seen in its ancient perspective as contemplation of the cosmos and the enhancement of understanding of self and creation.[52]

5. There is wisdom in the stability of natural processes unchanged by human intervention. Massive human-induced disruptions of ecosystems will be unethical and harmful to men. Design for human settlement should be with nature, not against nature.[53]

6. The quality and human existence and human welfare should not be measured only by quantity of products. Technology is returned to its ancient place as an appropriate tool for human welfare, not an end in itself.[54]

7. Optimal human carrying capacity should be determined for the planet as a biosphere and for specific islands, valleys, and continents. A drastic reduction of the rate of growth of population of *Homo sapiens* through humane birth-control programs is required.[55]

8. Treating the symptoms of man/nature conflict, such as air or water pollution, may divert attention from more important issues and thus be counterproductive to "solving" the problems. Economics must be subordinate to ecological-ethical criteria. Economics is to be treated as a small subbranch of ecology and will assume a rightfully minor role in the new paradigm.[56]

9. A new philosophical anthropology will draw on data of hunting/gathering societies for principles of healthy, ecologically viable societies. Industrial society is not the end toward which all societies should aim or try to aim.[57] Therefore, the notion of "reinhabiting the land" with hunting-gathering, and gardening as a goal and standard for postindustrial society should be seriously considered.[58]

10. Diversity is inherently desirable both culturally and as a principle of health and stability of ecosystems.[59]

11. There should be a rapid movement toward "soft" energy paths and "appropriate technology" and toward life-styles which will result in a drastic decrease in per capita energy consumption in advanced industrial societies while increasing appropriate energy in decentralized villages in so-called Third-World nations.[60] Deep ecologists are committed to rapid movement to a "steady-state" or "conservor society" both from ethical principles of harmonious integration of humans with nature and from appreciation of ecological realities.[61] Integration of sophisticated, elegant, unobtrusive, ecologically sound, appropriate technology with greatly scaled down, diversified, organic, labor-intensive agriculture, hunting, and gathering is another goal.[62]

12. Education should have as its goal encouraging the spiritual development and personhood development of the members of a community, not just training them in occupations appropriate for oligarchic bureaucracies and for consumerism in advanced industrial societies.[63]

13. More leisure as contemplation in art, dance, music, and physical skills will return play to its place as the nursery of individual fulfillment and cultural achievement.[64]

14. Local autonomy and decentralization of power is preferred over centralized political control through oligarchic bureaucracies. Even if bureaucratic modes of organization are more "efficient," other modes of organization for small-scale human communities are more "effective" in terms of the principles of Deep Ecology.[65]

15. In the interim, before the steady-state economy and radically changed social structure are instituted, vast areas of the planet biospheres will be zoned "off limits" to further industrial exploitation and large-scale human settlement; these should be protected by defensive groups of people. . . .

Deep Ecology is liberating ecological consciousness. The writers I have cited in this paper provide radical critiques of modern society and of the dominant values of this society. They also provide, or some of them do, a profound utopian alternative. The elaborating of or deepening of ecological consciousness is a continuing process. The goal is to have action and consciousness as one. But the development of ecological consciousness is seen as prior to ecological resistance in many of the writings cited. This ecological consciousness may not be very well articulated except by intellectuals who are in the business of verbalizing. But they, as much as anyone, realize the limitation of just verbalizing. Consciousness is *knowing*. From the perspective of Deep Ecology, ecological resistance will naturally flow from and with a developing ecological consciousness.

Notes

1. Arne Naess, "The Shallow and the Deep, Long-Range Ecology Movement," *Inquiry* 16 (1973): 95–100.
2. John Rodman, "Four Forms of Ecological Consciousness: Beyond Economics, Resources, and Conservation" (1977) Pitzer College; [revised version published as "Four Forms of Ecological Consciousness Reconsidered," in *Ethics and the Environment*, ed. Donald Scherer and Thomas Attig (Englewood Cliffs, N.J.: Prentice-Hall, 1983)].
3. Max Weber, *The Protestant Ethic and the Spirit of Capitalism* (New York: Scribner's, 1930).
4. Dennis C. Pirages and Paul Ehrlich, *Ark II: Social Response to Environmental Imperatives* (New York: Viking, 1974), 43. See also *The Future of the Great Plains*, H.R. Doc. 144, 75th Cong., 1st sess. (1937).
5. On the history of the paradigm, see Victor Ferkiss, *The Future of Technological Civilization* (New York: George Braziller, 1974). For a critique of the "me-now" consumerism of the 1970s, see Christopher Lasch, *The Culture of Narcissism: American Life in an Age of Diminishing Expectations* (New York: Warner Books, 1979). See also Manager's Journal, "Monitoring America, Values of Americans," *Wall Street Journal*, 2 Oct. 1978.
6. Lynn White, Jr., "The Historical Roots of Our Ecologic Crisis," *Science* 155 (1967): 1203.

7. Barry Weisberg, *Beyond Repair: The Ecology of Capitalism* (Boston: Beacon, 1971); England and Bluestone, "Ecology and Class Conflict," *Review of Radical Political Economics* 3 (1971): 21. On Locke's view of "property," see Ferkiss, *Future of Technological Civilization*.

8. Leo Marx, *The Machine in the Garden: Technology and the Pastoral Ideal in America* (New York: Oxford, 1967); and Lewis Mumford, *The Myth of the Machine*, vol. 2, *The Pentagon of Power* (New York: Harcourt Brace, 1970).

9. Thomas Kuhn, *The Structure of Scientific Revolutions*, 2d ed. (Chicago: University of Chicago Press, 1970). For criticism of Kuhn, see Imre Lakatos and Alan Musgrave, eds., *Criticism and the Growth of Knowledge* (Cambridge: Cambridge University Press, 1970).

10. Roderick Nash, *Wilderness and the American Mind*, 2d. ed. (New Haven: Yale University Press, 1973); Joseph Sax, "America's National Parks: Their Principles, Purposes and their Prospects," *Natural History* 35 (1976): 57.

11. Barry Commoner, *The Closing Circle* (New York: Knopf, 1971); James Ridgeway, *The Politics of Ecology* (New York: Dutton, 1970).

12. Natural Resources Defense Council, *Land Use Controls in the United States: A Handbook of Legal Rights of Citizens* (New York: Dial Press, 1977); Ian McHarg, *Design with Nature* (Garden City, N.Y.: Doubleday, 1971).

13. Rodman, "Four Forms of Ecological Consciousness"; Samuel Hays, *Conservation and the Gospel of Efficiency* (Cambridge: Harvard University Press, 1959); Gifford Pinchot, *Breaking New Ground* (New York: Harcourt Brace, 1947).

14. Paul Ehrlich, *The Population Bomb* (New York: Ballantine, 1968). See, generally, publications of the organization Zero Population Growth. Also, United States Commission on Population and the American Future, *The Report of the Commission on Population Growth and the American Future* (Washington, D.C.: Government Printing Office, 1972).

15. Tom Regan and Paul Singer, *Animal Rights and Human Obligations* (Englewood Cliffs, N.J.: Prentice-Hall, 1976); Peter Singer, *Animal Liberation* (New York: Random House, 1977).

16. Donella H. Meadows, Dennis L. Meadows, Jorgen Randers, and William W. Behrens III, *The Limits to Growth*, 2d ed. (New York: Universe Books, 1974); Mihajlo Mesarovic and Edward Pestel, *Mankind at the Turning Point: Second Report of the Club of Rome* (New York: Dutton, 1974); Donella H. Meadows, *Alternatives to Growth* (Cambridge, Mass.: Ballinger, 1977). For a critique of the limits to growth model, see H. S. D. Cole, Christopher Freeman, Marie Johoda, and K. L. R. Pavitt, eds., *Models of Doom: A Critique of the Limits to Growth* (New York: Universe, 1973); Herman Daly, ed., *Toward a Steady-State Economy* (San Francisco: W. H. Freeman, 1973).

17. White, "Historical Roots."

18. Alan Watts, *Psychotherapy East and West* (New York: Pantheon, 1975); Alan Watts, *Nature, Man and Woman* (New York: Pantheon, 1970); Alan Watts, *The Spirit of Zen: A Way of Life, Work and Art in the Far East* (London: J. Murray, 1936); Alan Watts, *The Essence of Alan Watts* (Millbrae, Calif.: Celestial Arts, 1977).

19. Daisetz Suzuki, *Essays in Zen Buddhism* (New York: Grove, 1961).

20. Fritjof Capra, *The Tao of Physics* (Berkeley, Calif.: Shambala, 1975).

21. Joseph Needham, *Science and Civilization in China*, 6 vols. (New York: Cambridge University Press, 1954–76).

22. Joseph Needham, "History and Human Values: A Chinese Perspective for World Science and Technology," *Centennial Review* 20 (1976): 1.

23. Houston Smith, *Forgotten Truth: The Perennial Tradition* (New York: Harper

and Row, 1976); H. Smith, "Tao Now: An Ecological Testament," in *Earth Might Be Fair*, ed. Ian Barbour (Englewood Cliffs, N.J.: Prentice-Hall, 1972).

24. Jacques Ellul, *The Technological Society* (New York: Vintage, 1964).

25. Paul Goodman, "Can Technology be Humane?" in *Western Man and Environmental Ethics*, ed. Ian Barbour (Reading, Mass.: Addison-Wesley, 1973), 225–42.

26. Herbert Marcuse, *One-Dimensional Man* (Boston: Beacon Press, 1964).

27. Theodore Roszak, *The Making of a Counter-Culture* (Garden City, N.Y.: Doubleday, 1969); T. Roszak, *Where the Wasteland Ends: Politics and Transcendence in Post-Industrial Society* (Garden City, N.Y.: Doubleday, 1972); T. Roszak, *Unfinished Animal: The Aquarian Frontier and the Evolution of Consciousness* (New York: Harper and Row, 1975); T. Roszak, *Person/Planet* (Garden City, N.Y.: Doubleday, 1978).

28. Robert Ornstein, *The Mind Field* (New York: Octagon Press, 1976), 105. The works of Carlos Castaneda have been influential. They include Castaneda, *The Teachings of Don Juan* (Berkeley and Los Angeles: University of California Press, 1968); *A Separate Reality* (New York: Simon and Schuster, 1971); *Journey to Ixtlan* (New York: Simon and Schuster, 1972); *Tales of Power* (New York: Simon and Schuster, 1974).

29. Luther Standing Bear, in *Touch the Earth*, ed. T. C. McLuhan (New York: Simon and Schuster, 1971). Among the most significant and original theories of Native Americans and nonhuman nature are: Vine Deloria, *God is Red* (New York: Grosset and Dunlap, 1975); Calvin Martin, *Keepers of the Game* (Berkeley and Los Angeles: University of California Press, 1978); Stan Steiner, *The Vanishing White Man* (New York: Harper and Row, 1976).

30. George Sessions, "Spinoza and Jeffers on Man in Nature," *Inquiry* 20 (1977): 481. See also George Sessions, *Spinoza, Perennial Philosophy and Deep Ecology* (unpublished MS, Sierra College, 1979).

31. Stuart Hampshire, *Two Theories of Morality* (Oxford: Oxford University Press, 1977); Stuart Hampshire, *Spinoza* (Harmondsworth, Eng.: Penguin Books, 1962); Arne Naess, "Spinoza and Ecology," *Philosophia* 7 (1977): 45.

32. Sessions, "Spinoza and Jeffers on Man in Nature." See also Arthur Coffin, *Robinson Jeffers: Poet of Inhumanism* (Madison: University of Wisconsin Press, 1971); Bill Hotchkiss, *Jeffers: The Siviastic Vision* (Auburn, Calif.: Blue Oak Press, 1975); R. Brophy, "Robinson Jeffers, Metaphysician of the West" (unpublished MS, Dept. of English, Long Beach State University, Long Beach, California).

33. John Cobb, Jr., *Is It Too Late? A Theology of Ecology* (1972); Charles Hartshorne, *Beyond Humanism: Essays in the Philosophy of Nature* (Chicago: Willett Clark, 1937); Alfred North Whitehead, *Science and the Modern World* (New York: Macmillan, 1925), chaps. 5, 13; Griffin, "Whitehead's Contribution to the Theology of Nature," *Bucknell Review* 20 (1972): 95.

34. Martin Heidegger, *The Question Concerning Technology and Other Essays*, trans. William Lovitt (New York: Harper and Row, 1977); George Steiner, *Martin Heidegger* (Hassocks, Eng.: Harvester Press, 1978); Vincent Vycinas, *Earth and Gods: An Introduction to the Philosophy of Martin Heidegger* (The Hague: M. Nijhoff, 1961). On the approach taken by Heidegger and the contemporary ecological consciousness, see Delores LaChapelle, *Earth Wisdom* (Los Angeles: Guild of Tutors Press, 1978), chap. 9. The writings of Michael Zimmerman on Heidegger are also useful, including his "Beyond 'Humanism': Heidegger's Understanding of Technology," *Listening* 12 (1977): 74. See also, M. Zimmerman, "Marx and Heidegger on the Technological Domination of Nature," *Philosophy Today* 12 (Summer 1979): 99.

35. William Murdoch and Joseph Connell, "All About Ecology," in *Western Man and Environmental Ethics*, ed. Ian Barbour (Reading, Mass.: Addison-Wesley, 1973), 156–70.

36. Donald Worster, "Epilogue," in *Nature's Economy* (San Francisco: Sierra Club Books, 1977). The importance of the thinking of ecologist Aldo Leopold should be emphasized. There are many articles interpreting Leopold's message. See, for example, Hwa Yol and Petee Jung, "The Splendor of the Wild: Zen and Aldo Leopold," *Atlantic Naturalist* 5 (1974): 29.

37. Paul Shepard, "Introduction: Man and Ecology," in *The Subversive Science*, ed. Paul Shepard and Daniel McKinley (Boston: Houghton Mifflin, 1969), 1. See also Neil Everndon, "Beyond Ecology," *North American Review* 263 (1978): 16–20.

38. Worster, Nature's Economy, 338.

39. Hans Huth, "Wilderness and Art," in *Wilderness, America's Living Heritage*, ed. David Brower (San Francisco: Sierra Club Books, 1961), 60; Paul Shepard, "A Sense of Place," *North American Review* 262 (1977): 22.

40. Ansel Adams, "The Artist and the Ideals of Wilderness," in *Wilderness, America's Living Heritage*, ed. David Brower (San Francisco: Sierra Club Books, 1961), 49.

41. Morris Graves, *The Drawings of Morris Graves*, ed. Ida E. Rubin (Boston: New York Graphic Society, 1974).

42. Mulford Quichert Sibley, *Nature and Civilization: Some Implications for Politics* (Ithaca, Ill.: F. E. Peacock, 1977), chap. 7, p. 251. Sibley makes a case for more utopian visions from contemporary intellectuals. Although many people have been revulsed by the visions of Marxism and fascist dictatorships, "the student of politics has an obligation not only to explain and criticize but also to propose and explicate ideals. We need more utopian visions, not fewer. For if politics be that activity through which man seeks consciously and deliberately to order and control his collective life, then one of the salient questions in all politics must be: Order and control for what ends? Without utopian visions these ends cannot be stated as wholes; and even a discussion of means and strategies will be clouded unless ends are at least relatively clear" (p. 47).

43. Hwa Yol and Petee Jung, "To Save the Earth," *Philosophy Today* 8 (1975): 108.

44. Sessions, "Spinoza and Jeffers on Man and Nature."

45. Henryk Skolimowski, "The Ecology Movement Re-examined," *Ecologist* 6 (1976): 298; Skolimowski, "Options for the Ecology Movement," *Ecologist* 7 (1977): 318.

46. Roderick Nash, "Do Rocks Have Rights," *Center Magazine* 10 (1977): 2.

47. Skolimowski, "Ecology Movement Re-examined."

48. Jacob Needleman, *A Sense of the Cosmos* (Garden City, N.Y.: Doubleday, 1975), 76–7, 100–102.

49. Sessions, "Spinoza and Jeffers on Man and Nature." Spinoza is one of the important philosophers for Deep Ecology. The new translations of Spinoza's work are absolutely essential for understanding his thought. See Paul Weinpahl, *The Radical Spinoza* (New York: New York University Press, 1979).

50. Ornstein, *Mind Field*.

51. Sessions, "Spinoza and Jeffers on Man and Nature"; Gary Snyder, *The Old Ways* (San Francisco: City Lights Books, 1977).

52. Capra, *Tao of Physics*; Sessions, "Spinoza and Jeffers on Man and Nature"; Needleman, *Sense of the Cosmos*.

53. Commoner, *Closing Circle*; McHarg, *Design with Nature*.

54. Needleman, *Sense of the Cosmos*; Sessions, "Spinoza and Jeffers on Man and Nature."

55. E. F. Schumacher, *Small Is Beautiful, Economics as if People Mattered* (New York: Harper and Row, 1973). As an example of this argument, see Paul Ehrlich, Anne Ehrlich, and John Holdren, *Ecoscience* (San Francisco: W. H. Freeman, 1977).

56. William Ophuls, *Ecology and the Politics of Scarcity* (San Francisco: W. H. Freeman, 1977). Schumacher, *Small Is Beautiful*.

57. Steiner, *Vanishing White Man*; Snyder, *Old Ways*; Paul Shepard, *The Tender Carnivore and the Sacred Game* (New York: Scribner's, 1974).

58. Peter Berg and Raymond Dasmann, *Reinhabiting A Separate Country* (San Francisco: Planet Drum, 1978).

59. Raymond Dasmann, *A Different Kind of Country* (New York: John Wiley, 1968); Norman Myers, *The Sinking Ark: A New Look at the Problems of Disappearing Species* (New York: Pergamon, 1979).

60. Raymond Dasmann, *Ecological Principles for Economic Development* (New York: John Wiley, 1973).

61. Ophuls, *Politics of Scarcity*. On the "steady-state," see Daly, *Toward a Steady-State Economy*.

62. Schumacher, *Small Is Beautiful*.

63. Roszak, *Making of a Counter-Culture*.

64. Johan Huizinga, *Homo Ludens: A Study of the Play Element in Culture*, trans. R. F. C. Hull (London: Routledge, 1949). John Collier, "The Fullness of Life Through Leisure," in *The Subversive Science*, ed. Paul Shepard and Daniel McKinley (Boston: Houghton Mifflin, 1969), 416–36.

65. Ophuls, *Politics of Scarcity*; Dennis C. Pirages, ed., *The Sustainable Society* (New York: Praeger, 1977); Peter Berg and Raymond Dasmann, *Reinhabitating a Separate Country* (San Francisco: Planet Drum, 1978).

Ecocentrism and the Anthropocentric Detour

GEORGE SESSIONS

WARWICK FOX POINTS OUT that philosophical Deep Ecology has two main tasks: "A positive or constructive task of encouraging an egalitarian [or ecocentric] attitude on the part of humans towards all entities in the ecosphere, [and] a negative or critical task of dismantling anthropocentrism."[1] In hopes of helping contribute to these tasks, I offer a brief summary of the history of the twists and turns of ecocentrism and anthropocentrism in the West. This summary discusses only the main highlights; no attempt is made to be all inclusive.

I. ECOCENTRISM AND PRIMAL CULTURES

Although this issue has been hotly debated in the literature over the last twenty years, it seems accurate to say that the cultures of most primal (hunting/ gathering) societies throughout the world were permeated with nature-oriented religions that expressed the ecocentric perspective. These cosmologies, involving a sacred sense of the earth and all its inhabitants, helped order their lives and determine their values. For example, anthropologist Stan Steiner describes the traditional American Indian philosophy of the sacred "circle of life": "In the Circle of Life, every being is no more, or less, than any other. We are all sisters and brothers. Life is shared with the bird, bear, insects, plants, mountains, clouds, stars, sun."[2] Countless expressions of this kind can be cited from primal cultures all over the globe.[3] Given that the vast majority of humans who have lived on earth over the millennia have been hunter/gatherers, it is clear that ecocentrism has been the dominant human religious/philosophical perspective through time.

From: *ReVISION* 13, no. 3 (Winter 1991): 109–15.

With the beginning of agriculture, most ecocentric cultures (and religions) were gradually replaced or driven off into remote corners of the earth by pastoral and, eventually, "civilized" cultures (Latin: *civitas*, cities).[4] It seems likely that one of the functions of the Garden of Eden story, for instance, was to provide a moral justification for this process. Thus, the environmental crisis, in Paul Shepard's view, has been ten thousand years in the making: "As agriculture replaced hunting and gathering it was accompanied by radical changes in the way men saw and responded to their natural surroundings. . . . [Agriculturalists] all shared the aim of completely humanizing the earth's surface, replacing wild with domestic, and creating landscapes from habitat."[5] Whereas Taoism and certain other Eastern religions retained elements of the ancient shamanistic nature religions, the Western religious tradition radically distanced itself from wild nature and, in the process, became increasingly anthropocentric. Henri Frankfort claims, for example, that Judaism sacrificed "the greatest good ancient Near East religion could bestow—the harmonious integration of man's life with the life of nature—Man remained outside nature, exploiting it for a livelihood . . . using its imagery for the expression of his moods, but never sharing its mysterious life."[6] This overall analysis is also supported by Loren Eiseley, who points out that

> primitive man existed in close interdependence with his first world . . . he was still inside that world; he had not turned her into an instrument or a mere source of materials. Christian man in the West strove to escape this lingering illusion the primitives had projected upon nature. Intent upon the destiny of his own soul, and increasingly urban, man drew back from too great an intimacy with the natural. . . . If the new religion was to survive, Pan had to be driven from his hillside or rendered powerless by incorporating him into Christianity.[7]

Similar conclusions were arrived at by D. H. Lawrence in his extraordinary 1924 essay, "Pan in America": "Gradually men moved into cities. And they loved the display of people better than the display of a tree. They liked the glory they got out of overpowering one another in war. And, above all, they loved the vainglory of their own words, the pomp of argument and the vanity of ideas. . . . Til at last the old Pan died and was turned into the devil of the Christians."[8]

The intellectual Greek strand in Western culture also exhibits a similar development from early ecocentric nature religions, and the nature-oriented cosmological speculations of the pre-Socratics, to the anthropocentrism of the classical Athenian philosophers. The Milesian cosmologists, according to Karl Popper, "envisaged the world as a kind of a house, the home of all creatures—our home."[9] Beginning with Socrates, however, philosophical speculation was characterized by "an undue emphasis upon man as compared with the universe," as Bertrand Russell and certain other historians of Western philosophy have observed.

With the culmination of Athenian philosophy in Aristotle, an anthropocentric system of philosophy and science was set in place that was to play a major role in shaping Western thought until the seventeenth century. Aristotle rejected the pre-Socratic ideas of an infinite universe, cosmological and biological evolution, and heliocentrism, and proposed instead an earth-centered finite universe wherein humans were differentiated from, and seen as superior to, animals and plants by virtue of their rationality. Although Aristotle's philosophy was biologically inspired, nevertheless he arrived at a hierarchical concept of the "great chain of being" in which nature *made* plants for the use of animals, and animals were *made* for the sake of humans (*Politics* 1).

In the great medieval Christian synthesis of Saint Thomas Aquinas, there were problems reconciling Aristotle's naturalism with Christian otherworldliness (including the idea of an immaterial soul), but Aristotle's anthropocentric cosmology turned out to be quite compatible with Judeo-Christian anthropocentrism. In the Christian version of the "great chain of being," the hierarchical ladder led from a transcendent God, angels, men, women, and children, down to animals, plants, and the inanimate realm.[10] By way of summarizing this medieval culmination and synthesis of Greek and Christian thought, philosopher Kurt Baier remarked: "The medieval Christian world picture assigned to man [humans] a highly significant, indeed the central part in the grand scheme of things. The universe was made for the express purpose of providing a stage on which to enact a drama starring Man in the title role."[11] The West has had several decisive historical opportunities to leave the path of the narrow and ecologically destructive "anthropocentric detour" and return to ecocentrism, but the dominant culture has not done so. In his classic paper of the 1960s that linked anthropocentrism with the environmental crisis, historian Lynn White, Jr., claimed that, in the thirteenth century, Saint Francis of Assisi tried to undermine the Christian views of human dominance over, and separation from, the rest of nature. According to White: "Francis tried to depose man from his monarchy over creation and set up a democracy of all God's creatures. . . . The greatest spiritual revolutionary in Western history, Saint Francis, proposed what he thought was an alternative Christian view of nature and man's relation to it: he tried to substitute the idea of the equality of all creatures, including man, for the idea of man's limitless rule of creation. He failed."[12]

II. THE RISE OF THE ANTHROPOCENTRIC MODERN WORLD

The second opportunity to abandon the "anthropocentric detour" came in the seventeenth century with the development of the nonanthropocentric philosophical system of the Dutch philosopher, Baruch Spinoza. But even as the West underwent the major intellectual/social paradigm shift from the medieval to the modern world in the seventeenth and eighteenth centuries, the anthropocentrism of the Greek and Judeo-Christian traditions continued to dominate the major theorists of this period. For example, the leading philosophical spokesmen for the scientific revolution (Francis Bacon, René Descartes, and Liebniz) were all strongly influenced by Christian anthropocentric theology. Bacon claimed that modern science would allow humans to regain a command over nature, which had been lost with Adam's fall in the garden. Descartes, considered to be the "father" of modern Western philosophy, argued that the new science would make humans the "masters and possessors of nature." Also in keeping with his Christian background, Descartes's famous "mind-body dualism" resulted in the view that only humans had minds (or souls); all other creatures were merely bodies (machines); they had no sentience (mental life) and, as a result, could feel no pain. And so, in the middle of the nineteenth century, Darwin had to argue, against prevailing opinion, that at least the great apes experienced various feelings and emotions! Descartes set Western academic philosophy on a subjectivist, "inside-out" epistemological path that, among other things, made the reality of the world, apart from human consciousness, derivative and thus highly suspect. It was certainly not possible, in this view, to recognize that nature itself had intrinsic value or was sacred.[13]

These Christian anthropocentric attitudes combined with, and reinforced, Renaissance anthropocentric humanism, which arose prior to the scientific revolution (from classical Greek and Roman sources), in, for example, the fifteenth-century pronouncements of Pico della Mirandola and continued with the Enlightenment philosophers into the twentieth century with Karl Marx, John Dewey, and the humanistic existentialism of Jean-Paul Sartre. Like medieval Christianity, Renaissance humanism portrayed humans as the central fact in the universe while also supporting the exalted view that humans had unlimited powers, potential, and freedom (what eco-philosopher Peter Gunter refers to as "man-infinite").[14]

Modern science, however, turned out to be a two-edged sword. As we have seen, seventeenth-century science, on the one hand, was conceived within a Christian matrix with the avowed anthropocentric purpose of conquering and dominating nature. This led Lynn White to claim that both modern science and the more recent scientifically based technology "is permeated with Christian arrogance toward nature."[15]

On the other hand, the development of modern theoretical science over the last three hundred years, has resulted in the replacement of the Aristotelian anthropocentric cosmology with essentially the original non-anthropocentric cosmological worldview of the pre-Socratics, first in astronomy with heliocentrism, the infinity of the universe, and cosmic evolution, then in biology with Darwinian evolution. Ecology, as the "subversive science," has, even more than Darwinian biology, stepped across the anthropocentric threshold, so to speak, and implied an ecocentric orientation to the world. Thus, as the modern West has tried to cling to its anthropocentric illusions, each major theoretical scientific development since the seventeenth century has served to "decentralize" humans from their preeminent place in the Aristotelian-Christian cosmology. Lynn White also recognized this aspect of the development of modern science when he pointed out: "Despite Copernicus, all the cosmos rotates around our little globe. Despite Darwin, we are not, in our hearts, part of the natural process. We are superior to nature, contemptuous of it, willing to use it for our slightest whim."[16]

Carrying this a step further, the physicist Fritjof Capra claims that twentieth-century developments in Einsteinian relativity and quantum physics have undermined the atomistic "discrete entity" mechanistic paradigm of seventeenth-century science and now point instead to a process metaphysics of interrelatedness more typical of Eastern cosmologies such as Hinduism. Capra finds the parallels between the "new physics," Eastern cosmologies, and the ecological sense of interrelatedness so striking that he has said: "I think what physics can do is help generate ecological awareness."[17] Thomas Berry, and the physicist Brian Swimme, claim that the unfolding of twentieth-century science actually provides us with a new, non-anthropocentric "cosmic story" within which humans can now metaphysically find their place and values in an evolutionary/ecological context.[18]

From the vantage point of the late twentieth century and the deepening environmental crisis, it is now possible to see that the transition from the medieval to the modern world was not as radical as many Enlightenment-inspired historians have thought. The key was not so much a shift to reason, proper empirical scientific method, or the theories of the new sciences (which in many cases were a rediscovery of the theories of the pre-Socratics). Rather, the transition amounted to an increasingly unfettered playing out of the full implications of a radical anthropocentrism with roots deep in ancient Western culture: the discovery of new worlds to plunder by a resource-depleted, ecologically damaged Europe; the development of a new science, science-driven technology, and industrialization capable of inflicting great damage on the earth; the exponential growth of human populations; the crowding out and extinction of other species through habitat degradation and loss; but ultimately the desanctification of Nature and the concept of private property giving rise to economic systems designed to treat the earth exclusively as a human resource and as a commodity, together with the reconceptualization and reinforcement of humans as basically Hobbesian, selfish exploiters and consumers. All this was to be the new freedom under democratic regimes for humans.

III. SPINOZA'S INFLUENCE UPON MODERN ECOCENTRISM

Spinoza's seventeenth-century philosophical system provided the second intellectual opportunity for Western culture to abandon the anthropocentric detour. Drawing upon ancient Jewish pantheistic roots, Spinoza seemed to foresee the non-anthropocentric development of modern science. Spinoza attempted to resanctify the world by identifying God with nature, which was conceived of as each and every existing entity—human and nonhuman. Unlike Descartes, mind (or the mental attribute), for Spinoza, was found throughout nature. Through criticism of Hobbes and Descartes, Spinoza developed a philosophy that would channel the new scientific understanding of nature primarily into spiritual human self-realization and into an appreciation of God/nature, rather than into the misguided attempt to dominate and control nature. Some contemporary philosophers believe that Spinoza's system is the most sophisticated ever developed in the West and that it provides a fruitful guide to ecological understanding and self-realization. Others point out that it is the Western system most similar to Eastern thought, especially Zen Buddhism.[19]

Spinoza's pantheistic vision did not derail the dominant Western philosophic and religious anthropocentrism and the dream of the human conquest of nature in the seventeenth century, but it influenced many who questioned and resisted this trend. Some of the leading figures of the eighteenth-century European romantic movement (the main Western countercultural force speaking on behalf of nature and against the uncritical and unbridled enthusiasm of the scientific and Industrial Revolutions) were inspired by Spinoza's pantheism.

Two of the leading intellects of the early twentieth century, Bertrand Russell and Albert Einstein, were deeply influenced by Spinoza. Einstein called himself a "disciple of Spinoza," expressed his admiration as well for Saint Francis, and held that "cosmic religious feeling" was the highest form of religious life.[20]

Recent scholarship has revealed that Russell's cosmology, ethics, and religious orientation were essentially Spinozistic. At the end of World War II and long before there was any public awareness of a pending environmental crisis, Russell was a lone philosopher (with the possible exception of Heidegger) speaking out against anthropocentrism. The philosophies of Marx and Dewey are anthropocentric, Russell claimed; they place humans in the center of things and "have been inspired by scientific *technique*." Further, they are "power philosophies, and tend to regard everything non-human as mere raw material." Prophetically, Russell warned that the desire of Dewey (and Marx) for social power over nature "contributes to the increasing danger of vast social disaster."[21]

Einstein and Russell were drawn to Spinoza largely from the perspective of cosmology and astronomy (although Russell was beginning to see the implications of Spinoza's pantheism for the human exploitation of the earth).[22] An explicitly ecological expression of Spinoza's pantheism had to await the California poet Robinson Jeffers and the Norwegian philosopher Arne Naess.

From his perch on the Hawk Tower on the Carmel coast beginning in the 1920s, Jeffers developed a pantheistic philosophy, expressed through his poetry, which he called "Inhumanism," as a counterpoint to Western anthropocentrism. One commentator claimed that Jeffers was "Spinoza's twentieth century evangelist." And Jeffers's ecocentric philosophy and poetry has inspired David Brower, Ansel Adams, Nancy Newhall, and other prominent contemporary environmentalists.[23]

Arne Naess lived in wild mountain and coastal areas of Norway as a young boy and, at the age of seventeen, Spinoza became his favorite philosopher. Two years later, while a young student in Paris, Naess became "hooked" on Gandhi as well, as a result of Gandhi's 1931 "salt march." As the age of ecology dawned in the 1960s, Naess was influenced by Rachel Carson's great efforts and began to see the relevance of the philosophies of Spinoza and Gandhi for ecological understanding and action. He formed Spinoza study groups at the University of Oslo that produced important and original Spinoza scholarship. Naess has claimed that "no great philosopher has so much to offer in the way of clarification and articulation of basic ecological attitudes as Baruch Spinoza."[24]

IV. The Ecocentrism of Santayana and Muir

While Spinoza's path to ecocentrism was not taken in the seventeenth and eighteenth centuries by the dominant Western culture, yet a third opportunity occurred in America at the beginning of the twentieth century. Echoing the spirit of Thoreau, Harvard University philosopher George Santayana had grown increasingly disillusioned with his anthropocentric, pragmatist, and idealist colleagues, with the direction of urbanization, and with the economic-technological domination-over-nature path of late nineteenth-century America. Upon his retirement, Santayana came to the University of California, Berkeley, in 1911 to deliver a parting shot at the prevailing anthropocentric American philosophy and religion. In his lecture, Santayana pointed out that

> a Californian whom I had recently the pleasure of meeting observed that, if the philosophers had lived among your mountains their systems would have been different ... from what those systems are which the European genteel tradition has handed down since Socrates; for these systems are egotistical; directly or indirectly they are anthropocentric, and inspired by the conceited notion that man, or human reason, is the center and pivot of the universe. That is what the mountains and the woods should make you at last ashamed to assert.[25]

Santayana claimed that only one American writer, Walt Whitman, had escaped anthropocentrism by extending the democratic principle "to the animals, to inanimate nature, to the cosmos as a whole. Whitman was a pantheist; but his pantheism, unlike that of Spinoza, was unintellectual, lazy, and self-indulgent." Santayana looked forward to an ecocentric revolution in philosophy.[26]

It is not known whether Santayana was aware of John Muir. But Muir, still alive in California at the time of Santayana's Berkeley lecture, was the one American who (with the possible exception of Thoreau) preeminently exemplified Santayana's ecocentric revolution. Muir's personal papers were unavailable to scholars until the mid-1970s, and so the full extent of his philosophic ecocentric achievement was unknown until studies on Muir began to appear in the 1980s.[27] As a result of his deep personal experiences in nature (and independently of the Transcendentalists), Muir had rejected the anthropocentrism of his strict Calvinist upbringing ("Lord Man"), arriving at a basically ecocentric perspective by the age of twenty-nine during his one-thousand-mile walk to the Gulf of Mexico in 1867. For the next ten years, Muir wandered through Yosemite and the High Sierra, climbing mountains, studying glaciers, and further developing his ecological consciousness. He arrived at the major generalizations of ecology by direct observation and intuitive experiencing of nature. In 1892, Muir was drafted to be the first president of the Sierra Club, a position he held until his death in 1914. Only now is he belatedly recognized by historians as the founder of the American conservation (environmental) movement.[28]

The turning point for early twentieth-century ecocentrism occurred in the confrontation between Muir, Gifford Pinchot, and President Theodore Roosevelt. Muir camped with Roosevelt in Yosemite and tried to influence his philosophical outlook but, by 1908, Roosevelt had turned away from Muir's ecocentrism and adopted Pinchot's anthropocentric policies of the scientific-technological management and development of nature as a human resource and commodity. A unique opportunity to set America on an ecocentric ecological path was lost, and Santayana's lecture had largely been in vain.[29]

V. ECOCENTRISM AND THE FUTURE

While America and the rest of Western culture continued with the anthropocentric detour, now under the resource conservation and development policies of Pinchot, ecocentrism remained alive during this period in the writings of D. H. Lawrence, Robinson Jeffers, Joseph Wood Krutch, and various professional ecologists, including Aldo Leopold. Ecocentrism was to reemerge as an intellectual force in American life during the age of ecology of the 1960s and 1970s (the transition "from conservation to ecology") in the writings of Aldous Huxley, Paul Sears, Loren Eiseley, Rachel Carson, Lynn White, Jr., Paul Shepard, and in Gary Snyder, Edward Abbey, Arne Naess, and the Deep Ecology movement. John Muir's philosophy again came to the fore as a guiding beacon.

In the 1980s, as the philosophical differences between ecocentrism and anthropocentrism loomed more dramatically into view, and as Earth First! appeared and green political parties tried to sort out their philosophical underpinnings, debates broke out among the supporters of the traditions of Muir, Pinchot, and Marx.[30] From an ecological standpoint, which constitutes the genuinely radical tradition? Scholars are now in agreement that Marx's environmental views are anthropocentric and most closely resemble Pinchot's.[31] As environmental historian Stephen Fox assesses the situation: "The conservation movement, the most successful exercise in anti-modernism, corresponded to the Russian Revolution. Muir was its Lenin. Pinchot was its Stalin."[32]

As professional ecologists have been documenting since the 1960s, humans are seriously out of balance with the rest of the earth. A human population of one to two billion living lightly on the planet would probably be sustainable given the ecological requirements of carrying capacity for all the earth's species. As things now stand, according to Michael Soulé and the new field of conservation biology, almost all currently protected wilderness areas and wildlife preserves around the world are too small, and the boundaries are not ecologically drawn. As a result, natural evolution and continued speciation for many species on the planet have ground to a halt, and massive species extinctions and wild ecosystem degradation are inevitable. To correct this situation, Arne Naess once suggested that an ideal ecological balance on the planet might consist of one-third wilderness (wild species habitat), one-third "free nature" (where there are mixed communities of human and wild species living in largely nondomesticated ecosystems), and one-third cities, roads, agriculture, etcetera, for intensive human inhabitation of the planet.[33]

As the "social justice" movement (with its anthropocentric lineage) attempts to join forces with the ecocentric ecological movement, what will be the outcome?[34] Will the very real need to secure human livelihood and equality, a toxin-free human environment, and "jobs" in the short-run in a seriously overcrowded world overshadow the necessity to protect and restore the long-range ecological integrity of the planet? Have most humans now become so thoroughly domesticated in their urban environments that they are unable to perceive the need for a wild planet, wild species and eco-systems, and the wild in themselves? Murray Bookchin and the "social ecologists" have, for the most part, demonstrated an unwillingness or inability to transcend a narrow anthropocentric perspective and consider the necessity for human population stabilization and reduction, and the high priority of protection and restoration of wild species and ecosystems.[35] The recent upheaval in Earth First! has resulted in the ouster of ecocentrically oriented leaders and their replacement by less-imaginative leaders with a "social justice" background who want to emphasize urban pollution problems and deemphasize wild ecosystem protection.

It appears that the age of ecology stands today at a major crossroads. One path follows the crucial philosophical and ecological insights of Thoreau and Muir and Naess, while incorporating, to the extent possible in this radically out-of-balance world, the concerns of social justice. It involves what Gary Snyder calls "the practice of the wild."[36] It means recapturing for humanity what philosopher Max Oelschlaeger refers to as "Paleolithic consciousness."[37] All of the ecological evidence and wisdom suggest that the other path (a continuation of the "anthropocentric detour") will lead inexorably, perhaps with the best of intentions, to an accelerating decline of the earth and all its inhabitants.

Notes

1. Warwick Fox, "The Deep Ecology-Ecofeminism Debate and Its Parallels, *Environmental Ethics* 11, no. 1 (1989): 5–25, quotation on 5.
2. Stan Steiner, *The Vanishing White Man* (New York: Harper and Row, 1976), 113.
3. See also J. Baird Callicott, "Traditional American Indian and Western European Attitudes Toward Nature," *Environmental Ethics* 4 (Winter 1982): 293; J. Donald Hughes, *American Indian Ecology* (El Paso: Texas Western Press, 1982); Max Oelschlaeger, *The Idea of Wilderness: From Prehistory to the Age of Ecology* (New Haven, Conn.: Yale University Press, 1991), 1–30.
4. See also Oelschlaeger, *Idea of Wilderness*.
5. Paul Shepard, *The Tender Carnivore and the Sacred Game* (New York: Scribner's, 1973), 237.
6. H. Frankfort, *Kingship and the Gods* (Chicago: University of Chicago Press, 1948), 342.
7. Loren Eiseley, "The Last Magician," in *The Invisible Pyramid*, ed. L. Eiseley (New York: Scribner's, 1970), 154.
8. D. H. Lawrence, "Pan in America" (1926), reprinted in *The Everlasting Universe*, ed. L. Forstner and J. Todd (Lexington, Mass.: D. C. Heath, 1971), 221; for discussions of Lawrence's ecological views, see Del Ivan Janik, "Environmental Consciousness in Modern Literature," in *Ecological Consciousness*, ed. J. Hughes and R. Schwartz (Washington, D.C.: University Press of America, 1981); and Dolores LaChapelle, "D. H. Lawrence and Deep Ecology," *Trumpeter* 7 (Spring 1990): 26–30.
9. Karl Popper, "Back to the Presocratics," in *Conjectures and Refutations*, ed. Karl Popper (London: Routledge and Kegan Paul, 1965), 141.
10. For a critical discussion of the anthropocentric Christian "great chain of being" from an ecofeminist perspective, see Elizabeth Gray, *Green Paradise Lost* (Wellesley, Mass.: Roundtable Press, 1982).
11. K. Baier, "The Meaning of Life: Christianity versus Science," in *Philosophy for a New Generation*, ed. A. Bierman and J. Gould (New York: Macmillan, 1973), 596.
12. L. White, Jr., "Historical Roots of Our Ecologic Crisis," in Forstner and Todd, *Everlasting Universe*, 16.
13. For an ecological critique of Bacon, Descartes, and Leibniz, see Clarence J. Glacken, "Man against Nature: An Outmoded Concept," in *The Environmental Crisis*, ed. H. Helfrich, Jr. (New Haven, Conn.: Yale University Press, 1970);

William Leiss, *The Domination of Nature* (New York: Braziller, 1970). For critiques of the anthropocentric development of Western academic philosophy, see John B. Cobb, Jr., "The Population Explosion and the Rights of the Subhuman World," in *Environment and Society: A Book of Readings on Environmental Policy, Attitudes, and Values,* ed. Robert T. Roelofs (Englewood Cliffs, N.J.: Prentice-Hall, 1974); George Sessions, "Anthropocentrism and the Environmental Crisis," *Humboldt Journal of Social Relations* 2 (Spring 1974): 71–81; Eugene Hargrove, *Foundation of Environmental Ethics* (Englewood Cliffs, N.J.: Prentice-Hall, 1989), 14–47.

14. For ecological critiques of humanism, see Roderick French, "Is Ecological Humanism a Contradiction in Terms?: The Philosophical Foundations of the Humanities Under Attack," in Hughes and Schwartz, *Ecological Consciousness,* 43–66; David Ehrenfeld, *The Arrogance of Humanism* (Oxford: Oxford University Press, 1978).

15. White, "Historical Roots," 15.

16. White, "Historical Roots," 15.

17. Fritjof Capra, *The Tao of Physics* (Berkeley, Calif.: Shambala, 1975). See also J. Baird Callicott, "Intrinsic Value, Quantum Theory, and Environmental Ethics," *Environmental Ethics* 7 (Fall 1985): 257; Michael Zimmerman, "Quantum Theory, Intrinsic Value, and Panentheism," *Environmental Ethics* 10 (Spring 1988): 3; Andrew McLaughlin, "Images and Ethics of Nature," *Environmental Ethics* 7 (Winter 1985): 293.

18. Thomas Berry, *The Dream of the Earth* (San Francisco: Sierra Club Books, 1989); Thomas Berry and Brian Swimme, *The Universe Story* (Clinton, Wash.: New Story Productions, 1989).

19. George Sessions, "Western Process Metaphysics: Heraclitus, Whitehead, and Spinoza," in *Deep Ecology: Living as if Nature Mattered* ed. Bill Devall and George Sessions (Salt Lake City: Peregrine Smith, 1985), 236–42; Paul Wienpahl, *The Radical Spinoza* (New York: New York University Press, 1979); Arne Naess, "Through Spinoza to Mahayana Buddhism, or Through Mahayana Buddhism to Spinoza?" in *Spinoza's Philosophy of Man,* ed. Jon Wetlesen (Oslo: University Press, 1978).

20. Albert Einstein, *The World as I See It* (New York: Philosophical Library, 1942), 14.

21. Bertrand Russell, *A History of Western Philosophy* (New York: Simon and Schuster, 1945), 494.

22. B. Hoffman and H. Dukas, *Albert Einstein: Creator and Rebel* (New York: New American Library, 1972), 94–95; Arne Naess, "Einstein, Spinoza, and God," in *Old and New Questions in Physics, Cosmology, Philosophy and Theoretical Biology,* ed. Alwyn van der Merwe (Holland: Plenum, 1983), 683–87; Kenneth Blackwell, *The Spinozistic Ethics of Bertrand Russell* (London: Allen and Unwin, 1985).

23. David Brower, ed., *Not Man Apart: Lines from Robinson Jeffers* (San Francisco: Sierra Club Books, 1965): Ansel Adams, *Ansel Adams: An Autobiography* (Boston: Little, Brown, 1985), 84–87; George Sessions, "Spinoza and Jeffers on Man in Nature," *Inquiry* 20 (1977): 481; for "Jeffers as Spinoza's Evangelist," see Arthur Coffin, *Robinson Jeffers: Poet of Inhumanism* (Madison: University of Wisconsin Press, 1971), 255.

24. Arne Naess, *Freedom, Emotion, and Self-Subsistence: The Structure of a Central Part of Spinoza's Ethics* (Oslo: University Press, 1975), 118–119; Arne Naess, "Spinoza and Ecology," in *Speculum Spinozanum,* ed. S. Hessing (Boston: Routledge and Kegan Paul, 1978); Arne Naess, *Ghandi and Group Conflict* (Oslo: University Press, 1974).

25. George Santayana, "The Genteel Tradition in American Philosophy," in *Winds of Doctrine* (New York: Scribner's, 1926).

26. Ibid., 203. For a discussion of Santayana's address in an ecological context, see William Everson, *Archetype West* (Berkeley, Calif.: Oyez Press, 1976), 54–60.

27. For the new Muir scholarship and discussions of his ecocentric philosophy, see Frederick Turner, *Rediscovering America: John Muir in His Time and Ours* (San Francisco: Sierra Club Books, 1985); Stephen Fox, *John Muir and His Legacy: The American Conservation Movement* (Boston: Little, Brown, 1981), 43–53, 59, 79–81, 289–91, 350–55, 361; and especially, Michael Cohen, *The Pathless Way: John Muir and American Wilderness* (Madison: University of Wisconsin Press, 1984), chaps. 1, 6, 7.

28. See Fox, *John Muir and His Legacy*; Michael Cohen, *The History of the Sierra Club 1892–1970* (San Francisco: Sierra Club Books, 1988).

29. For the split between Muir, Pinchot, and Roosevelt, see Roderick Nash, *Wilderness and the American Mind*, 3d ed. (New Haven, Conn.: Yale University Press, 1982), 129–40; Fox, *John Muir and His Legacy*, 109–30; Cohen, *Pathless Way*, 160–61, 292ff; Bill Devall and George Sessions, "The Development of Natural Resources and the Integrity of Nature," *Environmental Ethics* 6 (Winter 1984): 293.

30. See, for example, Robyn Eckersley, "The Road to Ecotopia?: Socialism versus Environmentalism," *Trumpeter* 5 (Summer 1988): 60; Francis Moore Lappé and J. Baird Callicott, "Marx Meets Muir: Toward a Synthesis of the Progressive Political and Ecological Visions," *Tikkun* 2 (Winter 1987): 16.

31. Howard L. Parsons, ed., *Marx and Engels on Ecology* (New Haven, Conn.: Greenwood Press, 1977); Val Routley, "On Karl Marx as an Environmental Hero," *Environmental Ethics* 3 (Spring 1981): 237; Donald Lee, "Toward a Marxian Ecological Ethic," *Environmental Ethics* 4 (Winter 1982): 339–43; John Clark, "Marx's Inorganic Body," *Environmental Ethics* 11 (Fall 1989): 243.

32. Fox, *John Muir and His Legacy* (Boston: Little, Brown, 1981).

33. For a discussion of the findings of Soulé and the conservation biologists, see Christopher Manes, *Green Rage: Radical Environmentalism and the Unmaking of Civilization* (Boston: Little, Brown, 1990), 34–35; George Sessions, "Ecocentrism, Wilderness and Global Ecosystem Protection," in *The Wilderness Condition: Essays on Environment and Civilization*, ed. Max Oelschlaeger (San Francisco: Sierra Club Books, 1991); Arne Naess, "Ecosophy, Population, and Free Nature," *Trumpeter* 5 (Fall 1988): 113.

34. Lappé and Callicott, "Marx Meets Muir."

35. See Murray Bookchin, *The Philosophy of Social Ecology: Essays on Dialectical Naturalism* (New York: Black Rose Books, 1990).

36. Gary Snyder, *The Practice of the Wild* (San Francisco: North Point Press, 1990).

37. Oelschlaeger, *Idea of Wilderness*.

The Concept of Social Ecology

MURRAY BOOKCHIN

THE RECONSTRUCTIVE AND DESTRUCTIVE tendencies in our time are too much at odds with each other to admit of reconciliation. The social horizon presents the starkly conflicting prospects of a harmonized world and an ecological sensibility based on a rich commitment to community, mutual aid, and new technologies on the one hand, and the terrifying prospect of some sort of thermonuclear disaster on the other. Our world, it would appear, will either undergo revolutionary changes, so far-reaching in character that humanity will totally transform its social relations and its very conception of life, or it will suffer an apocalypse that may well end humanity's tenure on the planet.

The tension between these two prospects has already subverted the morale of the traditional social order. We have entered an era that consists no longer of institutional stabilization but of institutional decay. A widespread alienation is developing toward the forms, the aspirations, the demands, and above all the institutions of the established order. The most exuberant, in fact, theatrical evidence of this alienation occurred in the sixties, when the "youth revolt" exploded into what seemed to be a counterculture. Considerably more than protest and adolescent nihilism marked the period. Almost intuitively, new values of sensuousness, new forms of communal life-style, changes in dress, language, and music, all borne on the wave of a deep sense of impending social change, rolled over a sizable section of an entire generation. We still do not know in what sense this wave began to ebb: whether as a historic retreat or as a transformation into a serious project for inner and social development. That the symbols of this movement eventually became the artifacts for a new culture industry does not alter the movement's far-reaching effects. Western society will never be the same again. . . .

From: *CoEvolution Quarterly* (Winter 1981): 15–22.

Crucial as this decay of institutions and values may be, it by no means exhausts the problems that confront the existing society. Intertwined with the social crisis is a crisis that has emerged directly from man's exploitation of the planet.[1] Established society is faced with a breakdown not only of its values and institutions, but also of its natural environment. This problem is not unique to our times: the dessicated wastelands of the Near East, the areas where the arts of agriculture and urbanism had their beginnings, are evidence of ancient human despoliation. But this example pales before the massive destruction of the environment that has occurred since the days of the Industrial Revolution, and especially since the end of the Second World War. The damage inflicted on the environment by contemporary society encompasses the entire earth. The exploitation and pollution of the earth have damaged not only the integrity of the atmosphere, climate, water resources, soil, flora and fauna of specific regions, but also the basic natural cycles on which all living things are dependent.

Yet modern man's capacity for destruction is quixotic evidence of his capacity for reconstruction. The powerful technological agents we have unleashed against the environment include many of the very agents we require for its reconstruction. What we crucially lack is the consciousness and sensibility that will help us achieve such eminently desirable goals—a consciousness and sensibility that is far broader than we customarily mean by these terms. Our definitions must include not only the ability to reason logically and respond emotionally in a humanistic fashion; they must also include a fresh awareness of the relatedness between things and an imaginative insight into the possible. On this score, Marx was entirely correct to emphasize that the revolution required by our time must draw its poetry not from the past but from the future, from the humanistic potentialities that lie on the horizons of social life.

The new consciousness and sensibility cannot be poetic alone; they must also be scientific. Indeed, there is a level at which our consciousness must be neither poetry nor science, but a transcendence of both into a new realm of theory and practice, an artfulness that combines fancy with reason, imagination with logic, vision with technique. We cannot shed our scientific heritage without returning to a rudimentary technology, with its shackles of material insecurity, toil, and renunciation. By the same token, we cannot allow ourselves to be imprisoned within a mechanistic outlook and a dehumanizing technology—with its shackles of alienation, competition, and brute denial of humanity's potentialities. Poetry and imagination must be integrated with science and technology, for we have evolved beyond an innocence that can be nourished exclusively by myths and dreams.

Is there a scientific discipline that allows for the indiscipline of fancy, imagination, and artfulness? Can it encompass problems created by the social and environmental crises of our time? Can it integrate critique with reconstruction, theory with practice, vision with technique? In view of the enormous dislocations that now confront us, our own era raises the need for a more sweeping and insightful body of knowledge—scientific as well as social—to deal with our problems. Without renouncing the gains of earlier scientific and social theories, we are obliged to develop a more rounded critical analysis of our relationship with the natural world. We must seek the foundations for a more reconstructive approach to the grave problems posed by the apparent "contradictions" between nature and society. We can no longer afford to remain captives to the tendency of the more traditional sciences to dissect phenomena and examine their fragments. We must combine them, relate them, and see them in their totality as well as their specificity.

In response to these needs, we have formulated a discipline unique to our age: "social ecology." The more well-known term "ecology" was coined by Ernst Haeckel a century ago to denote the investigation of the inter-relationships between animals, plants, and their inorganic environment. Since Haeckel's day, the term has been expanded to include ecologies of cities, of health, and of the mind. This proliferation of a word into widely disparate areas may seem particularly desirable to an age that fervently seeks some kind of intellectual coherence and unity of perception. But it can also prove to be extremely treacherous. Like such newly arrived words as "holism," "decentralization," and "dialectics," the term "ecology" runs the peril of merely hanging in the air without any roots, context, or texture. Often it is used as a metaphor, an alluring catchword, that loses the potentially compelling internal logic of its premises.

Accordingly, the radical thrust of these words is easily neutralized. "Holism" evaporates into a mystical sigh, a rhetorical expression for ecological fellowship and community that ends with such in-group greetings and salutations as "holistically yours." What was once a serious philosophical stance has been reduced to environmentalist kitsch. "Decentralization" commonly means logistical alternatives to gigantism, not the human scale that would make an intimate and direct democracy possible. "Ecology" fares even worse. All too often it becomes a metaphor, like the word "dialectics," for any kind of integration and development. Perhaps even more troubling, the word in recent years has been identified with a very crude form of natural engineering which might well be called "environmentalism."

I am mindful that many ecologically oriented individuals use "ecology" and "environmentalism" interchangeably. Here, I would like to draw a semantically convenient distinction. By "environmentalism" I propose to designate a mechanistic, instrumental outlook that sees nature as a passive habitat composed of "objects" such as animals, plants, minerals, and the like which must merely be rendered more serviceable for human use. Given my use of the term, "environmentalism" tends to reduce nature to a storage bin of "natural resources" or "raw materials." Within this context, very little of a social nature is spared from the environmentalist's vocabulary: cities become "urban resources" and their inhabitants "human resources." If the word "resources" leaps out so frequently from environmentalistic discussions of nature, cities, and people, an issue more important than mere word play is at stake. Environmentalism, as I use this term, tends to view the ecological project for attaining a harmonious relationship between humanity and nature as a truce rather than a lasting equilibrium. The harmony of the environmentalist centers around the development of new techniques for plundering the natural world with a minimal disruption of the human habitat. Environmentalism does not question the most basic premise of the present society, notably, that humanity must dominate nature; rather, it seeks to *facilitate* that notion by developing techniques for diminishing the hazards caused by the reckless despoliation of the environment.

To distinguish ecology from environmentalism and from abstract, often obfuscatory definitions of the term, I must return to its original usage and explore its direct relevance to society. Put quite simply: ecology deals with the dynamic balance of nature, with the interdependence of living and nonliving things. Since nature also includes human beings, the science must include humanity's role in the natural world—specifically, the character, form, and structure of humanity's relationship with other species and with the inorganic substrate of the biotic environment. From a critical viewpoint, ecology opens to wide purview the vast disequilibrium that has emerged from humanity's split with the natural world. One of nature's very unique species, *Homo sapiens*, has slowly and painstakingly developed from the natural world into a unique social world of its own. As both worlds interact with each other through highly complex phases of evolution, it has become as important to speak of a *social ecology* as to speak of a natural ecology.

Let me emphasize that the failure to explore these phases of human evolution—which have yielded a succession of hierarchies, classes, cities, and finally states—is to make a mockery of the term "social ecology." Unfortunately, the discipline has been beleaguered by self-professed acolytes who continually try to collapse all the phases of natural and human development into a universal "oneness" (not wholeness), a yawning "night in which all cows are black," to apply one of Hegel's caustic phrases to a widely accepted pop mysticism that clothes itself in ecological verbiage. If nothing else, our common use of the word "species" to denote the wealth of life around us should alert us to the fact of *specificity*, of *particularity*—the rich abundance of *differentiated* beings and things that enter into the very subject matter of natural ecology. To explore these differentia, to examine the phases and interfaces that enter into their making and into humanity's long development from animality to society—a development latent with problems and possibilities—is to make social ecology one of the most powerful disciplines from which to draw our critique of the present social order.

But social ecology not only provides a critique of the split between humanity and nature; it also poses the need to heal them. Indeed, it poses the need to radically transcend them. As E. A. Gutkind pointed out, "The goal of Social Ecology is wholeness, and not mere adding together of innumerable details collected at random and interpreted subjectively and insufficiently." The science deals with social and natural relationships in communities or "ecosystems."[2] In conceiving them holistically, that is to say, in terms of their mutual interdependence, social ecology seeks to unravel the *forms* and *patterns* of interrelationships that give intelligibility to a community, be it natural or social. Holism, here, is the result of a conscious effort to discern how the particulars of a community are arranged, how its geometry (as the Greeks might have put it) makes the whole more than the sum of its parts. Hence, the wholeness to which Gutkind refers is not to be mistaken for a spectral oneness that yields cosmic dissolution in a structureless nirvana; it is a richly articulated structure that has a history and an internal logic of its own.

History, in fact, is as important as form or structure. To a large extent, the history of a phenomenon *is* the phenomenon itself. We are, in a real sense, everything that existed before us and, in turn, we can eventually become vastly more than we are. Surprisingly, *very* little in the evolution of life forms has been lost in natural and social evolution, as the embryonic development in our very bodies attests. Evolution lies within us (as well as around us) as parts of the very nature of our beings. For the present, it suffices to point out that wholeness is not a bleak undifferentiated universality that involves the reduction of a phenomenon to what it has in common with everything else. Nor is it a celestial, omnipresent energy that replaces the vast material differentia of which the natural and social realms are composed. To the contrary, wholeness comprises the variegated structures, articulations, and mediations that impart to the whole a rich variety of forms and thereby add unique qualitative properties to what a strictly analytic mind often reduces to "innumerable" and "random" details.

Terms like "wholeness," "totality," and even "community" have perilous nuances for a generation that has known fascism and other totalitarian ideologies. The words evoke images of a "wholeness" achieved through homogenization, standardization, and a repressive coordination of human beings. These fears are reinforced by a "wholeness" that seems to provide an inexorable finality to the course of human history—one that implies a suprahuman, narrowly teleological concept of "social law" which denies the ability of human will and individual choice to shape the course of social events. Such notions of social law and teleology have been used to achieve a ruthless subjugation of the individual to suprahuman forces beyond human control. Our century has been afflicted by a plethora of totalitarian ideologies that, placing human beings in the service of "history," have denied them a place in the service of their own humanity.

Actually, such a totalitarian concept of wholeness stands sharply at odds with what ecologists denote by the term. In addition to comprehending its heightened awareness of form and structure, we now come to a very important tenet of ecology: ecological wholeness means not an immutable homogeneity but rather the very opposite—a dynamic *unity of diversity*. In nature, balance and harmony are achieved by ever-changing differentiation, by ever-expanding diversity. Ecological stability, in effect, is a function not of simplicity and homogeneity but of complexity and variety. The capacity of an ecosystem to retain its integrity depends not upon the uniformity of the environment but upon its diversity.

If we assume that the thrust of natural evolution has been toward increasing complexity, that the colonization of the planet by life has been possible only as result of biotic variety, a prudent rescaling of man's hubris should call for caution in disturbing natural processes. To assume that science commands life's vast nexus of organic and inorganic interrelationships in all its details is worse than arrogance: it is sheer stupidity. If unity in diversity forms one of the cardinal tenets of ecology, the wealth of biota that exists in a single acre of soil leads us to still another basic ecological tenet: the need to allow for a high degree of natural spontaneity. The compelling dictum "respect for nature" has concrete implications.

Thus a considerable amount of leeway must be permitted for natural spontaneity, for the diverse biological forces that yield a variegated ecological situation. "Working with nature" largely means that we must foster the biotic variety that emerges from a spontaneous development of natural phenomena. I hardly mean that we must surrender ourselves to a mythical nature that is beyond all human comprehension and intervention, a nature that demands human awe and subservience. Perhaps the most obvious conclusion we can draw from these ecological tenets is Charles Elton's sensitive observation: "The world's future has to be managed, but this management would not be just like a game of chess—[but] more like steering a boat." What ecology, both natural and social, can hope to teach us is the way to find the current and understand the direction of the stream.

What ultimately distinguishes an ecological outlook as uniquely liberatory is the challenge it raises to conventional notions of hierarchy. Let me emphasize, however, that this challenge is implicit: it must be painstakingly elicited from the discipline, which is permeated by conventional scientistic biases. Ecologists are rarely aware that their science provides strong philosophical underpinnings to a nonhierarhical view of reality. Like many natural scientists, they resist philosophical generalizations as alien to their research and conclusions—a prejudice that is itself a philosophy rooted in the Anglo-American empirical tradition. Moreover, they follow their colleagues in other disciplines who model their notions of science on physics. This prejudice, which goes back to Galileo's day, has led to a widespread acceptance of systems theory in ecological circles. While systems theory has its place in the repertoire of science, it can easily become an all-encompassing, quantitative, reductionist theory of energetics if it acquires preeminence over *qualitative* descriptions of ecosystems, that is, descriptions rooted in organic evolution, variety, and holism. Whatever the merits of systems theory as an account of energy flow through an ecosystem, the primacy it gives to this quantitative aspect of ecosystem analysis fails to take adequate account of life forms as more than consumers and producers of calories.

If we recognize that every ecosystem can also be viewed as a food web, we can think of it as a circular, interlacing nexus of plant-animal relationships (rather than a stratified pyramid with man at its apex) that includes such widely varying creatures as microorganisms and large mammals. What ordinarily puzzles anyone who sees food-web diagrams for the first time is the impossibility of discerning a point of entry into the nexus. The web can be entered at any point and leads back to its point of departure without any apparent exit. Aside from the energy provided by sunlight (and dissipated by radiation) the system to all appearances is closed. Each species, be it a form of bacteria or deer, is knitted together in a network of interdependence with each other, however indirect the links may be. A predator in the web is also prey, even if the "lowliest" of organisms merely makes it ill or helps to consume it after death.

Nor is predation the sole link that unites one species with another. A resplendent literature now exists that reveals the enormous extent to which symbiotic mutualism is a major factor in fostering ecological stability and organic evolution. That plants and animals continually adapt to unwittingly aid each other, be it by an exchange of biochemical functions that are mutually beneficial or even dramatic instances of physical assistance and succor, has opened a whole new perspective on the nature of ecosystem stability and development.

We must not get caught up in direct comparisons between plants, animals, and human beings or in direct comparisons between plant-animal ecosystems and human communities. None of these is completely congruent with another. It is not in the *particulars* of differentiation that plant-animal communities are ecologically united with human communities but rather in their logic of differentiation. Wholeness, in fact, is completeness. The dynamic stability of the whole derives from a visible level of completeness in human communities as in climax ecosystems. What unites these modes of wholeness and completeness, however different they are in their specificity and their qualitative distinctness, is the *logic of development* itself. A climax forest is whole and complete as a result of the same unifying process—the same *dialectic* that makes a particular social form whole and complete.

When wholeness and completeness are viewed as the result of an immanent dialectic within phenomena, we do no more violence to the uniqueness of these phenomena than the principle of gravity does violence to the uniqueness of objects that fall within its "lawfulness." In this sense, the ideal of human roundedness, a product of the rounded community, is the legitimate heir to the ideal of a stabilized nature, a product of the rounded natural environment. Marx tried to root humanity's identity and self-discovery in its productive interaction with nature. But I must add that not only does humanity place its imprint on the natural world and transform it, but also nature places its imprint on the human world and transforms it. To use the language of hierarchy against itself: it is not only we who "tame" nature but also nature that "tames" us.

These turns of phrase should be taken as more than metaphors. Lest it seem that I have rarefied the concept of wholeness into an abstract dialectical principle, let me note that natural ecosystems and human communities interact with each other in very existential ways. Our animal nature is never so distant from our social nature that we can remove ourselves from the organic world outside us and from the one within us. From our embryonic development to our layered brain, we recapitulate the totality of our natural evolution. We are not so remote from our primate ancestry that we can ignore its physical legacy in our stereoscopic vision, acuity of intelligence, and grasping fingers. We phase into society as individuals in the same way that society, phasing out of nature, comes into itself.

These continuities, to be sure, are obvious enough. What is often less obvious is the extent to which nature itself is a realm of potentiality for the emergence of *social* differentia. Nature is as much a precondition for the *development* of society—not merely its emergence—as technics, labor, language, and mind. And it is a precondition not merely in William Petty's sense—that if labor is the "father" of wealth, nature is its "mother." This formula, so dear to Marx, actually slights nature by imparting to it the patriarchal notion of feminine passivity. The affinities between nature and society are more active than we care to admit. Very specific forms of nature, that is to say, very specific *ecosystems*, constitute the ground for very specific forms of society. At the risk of using a highly embattled phrase, I might say that a "historical materialism" of natural development could be written that would transform "passive nature"—the object of human labor—into "active nature," the creator of human labor. Labor's metabolism with nature cuts both ways, so that nature interacts *with* humanity to yield the actualization of their common potentialities in the natural and social worlds.

An interaction of this kind, in which terms like "father" and "mother" strike a false note, can be stated very concretely. The recent emphasis on bioregions as frameworks for various human communities provides a strong case for the need to readapt technics and work styles to accord with the requirements and possibilities of particular ecological areas. Bioregional requirements and possibilities place a heavy burden on humanity's claims of "sovereignty" over nature and "autonomy" from its needs. If it is true, as Marx wrote, that men make history but not under conditions of their own choosing, it is no less true that history makes society but not under conditions of its own choosing. The hidden dimension that lurks in this wordplay with Marx's famous formula is the natural history that enters into the making of social history—active, concrete, existential nature that emerges from stage to stage of its own evermore complex development in the form of equally complex and dynamic ecosystems. Our ecosystems, in turn, are interlinked into highly dynamic and complex bioregions. How concrete the hidden dimension of social development is—and how much humanity's claims to "sovereignty" must defer to it—has only recently become evident from our need to design an alternative technology that is as adaptive to a bioregion as it is productive to society. Hence our concept of wholeness is not a finished tapestry of natural and social relations that we can exhibit to the hungry eyes of sociologists. It is a fecund natural history, ever-active and ever-changing—the way childhood presses toward and is absorbed into youth, and youth into adulthood.

Within this highly complex context of ideas we must now try to transpose the nonhierarchical character of natural ecosystems to society. Sociobiology has made this project deceptively simple and crudely mechanistic. To refute the notion that savanna baboons are hierarchical, for example, we take refuge in the complete exclusivity of society—its immunity to natural principles. If nature is hierarchical, so the argument goes, need this be true of a human community guided by reason, love, and mutualism?

What renders social ecology so important is that it offers no case whatsoever for hierarchy in nature and society; it decisively challenges the very function of hierarchy as a stabilizing or "ordering" principle in *both* realms. The association of *order as such* with hierarchy is ruptured. And this association is ruptured without rupturing the association of nature with society—as sociology, in its well-meaning opposition to sociobiology, has been wont to do. In contrast to sociologists, we do not have to render the social world so supremely autonomous over nature that we are obliged to dissolve the continuum that phases nature into society. In short, we do not have to accept the brute tenets of sociobiology that link us crudely to nature at one extreme or the naive tenets of sociology that cleave us sharply from nature at the other extreme. Of course, the fact that hierarchy does exist in present-day society does not mean that it has to remain so—irrespective of its lack of meaning or reality for nature. But the case against hierarchy is not contingent on its uniqueness as a social phenomenon. That hierarchy threatens the existence of social life today certainly *does* mean that it cannot remain a social fact. And that it threatens the integrity of organic nature means that it *will* not remain so, given the harsh verdict of "mute" and "blind" nature. . . .

In concrete terms, what tantalizing issues does social ecology raise for our time and our future? In restoring a new and more advanced interface with nature than we have had previously, will it be possible to achieve a new balance between humanity and nature by sensitively tailoring our agricultural practices, urban areas, and technologies to the natural requirements of a region and the ecosystems of which it is composed? Can we hope to manage the natural environment by a drastic decentralization of agriculture which makes it possible to cultivate land as though it were a garden, balanced by a diversified fauna and flora? Will these changes require the decentralization of our cities into moderate-sized communities, creating a new balance between town and country? What technology will be required to achieve these goals— indeed, to avoid the further pollution of earth? What institutions will be required to create a new public sphere, what social relations to foster a new ecological sensibility, what forms of work to render human practice playful and creative, what sizes and populations of communities to scale life to human dimensions controllable by all? What kind of poetry? Concrete questions—ecological, social, political, and behavioral—rush in upon us like a flood that heretofore has been damned up by the constraints of traditional ideologies and habits of thought.

Let there be no mistake about it: the answers we provide to these questions have a direct bearing on whether or not humanity will be able to survive on the planet. The trends in our time are visibly directed against ecological diversity; in fact, they point toward a brute simplification of the entire biosphere. Complex food chains in the soil and on the earth's surface are being ruthlessly undermined by the fatuous application of industrial techniques to agriculture, with the result that soil has been reduced in many areas to a mere sponge for absorbing simple chemical nutrients. The cultivation of single crops over vast stretches of land is effacing natural, agricultural, and even physiographic variety. Immense urban belts are encroaching unrelentingly on the countryside, replacing flora and fauna with concrete, metals, and glass, and enveloping large regions with a haze of atmospheric pollutants. In this mass urban world, human experience itself becomes crude and elemental, subject to brute noisy stimuli and crass bureaucratic manipulation. A national division of labor, standardized along industrial lines, is replacing regional and local variety, reducing entire continents to immense smoking factories and cities to garish, plastic supermarkets.

Modern society, in effect, is disassembling the biotic complexity achieved by eons of organic evolution. The great movement of life from fairly simple to increasingly complex forms and relations is being ruthlessly reversed in the direction of an environment that will be able to support only simpler living things. To continue this reversal of biological evolution, to undermine the biotic food webs on which humanity depends for its means of life, places in question the very survival of the human species. If the reversal of the evolutionary process continues, there is good reason to believe—all control of other toxic agents aside—that the preconditions for complex forms of life will be irreparably destroyed and the earth will be incapable of supporting us as a viable species.

In this confluence of social and ecological crises, we can no longer afford to be unimaginative; we can no longer afford to do without utopian thinking. The crises are too serious and the possibilities too sweeping to be resolved by customary modes of thought—the very sensibilities that produced these crises in the first place. Years ago, the French students in the May–June uprising of 1968 expressed this sharp contrast of alternatives magnificently in their slogan: "Be practical! Do the impossible!" To this demand the generation that faces the next century can add the more solemn injunction: "If we don't do the impossible, we shall be faced with the unthinkable!"

Notes

1. I use the word "man," here, advisedly. The split between humanity and nature has been precisely the work of the male, who, in the memorable lines of Theodor Adorno and Max Horkheimer, "dreamed of acquiring absolute mastery over nature, of converting the cosmos into one immense hunting ground" (*Dialectic of Enlightenment* [New York: Herder and Herder, 1972], 248). For the words "one immense hunting ground," I would be disposed to substitute "one immense killing ground" to describe the male-oriented "civilization" of our era.
2. The term "ecosystem"—or ecological system—is often used very loosely in many ecological works. Here I employ it, as in natural ecology, to mean a fairly demarcatable animal-plant community and the abiotic or nonliving factors needed to sustain it. I also use it in social ecology to mean a distinct human and natural community, the social as well as organic factors that interrelate with each other to provide the basis for an ecologically rounded and balanced community.

Socialism and Ecology

JAMES O'CONNOR

THE PREMISE OF RED-green political action is that there is a global
ecological and economic crisis; that the ecological crisis cannot be resolved
without a radical transformation of capitalist production relationships; and
that the economic crisis cannot be resolved without an equally radical
transformation of capitalist productive forces. This means that solutions to
the ecological crisis presuppose solutions to the economic crisis and vice
versa. Another a priori of red-green politics is that both sets of solutions
presuppose an ecological socialism.

The problem is that socialism in theory and practice has been declared
"dead on arrival." In theory, post-Marxist theorists of radical democracy are
completing what they think is the final autopsy of socialism. In practice,
in the North, socialism has been banalized into a species of welfare capitalism. In Eastern Europe, the moment for democratic socialism seems to have
been missed over twenty years ago and socialism is being overthrown. In
the South, most socialist countries are introducing market incentives, re-
forming their tax structures, and taking other measures that they hope will
enable them to find their niches in the world market. Everywhere market
economy and liberal democratic ideas on the right, and radical democratic
ideas on the left, seem to be defeating socialism and socialist ideas.

Meanwhile, a powerful new force in world politics has appeared, an ecology
or green movement that puts the earth first and takes the preservation of
the ecological integrity of the planet as the primary issue. The simultaneous
rise of the free market and the greens together with the decline of socialism

From: *Capitalism, Nature, Socialism* 2, no. 3 (Oct. 1991): 1–12, and has been reprinted in
Our Generation (Canada), *Il Manifesto* (Italy), *Ecologia Politica* (Spain), *Nature and Society*
(Greece), *Making Sense* (Ireland), and *Salud y Cambio* (Chile). A revised version will appear
in James O'Connor, *Capitalism and Nature: Essays in the Political Economy and Politics of
Ecology*, to be published by Guilford Publications in association with *Capitalism, Nature,
Socialism*.

suggests that capitalism has an ally in its war against socialism. This turns out to be the case. Many or most greens dismiss socialism as irrelevant. Some or many greens attack it as dangerous. Especially are they quick to condemn those who they accuse of trying to appropriate ecology for Marxism.[1] The famous green slogan, "neither left nor right, but out front," speaks for itself.[2]

But most greens are not friends of capitalism, either, as the green slogan makes clear. The question then arises, who or what are the greens allied with? The crude answer is, the small farmers and independent business, that is, those who used to be called the "peasantry" and "petty bourgeoisie"; "liveable cities" visionaries and planners; "small is beautiful" technocrats; and artisans, cooperatives, and others engaged in ecologically friendly production. In the South, greens typically support decentralized production organized within village communal politics; in the North, greens are identified with municipal and local politics of all types.

By the way of contrast, mainstream environmentalists might be called "fictitious greens."[3] These environmentalists support environmental regulations consistent with profitability and the expansion of global capitalism, for example, resource conservation for long-run profitability and profit-oriented regulation or abolition of pollution. They are typically allied with national and international interests. In the United States, they are environmental reformers, lobbyists, lawyers, and others associated with the famous "Group of Ten."

As for ecology, everywhere it is at least tinged with populism, a politics of resentment against not only big corporations and the national state and central planning but also against environmentalism.

Ecology (in the present usage) is thus associated with "localism," which has always been opposed to the centralizing powers of capitalism. If we put two and two together, we can conclude that ecology and localism in all of their rich varieties have combined to oppose both capitalism and socialism. Localism uses the medium or vehicle of ecology and vice versa. They are both the content and context of one another. Decentralism is an expression of a certain type of social relationship, a certain social relation of production historically associated with small-scale enterprise. Ecology is an expression of a certain type of relationship between human beings and nature—a relationship which stresses the integrity of local and regional ecosystems. Together ecology and localism constitute the most visible political and economic critique of capitalism (and state socialism) today.

Besides the fact that both ecology and localism oppose capital and the national state, there are two main reasons why they appear to be natural allies. First, ecology stresses the site specificity of the interchange between human material activity and nature, hence opposes both the abstract valuation of nature made by capital and also the idea of central planning of production, and centralist approaches to global issues generally.[4] The concepts of site specificity of ecology, local subsistence or semi-autarkic economy, communal self-help principles, and direct forms of democracy all seem to be highly congruent.

Second, the socialist concept of the "masses" has been deconstructed and replaced by a new "politics of identity" in which cultural factors are given the place of honor. The idea of the specificity of cultural identities seems to meld easily with the site specificity of ecology in the context of a concept of social labor defined in narrow, geographic terms. The most dramatic examples today are the struggles of indigenous peoples to keep both their cultures and subsistence-type economies intact. In this case, the struggle to save local cultures and local ecosystems turns out to be two different sides of the same fight.

For their part, most of the traditional Left, as well as the unions, remain focused on enhanced productivity, growth, and international competitiveness, that is, jobs and wages, or more wage labor—not to abolish exploitation but to be exploited less. This part of the Left does not want to be caught any more defending any policies which can be identified with "economic austerity" or policies which labor leaders and others think would endanger past economic gains won by the working class (although union and worker struggles for healthy and safe conditions inside and outside of the workplace obviously connect in positive ways with broader ecological struggles). Most of those who oppose more growth and development are mainstream environmentalists from the urban middle classes who have the consumer goods that they want and also have the time and knowledge to oppose ecologically dangerous policies and practices. It would appear, therefore, that any effort to find a place for the working class in this equation, that is, any attempt to marry socialism and ecology, is doomed from the start.

But just because something has never happened does not mean that it cannot happen, or that it is not happening in various ways right now. In the developed capitalist countries, one can mention the green caucuses within Canada's NDP; the work of Barry Commoner, who calls for source reduction, the "social governance of technology," and economic planning based on a "deep scientific understanding of nature"; the antitoxic and worker and community health and safety movements which bring together labor, community, and ecological issues; various red-green Third-World solidarity movements, such as the Third World Network and Environmental Project on Central America; and the new emphasis on fighting ecological racism. One thinks of the Socialist party's struggle for control of the Upper House of the Diet against the long-entrenched Liberal Democrats, which reflects rising concern about both ecological and social issues in Japan. In Europe, we can see the greening of Labor, Social Democratic, and Communist parties, even if reluctantly and hesitatingly, as well as the rise of the green parties, some of which (as in Germany) are to the left of these parties with respect to some traditional demands of the labor movement. And in the subimperialist powers, which are taking the brunt of the world capitalist crisis, for example, Brazil, Mexico, and Argentina in Latin America, India, and perhaps Nigeria, Korea, and Taiwan, there are new ecological movements in which the traditional working class is engaged. And we cannot forget the Nicaraguan experiment which combined policies aimed at deep environmental reforms with socialism and populism.

There are good reasons to believe that these and other eco-socialist tendencies are no flash in the pan, which permits us to propose that ecology and socialism is not a contradiction in terms. Or, to put the point differently, there are good reasons to believe that world capitalism itself has created the conditions for an ecological socialist movement. These reasons can be collected under two general headings. The first pertains to the causes and effects of the world economic and ecological crisis from the mid-1970s to the present. The second pertains to the nature of the key ecological issues, most of which are national and international, as well as local, issues.

First, the vitality of Western capitalism since World War II has been based on the massive externalization of social and ecological costs of production. Since the slow-down of world economic growth in the mid-1970s, the concerns of both socialism and ecology have become more pressing than ever before in history. The accumulation of global capital through the modern crisis has produced even more devastating effects not only on wealth and income distribution, norms of social justice, and treatment of minorities, but also on the environment. An "accelerated imbalance of (humanized) nature" is a phrase that neatly sums this up. Socially, the crisis has led to more wrenching poverty and violence, rising misery in all parts of the world, especially the South, and, environmentally, to toxification of whole regions, the production of drought, the thinning of the ozone layer, the greenhouse effect, and the withering away of rain forests and wildlife. The issues of economic and social justice and ecological justice have surfaced as in no other period in history. It is increasingly clear that they are, in fact, two sides of the same historical process.

Given the relatively slow rate of growth of worldwide market demand since the mid-1970s, capitalist enterprises have been less able to defend or restore profits by expanding their markets and selling more commodities in booming markets. Instead, global capitalism has attempted to rescue itself from its deepening crisis by cutting costs, by raising the rate of exploitation of labor, and by depleting and exhausting resources. This "economic restructuring" is a two-sided process.

Cost cutting has led big and small capitals alike to externalize more social and environmental costs, or to pay less attention to the global environment, pollution, depletion of resources, worker health and safety, and product safety (meanwhile, increasing efficiency in energy and raw-material use in the factories). The modern ecological crisis is aggravated and deepened as a result of the way that capitalism has reorganized itself to get through its latest economic crisis.

In addition, new and deeper inequalities in the distribution of wealth and income are the result of a worldwide increase in the rate of exploitation of labor. In the United States during the 1980s, for example, property income increased three times as fast as wage and salary income. Higher rates of exploitation have also depended upon the ability to abuse undocumented workers and set back labor unions, social democratic parties, and struggles for social justice generally, especially in the South. It is no accident that in those parts of the world where ecological degradation is greatest—Central America, for example—there is greater poverty and heightened class struggles. The feminization of poverty is also a part of this trend of ecological destruction. It is the working class, oppressed minorities, women, and the rural and urban poor worldwide who suffer most from both economic and ecological exploitation. The burden of ecological destruction falls disproportionately on these groups.

Crisis-ridden and crisis-dependent capitalism has forced the traditional issues of socialism and the relatively new issues ("new" in terms of public awareness) of ecology to the top of the political agenda. Capitalism itself turns out to be a kind of marriage broker between socialism and ecology, or, to be more cautious, if there is not yet a prospect for marriage, there are at least openings for an engagement.

Second, the vast majority of economic and social and ecological problems worldwide cannot be adequately addressed at the local level. It is true that the degradation of local ecological systems often do have local solutions in terms of prevention and de-linking (although less so in terms of social transformation). Hence it comes as no surprise to find strong connections between the revival of municipal and village politics and local ecological destruction. But most ecological problems, as well as the economic problems which are both cause and effect of the ecological problems, cannot be solved at the local level alone. Regional, national, and international planning are also necessary. The heart of ecology is, after all, the interdependence of specific sites and the need to situate local responses in regional, national, and international contexts, that is, to sublate the "local" and the "central" into new political forms.

National and international priorities are needed to deal with the problem of energy supplies, and supplies of nonrenewable resources in general, not just for the present generation but especially for future generations. The availability of other natural resources, for example, water, is mainly a regional issue, but in many parts of the globe it is a national or international issue. The same is true of the destruction of forests. Or take the problem of soil depletion, which seems to be local or site specific. Insofar as there are problems of soil quantity and quality, or water quantity or quality, in the big food-exporting countries, for example, the United States, the food-importing countries are also effected. Further, industrial and agricultural pollution of all kinds spills over local, regional, and national boundaries.

North Sea pollution, acid rain, ozone depletion, and global warming are obvious examples.

Furthermore, if we broaden the concept of ecology to include urban environments, or what Marx called "general, communal conditions of production," problems of urban transport and congestion, high rents and housing, and drugs, which appear to be local issues amenable to local solutions, turn out to be global issues pertaining to the way that money capital is allocated worldwide; the loss of foreign markets for raw materials and foodstuffs in drug-producing countries; and the absence of regional, national, and international planning of infrastructures.

If we broaden the concept of ecology even more to include the relationship between human health and well-being and environmental factors (or what Marx called the "personal condition of production"), given the increased mobility of labor nationally and internationally, and greater emigration and immigration, partly thanks to the way capital has restructured itself to pull out of the economic crisis, we are also talking about problems with only or mainly national and international solutions.

Finally, if we address the question of technology and its transfer, and the relationship between new technologies and local, regional, and global ecologies, given that technology and its transfer are more or less monopolized by international corporations and nation-states, we have another national and international issue.

In sum, we have good reasons to believe that both the causes and consequences of, and also the solutions to, most ecological problems are national and international, hence that far from being incompatible, socialism and ecology presuppose one another. Socialism needs ecology because the latter stresses site specificity and reciprocity, as well as the central importance of the material interchanges within nature and between society and nature. Ecology needs socialism because the latter stresses democratic planning, and the key role of the social interchanges between human beings. By contrast, popular movements confined to the community, municipality, or village cannot by themselves deal effectively with most of both the economic and ecological aspects of the general destructiveness of global capitalism, not to speak of the destructive dialectic between economic and ecological crisis.

If we assume that ecology and socialism pressupose one another, the logical question is, why haven't they gotten together before now? Why is Marxism especially regarded as unfriendly to ecology and vice versa? To put the question another way, where did socialism go wrong, ecologically speaking?

The standard, and in my opinion correct, view is that socialism defined itself as a movement which would complete the historical tasks of fulfilling the promises of capitalism. This meant two things. First, socialism would put real social and political content into the formal claims of capitalism of equality, liberty, and fraternity. Second, socialism would realize the promise of material abundance which crisis-ridden capitalism was incapable of doing.

The first pertains to the ethical and political meanings of socialism; the second, to the economic meaning.

It has been clear for a long time to almost everyone that this construction of socialism failed on two counts. First, instead of an ethical, political society, in which the state is subordinated to civil society, we have the party bureaucratic state; and thus the post-Marxist attempt to reconcile social justice demands with liberalism.

Second, and related to the first point, in place of material abundance, we have the economic crisis of socialism; and thus the post-Marxist attempt to reconcile not only social justice demands and liberalism but also both of these with markets and market incentives.

However, putting the focus on these obvious failures obscures two other issues that have moved into the center of political debates in the past decade or two. The first is that the ethical and political construction of socialism borrowed from bourgeois society ruled out any ethical or political practice that is not more or less thoroughly human-centered, as well as downplaying or ignoring reciprocity and "discursive truth." The second is that the economic construction of abundance borrowed with only small modifications from capitalism ruled out any material practice that did not advance the productive forces, even when these practices were blind to nature's economy. Stalin's plan to green Siberia, which fortunately was never implemented, is perhaps the most grotesque example.

These two issues, or failures, one pertaining to politics and ethics, the other to the relationship between human economy and nature's economy, are connected to the failure of historical materialism itself. Hence they need to be addressed in methodological as well as theoretical and practical terms.

Historical materialism is flawed in two big ways. Marx tended to abstract his discussions of social labor, that is, the divisions of labor, from both culture and nature. A rich concept of social labor which includes both society's culture and nature's economy cannot be found in Marx or traditional historical materialism.

The first flaw is that the traditional conception of the productive forces ignores or plays down the fact that these forces are social in nature, and include the mode of cooperation, which is deeply inscribed by particular cultural norms and values.

The second flaw is that the traditional conception of the productive forces also plays down or ignores the fact that these forces are natural as well as social in character.

It is worth recalling that Engels himself called Marxism the "materialist conception of history," where "history" is the noun and "materialist" is the modifier. Marxists know the expression "in material life social relations between people are produced and reproduced" by heart, and much less well the expression "in social life the material relations between people and nature are produced and reproduced." Marxists are very familiar with the "labor process" in which human beings are active agents, and much less familiar with the "waiting process" or "tending process" characteristic of agriculture, forestry, and other nature-based activities in which human beings are more passive partners and, more generally, where both parties are "active" in complex, interactive ways.

Marx constantly hammered away on the theme that the material activity of human beings is two-sided, that is, a social relationship as well as a material relationship; in other words, that capitalist production produced and reproduced a specific mode of cooperation and exploitation and a particular class structure as well as the material basis of society. But in his determination to show that material life is also social life, Marx tended to neglect the opposite and equally important fact that social life is also material life. To put the same point differently, in the formulation "material life determines consciousness," Marx stressed that since material life is socially organized, the social relationships of production determine consciousness. He played down the equally true fact that since material life is also the interchange between human beings and nature, that these material or natural relationships also determine consciousness. These points have been made in weak and strong ways by a number of people, although they have never been integrated and developed into a revised version of the materialist conception of history.

It has also been suggested *why* Marx played up history (albeit to the exclusion of culture) and played down nature. The reason is that the problem facing Marx in his time was to show that capitalist property relationships were historical not natural. But so intent was Marx to criticize those who naturalized hence reified capitalist production relationships, competition, the world market, etc., that he forgot or downplayed the fact that the development of human-made forms of "second nature" does not make nature any less natural. This was the price he paid for inverting Feuerbach's passive materialism and Hegel's active idealism into his own brand of active materialism. As Kate Soper has written, "The fact is that in its zeal to escape the charge of biological reductionism, Marxism has tended to fall prey to an antiethical form of reductionism, which in arguing the dominance of social over natural factors literally spirits the biological out of existence altogether."[5] Soper then calls for a "social biology." We can equally call for a "social chemistry," "social hydrology," and so on, that is, a "social ecology," which for socialists means "socialist ecology."

The greens are forcing the reds to pay close attention to the material interchanges between people and nature and to the general issue of biological exploitation, including the biological exploitation of labor, and also to adopt an ecological sensibility. Some reds have been trying to teach the greens to pay closer attention to capitalist production relationships, competition, the world market, etc.—to sensitize the greens to the exploitation of labor and the themes of economic crisis and social labor. And feminists have been teaching both greens and reds to pay attention to the sphere of reproduction and women's labor.

What does a green socialism mean politically? Green consciousness would have us put "earth first," which can mean anything you want it to mean politically. As mentioned earlier, what most greens mean in practice most of the time is the politics of localism. By contrast, pure red theory and practice historically has privileged the "central."

To sublate socialism and ecology does not mean in the first instance defining a new category which contains elements of both socialism and ecology but which is in fact neither. What needs to be sublated politically is localism (or decentralism) and centralism, that is, self-determination and the overall planning, coordination, and control of production. To circle back to the main theme, localism per se won't work politically and centralism has self-destructed. To abolish the state will not work; to rely on the liberal democratic state in which "democracy" has merely a procedural or formal meaning will not work, either. The only political form that might work, that might be eminently suited to both ecological problems of site specificity and global issues, is a democratic state—a state in which the administration of the division of social labor is democratically organized.[6]

Finally, the only *ecological* form that might work is a sublation of two kinds of ecology, the "social biology" of the coastal plain, the plateau, the local hydrological cycle, etc., and the energy economics, the regional and international "social climatology," etc., of the globe—that is, in general, the sublation of nature's economy defined in local, regional, and international terms. To put the conclusion somewhat differently, we need "socialism" *at least* to make the social relations of production transparent, to end the rule of the market and commodity fetishism, to end the exploitation of human beings by other human beings; we need "ecology" *at least* to make the social productive forces transparent, to end the degradation and destruction of the earth.

Notes

1. This is a crude simplification of green thought and politics, which varies from country to country, and which are also undergoing internal changes. In the United States, for example, where Marxism historically has been relatively hostile to ecology, "Left green" is associated with anarchism or libertarian socialism.
2. This slogan was coined by a conservative cofounder of the German Greens and was popularized in the United States by antisocialist New Age greens Fritjof Capra and Charlene Spretnak. Needless to say, it was never accepted by Left greens of any variety.
3. "Mainstream environmentalists" is used to identify those who are trying to save capitalism from its ecologically self-destructive tendencies. Many individuals who call themselves "environmentalists" are alienated by, and hostile to, global capitalism, and also do not necessarily identify with the "local" (see below).

4. Martin O'Connor writes, "One of the striking ambivalencies of many writers on 'environmental' issues is their tendency to make recourse to authoritarian solutions, e.g., based on ethical elitism. An example is the uneasy posturings found in the collection by Herman Daly in 1973 on *Steady-State Economics*."
5. Quoted by Ken Post, "In Defense of Materialistic History," *Socialism in the World* 74/75 (1989): 67.
6. I realize that the idea of a "democratic state" seems to be a contradiction in terms, or at least immediately raises difficult questions about the desirability of the separation of powers; the problem of scale inherent in any coherent description of substantive democracy; and also the question of how to organize, much less plan, a nationally and internationally regulated division of social labor without a universal equivalent for measuring costs and productivity (however "costs" and "productivity" are defined) (courtesy of John Ely).

PART IV

Ecofeminism

The Time for Ecofeminism

FRANÇOISE D'EAUBONNE
Translated by Ruth Hottell

NEW PERSPECTIVES

IN SEPTEMBER 1973 A movement was born in France—a movement closer to the Belgian Unified Feminist party than to the French MLF [Movement de Libération de Femmes].[1] The new group, the Feminist Front, was formed by several women from an MLF group, others from the subgroup Evolution (founded in 1970 after "The Women's Estates General" and in reaction to it), and above all, by independent women from all parties and all movements. The group's statutes adhered to the 1901 law for associations, and accordingly, the new movement marks a tendency which is much more legalistic and even reformist.

In contrast with the Italian movement, which remains somewhat divided, there is no antagonism between this new front and the MLF, which it supports in some of its activities and invites to participate in the front's work. Its dream (rather utopian still, it must be admitted, given the average age, the settled nature, and the bourgeois status of its current members, notwithstanding the continuing possibility of an infusion of new blood) is to serve as a link between all the women's movements and associations, leading toward a massive "sorority" (the Americans say sisterhood) which would further women's causes and their liberation in a determinant fashion. The front believes in means as moderate as those espoused by Betty Friedan: parliamentary representation, obligatory home-economics courses for boys as well as girls, professional promotion, and equal salaries. A woman's right to choose abortion is supported without qualifications, the right to divorce is upheld against certain reactionary projects, but the sexual revolution still causes a great discomfort for these prudent ones—or for these prudes.

From: Françoise d'Eaubonne, "Le Temps de L'Ecoféminisme," *Le Féminisme ou la Mort* (Paris: Pierre Horay, 1974), 215–52.

Another group has arrived on the scene even more recently, formed from a group of the Revolutionary Feminists and also adhering the statutes of the law of 1901: "The League of Women's Rights." This new group is much more radical, although just as resolutely legalistic. Its goal is to fight on legal grounds, enclosing the male society in its own contradictions, with the help of a collective of women lawyers and judges, in order to have women proclaimed a group susceptible to racist treatment and, thus, able to protest against discrimination on constitutional grounds.

We do not yet know the future contributions of these groups which have so suddenly and so quickly emerged after three years of MLF existence. But thanks to the first group, an attempt at synthesis is possible between two struggles previously thought to be separated, feminism and ecology.

Even though Shulamith Firestone had already alluded to the ecological content inherent in feminism in *The Dialectic of Sex: The Case for Feminist Revolution*, this idea had remained at the embryo stage until 1973. It was picked up again by certain members of the Feminist Front, which first included it in its manifesto, then removed it. The authors then separated themselves from such a timorous new movement and founded an information center: the Ecology-Feminism Center, destined to become later, as a part of their project of melding an analysis of and the launching of a new action: *ecofeminism*.

In regards to this question, I would add the idea: "from revolution to mutation," which inspired the title of a portion of this work. Nor would it be superfluous to recall one of the questions posed earlier by the Revolutionary Feminists, a group within the MLF: "We need to know nonetheless if the movement will be a *mass movement to which all women will potentially belong as a specifically exploited group, or if it will be just another subgroup?*"

It is in keeping with this spirit that militants, as much from the Feminist Front as from the League of Women's Rights, concentrate their strength on mobilizing and sensitizing as many of their "sisters" as possible about the relatively restricted and immediate objectives, working toward reasonable goals which can seem "reassuring" (outside the acronym MLF, which has already become "frightening" for many), all the while not forgetting (at least for the "league," younger and more dynamic than the FF) the eventual targets: the disappearance of the salaried class (beyond equal salaries), the disappearance of competitive hierarchies (beyond access to promotion), the disappearance of the family (beyond the control of procreation), and, above and beyond all of these, a *new humanism* born with the irreversible end of the male society, and which by definition must work through the ecological problem (or rather the extreme ecological peril).

For the moment, certainly, the mobilization of women around "specifically feminine" issues can take, even on a legal level, a tone of exigency which widely surpasses the ancient demand for "rights."

We want, say these followers of the most recent feminism, to depart from what certain subversive German groups call the "anti-authoritarian quag-mire," without meanwhile sinking into bureaucracy or elitism; we also want to reach the workers and to throw out the bases in the provinces; but as immediate and concrete as these projects are, we know that, above all, we urgently need to remake the planet around a totally new model. This is not an ambition, this is a necessity; the planet is in danger of dying, and we along with it.

Besides the authoritarian socialists or leftists of various degrees whose drone I have situated here as "principal struggle and secondary struggle," there are analysts and agitators, possible companions in struggle much more enlightened than these neo-Stalinists; they never cease calling for the "totalization" of the combat and protesting that everything which is "fragmentary" compromises the final goal—to destroy the Carthage of the System. These activists do not place themselves in the minefield of the "class struggle"; rather, they emphasize the necessity of a global awareness. It is not a question of forwarding one's particular demands, but of introducing new areas of consciousness:

> It is logical that individuals begin with the real experience of their own alienation in order to define the movement of their revolt; but once that is defined, nothing other than integration into the cultural firmament of the system is possible for them. . . . Those of you who are normal, stop limiting yourselves to your normality; homosexuals, stop limiting your-selves to your ghettos; women, stop limiting yourselves to your feminin-ity or to your counter-femininity. Invade the world, exhaust your dreams.[2]

There is a lot of truth in this exhortation, for it was by limiting themselves to their supposed "counter-femininity," that is, the demands for such-and-such a right, that yesterday's feminists imagined themselves to be the magic transformers of their femininitude, that they buried themselves, having gained, for all practical purposes, nothing but air. I completely agree that women's struggle should be global, totalitarian—even when it presents itself from a modestly reformist aspect—or not be at all. But the authors of *Grand Soir* are committing the same error as the society they are fighting by conflating the category "women" with "homosexuals," "lunatics,"[3] and other minorities in revolt against their alienation. In so doing, they forget that the question for women is not that of some minority but that of a majority reduced to the status of a minority, and the only one to be so treated; and, what is more, the only one of the two sexes with the ability of accepting, refusing, slowing down, or accelerating the reproduction of humanity, with whom rests at the moment, even if they are not totally aware of it, the sentence of death or the salvation of humanity in its entirety.

The sole totalitarian combat capable of overturning the System, instead of simply changing it once again for another, and of shifting finally from the spent "revolution" to the mutation that our world calls for, is the women's combat, that of all women; and not only because they were placed in the situation described in the preceding pages, because the iniquity and absurdity outrage their very souls and demand the overthrow of unbearable excess; that is legitimate, but it remains sentimental. The point, quite simply, is that it is no longer a matter of well-being, but of necessity; not of a better life, but of escaping death; and not of a "fairer life," but of the sole possibility for the entire species *to still have a future.*

U-Thant, the French scholars from the Museum of Natural History, the European Council, the *Unesco Courier* have said it again and again to all the highest international authorities; Conrad Lorenz sums it up: "For the first time in human history, no society can pick up the reins." (Obviously he is thinking, like everyone else, of a male society; that is, a society composed of representation, competition, and industrialization, in short, aggression and sexual hierarchy.) The Futurologists Conference, which took place several months ago in Rome and received a great deal of press—more than the war in the Middle East—has repeated it, even if France is keeping silent.[4]

What were these Cassandras shouting from the rooftops? Quite frankly that the point of no return had practically been reached, that one cannot stop a vehicle careening at a hundred miles an hour toward a brick wall when one is only sixty feet away from it, and that everything could end with a very virile "Prepare to meet your maker!" or "Keep clear of industrialized zones!"

Why this flight? To what extent does this colossal declaration of failure coincide with the feminine wish to snatch the car's steering wheel from the hands of male society, with the intention not of driving in its place but of jumping from the car?

These new perspectives of feminism do not detach themselves from the MLF, but they stand apart within it, due less to the more classical and not underground[5] language or the acceptance of a fledgling organization and the concerns of the feminine "masses" than to the global objective, an answer to the critique of parcellization[6] referred to earlier, and which would even be a new humanism: *ecofeminism.*

The reasoning is simple. Practically everyone knows that the two most immediate threats of death today are overpopulation and the destruction of natural resources; fewer are aware of the entire responsibility of the male System—the System as male (and not capitalist or socialist)—in creating these two perilous situations; but very few have yet discovered that each one of the *two* menaces is the logical outcome of one of the *two* parallel discoveries which gave power to men fifty centuries ago: reproduction, and their capability of sowing the earth as they do women.

Until that time, women alone possessed a monopoly on agriculture, and the men believed women were impregnated by the gods. Upon discovering the two possibilities at once—agricultural and procreational—man launched what Lederer has called "the great upheaval" for his own benefit. Having seized control of the soil, thus of *fertility* (later, industry), and of woman's womb (thus of fecundity), it was logical that the overexploitation of the one and the other would result in this double peril, menacing and parallel: overpopulation (a glut of births) and destruction of the environment (a glut of products).

The only mutation which can save the world, therefore, is the "great upheaval" of male power which brought about, first, agricultural overexploitation, then lethal industrial expansion. Not a "matriarchy," of course, or "power to women," but the destruction of power by women, and finally, the way out of the tunnel: egalitarian administration of a world being reborn (and no longer "protected" as the soft, first-wave ecologists believed).

ECOLOGY AND FEMINISM

I had already defended this point of view, before the birth of the Feminist Front and before reading Shulamith Firestone, in 1972, in a letter to *Nouvel-Observateur*, after an ecologist conference where the absence of women was particularly blatant. This declaration brought me numerous, unexpected letters. Some advocated a demand for "women's power"[7] and accused me of reverse sexism. Pierre Samuel, in his otherwise just and honest book, *Ecologie: Cycle infernal ou détente*,[8] referred to my work as "dangerous exaggeration," and, in the end, aligned himself with the concept which was to become the one behind the Ecology-Feminism Center founded by the dissidents from the Feminist Front.

But, is worldwide catastrophe really at our doorstep? Stopping short of the noble remarks of somebody like Lartéguy, who, faced with the figures I was citing on television, shrugged his shoulders and said, "Oh, specialists are always mistaken in their predictions,"[9] nevertheless a lot of people say, "Someone will invent something." All this talk allows them to put the Horse of the Apocalypse in a bottle, like a goldfish.

Consider, however, this remark: "Ecology is going to replace Vietnam among the essential preoccupations for students." Who said this and when? The *New York Times* . . . in 1970. That year, the son of the woman physician Weill-Hall, who founded "family planning," rallied Americans in an effort to struggle against collective suicide, the correlation of which he saw clearly in the problem his mother attacked.

Ecology, the "science that studies the relations of living beings among themselves and the physical environment in which they are evolving," includes, by definition, the relations between the sexes and the ensuing birthrate. Because of the horrors which menace us, the most intense interest is oriented toward the exhaustion of resources and the destruction of the environment—which is why it is time to recall that other element, the one which so closely ties together the question of women and of their combat.

Shall I give some technical details?

In America, the alarm was sounded in 1970 by the anti-establishment youth of Earth Day, who moved immediately to eco-terrorism—they bury cars and physically fight with loggers. Celestine Ward, in her feminist book *Women Power*, says that "they want to breathe in accordance with the cycle of the cosmos." No—they want to breathe, period. Judge for yourselves:

Each year, Irène Chédeaux[10] tells us, America must eliminate 142 tons of smoke, 7 million old cars, 30 million tons of paper, 28,000 million bottles, 48,000 million tin cans. "Each one of the 8 million New Yorkers breathes as much noxious material as if he had smoked 37 cigarettes, and he knows now that there are more rats in the dumps than humans in the city."

Shall I give you other details on the state of affairs? Fish are dying en masse in the Delaware River (oh, Westerns and Fenimore Cooper!) because Sun Oil, Scott Paper, and Dupont de Nemours dump their waste; but that undoubtedly is nothing next to the fact that Lake Zurich is dead, suffocated by human waste, and that Lake Léman will not last much longer; still nothing next to the fact that the Mediterranean, called the Great Pure One by the ancient Greeks, has become such a garbage can that in Marseille this summer, the author of this text had to drive along the coast for miles before finding a beach affording a little cleanliness. In America today numerous enterprises declare that 10 percent of their investments are absorbed by pollution control. A derisory effort given the catastrophic dimensions of the damage. We owe this modest result, and the more striking ban of a pipeline four feet in diameter which was to cross the Sahara, to the struggles of the various groups of "conservationists," the one best known for its radicalism being the Sierra Club. In January 1970, President Richard Nixon devoted a large part of his State of the Union Address to the ecological question. Soon after, a law was passed ordering the airlines to devise a plan by 1972 which would reduce the fumes from combustion by three-quarters. Jean-François Revel[11] relates how, in that very country, these endeavors, because of their anti-industrialism, have been labeled "communist maneuvers" by certain fanatics. That was the most beautiful consecration of a prophet: Kornblut, author of *Planète à Gogos*.[12]

A totally new sort of protester has just been born: the eco-guerrilla. At night, students saw down billboards that disfigure the countryside. The newspaper *Actuel* reminds us that in Berkeley in 1967 young engineers founded Ecology Action. They called themselves "apolitical"; we can imagine what political interest groups might react to their goals:

To suppress absurd practices: new model cars every year . . .

To control necessary but potentially dangerous practices: the treatment and distribution of food, the construction and organization of housing, the production and distribution of energy.

To eliminate destructive practices: social regimentation, covering the earth with concrete, asphalt and buildings, the dissemination of waste, *wars*.

It can quite simply be called a revolution, comments those signing the article. But *which one?*

In the meantime, the abomination is growing every day. The air is charged with 10 percent more carbon dioxide gas than at the beginning of this very recent era called industrial. There are "12 micromoles per cubic centimeter in the Alsace region, 88,000 on the Champs Elysées and 4 million in a department store."[13] A rat placed on the ground at the Place de la Concorde dies after ten or fifteen minutes, in the Paris of 1971. This leads us to assume that soon, since air pollution is rapidly increasing, a three-year-old child will have to wear a gas mask.

We have already mentioned demographic inflation; let us return to it.

If, according to the calculations of Edouard Bonnefous, world population multiplied by 2.6 between 1800 and 1950, whereas the total population of cities of more than 100,000 grew eight times faster, there was an insane jump from 1950 to 1960: this time, the world urban population felt a growth . . . of 35 percent. Deforestation follows the progress of technology: the Sunday edition of the *New York Times* devours seventy-seven hectares of forests!

When I was a child, the Garonne and the Seine were considered the only two truly polluted rivers in France. Today, six million tons of waste contaminate the hydrographic network of our gentle country, and the rate is increasing. In other words, there are as many kilos of refuse as there will be humans on the planet in the year 2000. That is as bad as anything in a horror film. *Le Figaro* may have called the Rhine "a big sewer," then, in a more prudent vein, a "great collector"; words change nothing about the thing (let's use the word "thing" as a discreet euphemism for the word "shit"). But we should refrain from accusing the camp of capital too quickly: if our lakes are agonizing or have become closed, dead seas (Léman and Annecy in Switzerland; Lakes Michigan, Erie, and Superior in America), we can savor this consoling factor—360,000 kilometers of running water in the Soviet Union are hardly in better condition. The underground layers of freshwater? That is capital on which we cannot write checks forever; for, with population growth, the growth of water needs increases. This was the subject of a recent science fiction novel by Jacques Sternberg: Tap water, throughout the world, had been replaced by a flow of revolting insects—a hazardous product had increased the size of every microbe or bacteria to monstrous proportions! How frightening is Dracula next to these images? A kindergarten horror story. And it is so true, the menace is so precise, that, in light of the imminent lack of fresh water, all industries are researching means of desalinating seawater. (But, to what avail if we have to remove the pollution from it too?) Alas! Where is the spirit of the Emperor Julien who said, in A.D. 358: "Water from the Seine is very good to drink"? He was not a "techno-logician."

Let's look at our own South. The Aquitaine National Petroleum Company has so harmed the surrounding farmland with its sulphur dioxide that recently it was ordered to pay damages and interest to the Béarn farmers who had taken it to court. This was not their only worry; they also had to defend themselves against the establishment of a large aluminium factory which, enclosed with Aquitaine Petroleum in a small valley, doubled the risks of atmospheric pollution.[14]

The catastrophic situation of urban agglomerations is becoming known everywhere. Paris holds the record for air pollution per square meter; every day it receives this celestial manna: 100,000 cubic meters of sulphurous acid and about ten tons of sulphuric acid. There is one city more poisoned— Tokyo, about which scholars have told this prophetic joke: in 1980, the Japanese John Doe will have to wear a gas mask to go about his daily business. I can picture the samurai removing his mask instead of performing hara-kiri.

Here again, pollution strictly speaking joins with the destruction of the natural environment. The air is barely more than a layer of smut in big cities, but their waste augments the aggression against human health. It sounds like a bad dream when we read that New York discards 250 million tons of nonnuclear waste every year: old cans, auto carcasses, plastic material, burned gas, disgusting refuse of all kinds. The image is in keeping with the habitual sadism of the heroine of the comic strip *Pravda la Survireuse*,[15] who contemplates a huge pit filled with old car hulks from atop her motorcycle and says, like some romantic beauty leaning over one of the lakes of yesteryear: "What a beautiful sight!" Filth in the air, thus approaching the atmospheric ceiling; filth on the earth, rising ever higher; we live between two layers of waste, like cave dwellers between the stalactites and stalagmites.

What solution can be found for such an absurd nightmare? Revolution? Yes, but not "our father's." The most antiquated quote from *Internationale* is doubtlessly:

Among those in control, who will contain Nature?

It should be replaced by:

In deadly infection, who will change Nature?

The answer cries out: It's you, it's us, it's me! But I, but you, but we are irresponsible: we cannot act otherwise, aside from infinitesimal behavior changes; foremost among the responsible parties is the technological civilization, super-urban and super-industrial, the insane race which launched the unstoppable wheel toward profit, in much the same way the Gauls sent burning wheels hurtling down steep slopes; but that was in order to fertilize by burn-beating, whereas our technological culture ruins and murders the nourishing soil under the wheel's fire. As the crime worsens, so does the madness: the overpopulation of the executioners who, at this moment, are engendering the future victims of their collective infamy.

The global population at the end of the Roman Empire is estimated to have been 250 to 300 million, less than the population of modern-day China.

Outside China and India, which always had the second highest birthrate, the rest of the human species was distributed mainly throughout the Mediterranean basin. Epidemics, famine, and natural disasters varied the demographic levels, and global growth was insignificant. It took sixteen centuries before the world population reached 500 to 550 million in the seventeenth century. Therefore, we can evaluate, *grosso modo*, the level of growth in the following way:

- 250 million in sixteen centuries (twice the original number). By 1750, we had reached 700 million. Thus:
- 200 million in one century only.
- Next, growth increases. The discoveries of civilization, work to decrease mortality rates, greater longevity gave us, in 1950, 2,500 million inhabitants. New figures:
- 1,800 million inhabitants in two centuries, that is, 900 million a century instead of the 200 million of preceding centuries.

"The situation is so grave that it demands predictions," says Elizabeth Draper in *Conscience et contrôle des naissances*.[16] But the expansion has only intensified; we can no longer count by centuries, but by decades; in 1960 we already numbered 3 billion (twenty years before the expected date, since experts predicted 3,000 to 3,600 million by 1980!) The rate of growth is five hundred million in ten years: in ten years, the world population increased, between 1950 and 1960, by the number of the total population in the seventeenth century!

The 3,600 million expected by 1980 had undoubtedly been exceeded in 1969. What is appalling is that we have been told that the current population will double in thirty-five years!

But overpopulation is not being produced everywhere at the same rate. Cities with more than 100,000 inhabitants are first: they are growing eight times faster than the others. The frightening jump from 1950–60 was even higher among the urban population; it increased by 35 percent.

"We have to build quickly, no matter what, no matter where, no matter how . . . Profits are fabulous and immediate. To make new factories work, we need workers on the spot. We build cities which are only cities by name: HLM,[17] shops, drugstores, supermarkets, cinemas. Advertisements are everywhere: the worker must consume. If he does not consume, production will drop. If production drops, he will lose his job. If he loses his job, he will not be able to pay the rent, the bill for the washing machine, etc.," explain Colette Saint-Cyr and Henri Gougeaud.

We find ourselves back in the demented universe of *Planète à Gogos*. Let's look at the American example; Robin Clarke, UNESCO expert, gives the following:

During the course of a lifetime an American baby will consume:

100 million liters of water,
28 tons of iron and steel,
25 tons of paper,
50 tons of food,
10,000 bottles, 17,000 cans, 27,000 capsules, (maybe) 2 or 3 cars, 35 tires,
will burn 1,200 barrels of petroleum,
will discard 126 tons of garbage,
and *will produce* 10 tons of radiant particles.

This means, the expert concludes, that *the birth of an American baby carries 25 times more significance for the ecology than that of a Hindu baby*.[18] It remains for us to evaluate, which Robin Clarke did not think of doing, the chances for survival of such a distressing world economy where, according to the region, one individual will consume during the course of sixty or seventy years twenty-five times more than another, for it is possible that the rebellion of the Third World, having enough of being the Büchenwald of the planet, will change the statisticians' figures. Even if there is no Third World rebellion, where will this American baby-turned-adult find to consume, burn, and throw away all that Robin Clarke predicts for him, since it has already been predicted that the earth will become a desert in twenty to thirty years at the rate we are going now? ("In 1985, announces Professor René Dubos, the entire planet will be nothing more than a barren desert.") Will massive death from malnutrition put an end to the exponential rate of demographic growth? And of urban concentration?

It was the Marquis de Sade, one of the greatest scientific geniuses of all time, who, on the eve of the Revolution, had already declared in *Justine* that the concentration of people in the cities contributed to the acceleration of mental health disorders and sexual obsessions. We have seen that this hyperconcentration is due to overpopulation, whose dangers Malthus, much later, would expose. (In a more muddled fashion, Restif de la Bretonne, in *Le paysan perverti*,[19] is saying exactly the same thing.)

One fine day in 1958, J. Calhoun, a researcher, had the idea of installing a colony of Norwegian rats in a barn in Maryland that he had set up for them. First observation: in a nine-hundred-square-meter space, five thousand rats lived in harmony. But the population did not exceed 150 new births.

Second stage of observation: the increase of adult rats in divided sectors brought about aberrations in both sexual and social conduct. Young rats set about practicing collective rape of an isolated female like vulgar gangs in Sarcelles or Billancourt.[20] Others were observed engaging in hypersexual behavior, practicing all the possible forms of the instinct, as if trying to overcome the anguish from overpopulation. Others finally became perfect gangsters, or other sorts of outlaws, isolating themselves from sexual as well as social contact in order to engage in nocturnal murders. Ordinary wars for the conquest of power multiplied into absurdity. The females finally stopped building their nests and suffered a *stress of fertility*.

We must, of course, mistrust analogies made too hastily. On the subject of does who lose interest in their fawn after a delivery under anesthesia, an argument brought up by opponents of painless childbirth, Simone de Beauvoir has answered in a decisive, albeit very simple, manner: "But women are not deer."[21] If female white Norwegian rats do not recognize their offspring following an intensive concentration of "urban" habitat, we can hardly compare this fact to that of the young Israeli mother growing accustomed to nursing not only her own child but also her comrades' babies in the day-care center of the kibbutz, in an interesting goal of disalienation of filial bonds. What takes place at the level of animal collectives can never be transposed directly to the human level without a careful examination of the differences in structure; it is the absence of this principle that renders so futile a sociological essay like Lionel Tiger's *Entre Hommes*.[22] But here, without falling into the same trap, we can envisage a definite similarity between the experience of the Norwegian rats and that of our urban concentrations. The German ethnologist Paul Leyhausen is of this opinion. Like Sade, he finds an important factor in urbanization contributing to the rise of mental illness. The number of suicides increases, as do alcoholism, mental deficiency, juvenile delinquency, "nervous breakdowns."[23] The growing difficulty of isolation and exhaustion due to overcrowding and transport contributes to a subsidence of nervous resistance which, so it seems, could be transmissible. We already know all that? Yes, but never before has something considered a trouble, worry, and imperfection of modern life reached such critical dimensions; the throbbing of a cavity in a tooth has become cancer of the jaw. Sexual crimes are starting up again in London; abominations previously beyond the scope of our experience—kidnapping, gang rape—are appearing in Paris; New York, to name only one city, protects itself against the madness by imitating it with a series of "little, insignificant murders."[24] To the destruction of the environment corresponds, at the current birthrate, the destruction of interior man. And also—finally a break in the clouds—the stress of the white female rat.

In an article entitled *Why We Abort*,[25] a woman who signs herself Christianne says:

> Rats, in cases of overpopulation, suffer fertility stress. Men have lost contact a little with their instincts and totally lost the power of spiritual command over their bodies. Women, thanks to their under-development, less so. Perhaps we can say they are beginning a kind of stress. And, what is more, that the demand for free abortion is a profoundly intuitive cry of alarm? Perhaps we should lend an ear.

It is no longer "lending an ear" that is necessary, but bellowing and roaring. The world is beginning to accept the idea of abortion for other reasons which make women violently demand their right to exercise control over their own bodies, over their future, over their procreation; it is thanks to worry over the exponential rate described and analyzed earlier that the male society is experiencing some tendency to question itself and to accept demands which are brought about by totally different motives. But what is remarkable about this situation is that this interest, if it is met, will lead to the realization of a situation more favorable for an oppressed caste—women—than for the caste of the oppressors; and the latter group knows it quite well. That is why they hesitate to accord something they themselves must desire: a stop to an insane growth in the birthrate which, along with the destruction of the environment, signs the death warrant for everyone. At the same time, obtaining this fortunate halt by giving women freedom of access to contraception and abortion is, for men, the conviction that *women won't stop there* and will begin to control their own lives; this is, recalling Fourier, a scandal of such violence that it is capable of undermining the bases of society. Hence the hesitations, the contradictions, the reforms and obstacles, the steps forward, the backlash; this comic grimace on the face of power conveys the extreme interior opposition which is ripping the male society apart, at all levels, in all countries.

"Making money, getting rich, exploiting man and nature in order to climb to the most expensive places on the social ladder.... As long as a society organizes its production for the goal of converting the resources of man and of nature into profits, no equitable and planned system of ecological balance can exist."[26]

It is evidence itself. At the base of the ecological problem are found the structures *of a certain power*. Like that of overpopulation, it is a problem of *men*; not only because it is men who hold world power and because for a century now they could have instituted radical contraception; but because power is, at the lower level, allocated in such a way to be exercised by men over women. In the domain of ecology as well as that of overpopulation, we see conflicts contradictory to capitalism harshly confronting one another, even though these problems far surpass the scope of capitalism and the socialist camp suffers from just as many, for the good reason that in both sexism still reigns. Under these conditions, where the devil can we find— even in the case of prolonged pacific coexistence of both economic camps— a possibility of implementing an "equitable and planned" system on a planetary scale, whether it be ecological or demographic? Lorenz is right: no (male) society can pick up the reins.

The first problem: stopping new births. The male society has begun to be afraid, and, in turn, Poland, Rumania, and Hungary have adopted the solution of "liberalizing" abortion; in the capitalist camp, England followed on 24 October 1967; Japan penalizes a surplus of births. Four American states, New York among them, "have followed in the path of the frank and massive yes" (*Guérir*,[27] August 1971). On the other hand, in the Latin Catholic countries, we find a sole timid overture in France; in Spain, even therapeutic abortion is forbidden: Let the mother die, so long as the fetus lives, even if it has only a minute chance of not following its mother in death. These examples represent some of the principal contradictions of a civilization panicked at the necessity of consenting to what will prolong humanity but also toll the first bell for its old putrid patriarchal forms.

If these problems are men's problems, it is because their origin is masculine: it is the male society, built by males and *for* males—we must repeat this tirelessly.

It would be derisory to play at the little game of historical "ifs": What would have happened if women had not lost the war of the sexes at the distant time when phallocracy was born—thanks to the passing of agriculture to the male sex? Without what August Bebel calls "the defeat of the feminine sex," and the modern Wolfgang Lederer, author of *Gynophobia*,[28] calls "the great upheaval," what would have happened? We would do well to keep ourselves from entering further into this fantasy (or else we will be in the domain of a charming science-fiction novel): "Ah! If the sky fell, a lot of sparrows would be caught," my grandmother used to say. A lone negative response emerges: perhaps humanity would have vegetated at an infantile stage, perhaps we would have never known either the jukebox or a spaceship landing on the moon, but the environment would have never known the current massacre, and even the word "ecology" would have remained in the little cervical box of *Homo sapiens*, like the word "kidney" or "liver," which never cross the lips or the pen if no suffering or pain is felt from these organs.

"Pollution," "destruction of the environment," "run-away demography" are all men's words, corresponding to men's problems: those of a male culture. These words would have no place in a female culture, directly linked to the ancient ancestry of the great mothers. That culture may have been simply a miserable chaos, much like the ones of the Orient which, as phallocratic as it may be, brings out much more anima than animus. It seems that neither of the two cultures would have been satisfactory, insofar as it would have been sexist also; but the ultimate negativity of a culture of women would have never been this, this extermination of nature, this systematic destruction—with maximum profit in mind—of all the nourishing resources.

It is interesting, returning to the famous stress of the white rat, to note that in the animal kingdom as in the human kingdom it is the female who tends to refuse procreation and not the male, although the instinct of preservation should be present throughout the species.

If we consider the behavior of the males in power in our society, what do we see? Conscious of the peril of overpopulation, they strive to make us believe that it is a "Third-World problem" and concentrate their efforts on the most disadvantaged spot of the planet, consequently the one which consumes the least. (Pierre Samuel, in *Ecologie, cycle infernal ou détente*,[29] notes that the United States alone, with 6 percent of the world population, consumes 45 percent of the global resources!)

The most material interest explains the attitude of the West: the caste of the masters profits most from growing itself and limiting the expansion of the disadvantaged, its future Spartacus; we saw this recently when Aimé Césaire stood up in protest against a proposed law designed to favor contraception further (already largely accessible) in Martinique; the black member of Parliament reproached the text for its manipulative intentions. He asked, Why Martinique, and not France? Certainly, the women in Martinique had every right to answer Aimé Césaire, telling him that it was up to them and not to a man to decide, much as was the case in Harlem a few years ago;[30] but the attitude of the political leader from Martinique was logical. For a long time now, the Third World has known that overpopulation is not a fact of wealth, of assurance that the old parents will be taken care of, etc., but that, on the contrary, it is an element of misery and of mortality; they continue to overprocreate through ignorance, material shortage, heightened oppression of women. It is understandable, additionally, that for the enlightened ones there is a legitimate distrust of the gifts of European family planning, whereas they would be welcomed with enthusiasm if the Occident began to benefit largely from its own discoveries in this respect.

The Catholic heterocops with their "right to life" or the ministers with their "moral order" are powerless in the face of this evidence: the scandalous revolt of women, regarding their bodies, is moving in the direction of the most immediate interest of humanity, of the future, of procreation itself. The true holders of power themselves, who are neither for Jesus Christ nor for Jehovah but for power, know it too; it pains them, but they know it. Neo-Malthusianism began to be a necessity well before what motivated the women's revolt and the manifesto of the 343.[31] When the arteries of capitalism age to the point where they need ever-greater doses of family planning like novocaine—oh blasphemous planning, oh economic antiliberalism!—the time is near.

The planet is going to overflow

Slogan shouted 5 March 1971, outside the Right to Life Conference in Maubert-Mutualité.

Before capitalism, the last arrival, aging and resistant, before feudalism, before phallocracy, feminine power, which never reached the dimension or the stature of matriarchy, was founded on the possession of agriculture; but it was an autonomous possession, probably accompanied by a sexual segregation; and that is why there was never a true matriarchy. Men controlled the pastorate and the hunt, women agriculture; each of these two armed groups confronted the other; such is the origin of the supposed legend of the Amazons.[32] When the family arrived on the scene, woman could still treat power from a position of power, as long as agricultural functions continued to make her sacred; the discovery of the process of fertilization—of the womb and of the soil—tolled the last bell. Thus began the iron age of the second sex. It has certainly not ended today. But the earth, symbol and former preserve of the womb of the great mothers, has had a harder life and has resisted longer; today, her conqueror has reduced her to agony. This is the price of phallocracy.

In a world, or simply a country, in which women (and not, as could be the case, *a* woman) found themselves truly in power, the first act would be to limit and space out births. For a long time, *well before overpopulation*, that is what they have been trying to do. The proof is the existence of anticonceptual folklore, where the most frighteningly dangerous procedures border on pure superstition. (Note: not only to avoid having a child, but *so that the husband would stay away from the bed*, oh Freud and the "feminine tendency to evade sexuality"!) These conjuration rites, obviously, were never cited by male scholars, whereas lists of opposite rites—those of fertility—exist everywhere. It is only on the planetary scale (not even national), that man deigns to notice the overpopulation; woman notices the rabbit effect at the family level. What methods did she have at her disposal to make it known?

The selfish handmaidens, so strongly denounced by the brilliant Henry Bordeaux and the Catholic novelists writing between the wars (from whom the Right to Life clique takes up the cause much to the benefit of Nestlé-Guigoz advertisements,[33] the selfish French mothers who declared they would only jeopardize their lives two or three times by foaling—we can bless them today for the feeble, very feeble, brakes they managed to apply: where would we be in a world peopled thirty years earlier and faster by the heroic devout women, readers of the *Bonne Presse* of Father Bethlehem?

Yes, we must repeat it to obsession, cry it, spread the word: just like the class struggle, demography is man's doing.[34] In areas of Catholic misogyny, it is doubly so: the transmission belt between male power at the top and us is, by definition, the men who control the exploitation of our fertility: the husband to whom we must submit, of course, but also the priest, who can only be a man, the doctor or the judge, who nine times out of ten is a man; all these civil servants of male power are males.

To summarize—if the class struggle, demography, ecology are all the problems and affairs of men, it is due to "the great defeat of the feminine sex" which took place throughout the planet in 3000 B.C. After the demise of the Amazons and agriculture, the guarantor of power, shared for a certain time between the sexes in the Hittite, Cretan, and Egyptian civilizations, little by little the wealth of the earth became masculine at the time when woman, tied to the family, no longer had recourse to vanished Amazonian ways. Patriarchal and masculine power peaked in the Bronze Age, with the discovery of what would later become industry. Women were then put under strict surveillance by the victorious sex, which still suffered from fear and distrust of them; they were exiled from all sectors other than the family ghetto; not only from power and from work outside the home, but even from areas in which man seemed to have no fear of competition: physical sports (ancient Greece, except Lacedaemonia), theater (feudal England, Japan), art and culture, higher education (practically the world over, apart from a few oft-cited exceptions). Pariah relegated to rabbitlike unions, and, in the best of situations, a luxurious ornament for the victors; finding the sole refuge, in all cultures, to be religious sequestration, escaping from man by devoting herself to the phantoms of the deity; today she sees, more and more profoundly and in more and more places, that only men control the "revolution" and profit from it, no matter how much help she gives them in hopes of escaping an oppression she believes to be caused by an economic system and *which is oppressive only due to the male characteristic of the system.* After the victory of a class or a category, *she* only benefits, at best, from slight improvements in conditions, reforms, advantages, a few coins which fall from the coffers of the upheaval, that is, if she happens to be on the right side; but power has only changed hands, never its structure; as we have seen, never do we find a deep challenge to the *rapport between the sexes*, a humanist and ecological question among others. "Seated, standing, lying, man is always his own tyrant." Céline would have been surprised at the connection of his thoughts to Marx: man's rapport with man is measured by man's rapport with woman. If some men, always and everywhere, end up the ultimate profiteers from a revolutionary change, it is precisely because they apply to others the measures of force that *all* men as a group apply to *all* women as a group—even if, on occasion, it happens that Mr. So-and-So capitulates to his own wife.

How, then, do we approach the problem of maximal profit which sacrifices the collective interest to private interest, or the race for power which takes the place of collective interest in revolutionary instances; how can this problem be resolved as long as mental structures remain as they are: that is to say, informed by fifty centuries of masculine planetary civilization, overexploitative and destructive of the resources?

The proof that no revolution directed and accomplished by men can achieve the necessary mutation is that none has ever gone further than replacing one regime by another, one system by another in accordance with the existing structures, and that none has ever even envisaged *the possibility of going further*, of departing from the infernal cycle of production-consumption which is the alibi for this enormous mass of work—useless, alienating, mystified and mystifying—the very base of the male society wherever it exists.

No, imagination is never in power. We keep falling into the same patterns, into the same fatal stereotypes; much like the sentiment expressed by the people when Louis XIII's favorite bird catcher replaced the Constable Concini: "Nothing but the cork has changed; the wine still tastes the same."

A few years ago, in a science-fiction collection entitled *Après*,[35] only one of several tales on the theme of a total change after a catastrophe[36] envisaged the following scenario: industrial production ended, scientific research limited to a small number of well-confined laboratories, collective efforts directed toward the areas of thought, art, and nonproductive activities. Astronauts returning to earth after long years away despaired at finding earth taken over by what they deemed an irremediable decay. Until the day they noticed that all the discoveries of the past had been carefully preserved and transferred to a restricted number of depositories, but were only used on rare occasions; for example, a very rare remedy was fabricated and a supersonic jet sent to get it in emergency cases, to heal a sick child; the rest of the time, transportation was by horse or bicycle. Importance, thus, was shifted from the speed of the transport to its motivation. The civilization of gadgets had disappeared and given way to a humanism, which all the while was not retrograde after dispensing with technical devices devoid of common sense. The author of this "utopia" was a woman, the only woman in the collection of short stories: Marion Zimmer.

In 1972, Gébé, an artist who seemed not to be aware of this tale translated from American English, produced his filmed utopia: *L'An 01*.[37] Work in general stops; the workers leave the factories, office workers leave their offices, propagandists shout in the streets: "Open your window and throw the key into the street." Then they start working again at a considerably slower pace, just enough to meet the immediate needs of the people; there was no need to kill the capitalists, for they took care of it themselves; the sky was darkened by company presidents throwing themselves from the skyscrapers. The answer to the inevitable question, But how do you occupy so much leisure time? was made with simplicity: You spend time reflecting, *and it's not a sad matter.* As long as humans shudder at such an unhappy prospect—the possibility of thinking about themselves—they are still stuck in Pascal's time: competition, aggression, and all the horrors of the male society will triumph.

Without a doubt, from the immediate feminist viewpoint, this new Village Seer that J. J. Rousseau-Gébé[38] offers us is hardly satisfying; everything therein is truly profoundly changed, except the rapport between the sexes; woman remains the sole object of erotic desire; she is still this White Woman who mends his pants for the former chief editor, this Black Woman who grinds the manioc while the men discuss the Year 01 in Africa; another finds she has not gained much by milking cows instead of typing. All the same, however, for the first time, and unknown even to the author, *a truly antimale* conception of the world was shown to us in the form of a public product, a film; the utopia (in the classical sense: yesterday's utopia, tomorrow's truth) that the Feminine alone can implement. That is to say, women above all, but not only women: their objective and natural allies, young people, these youths from both sexes which will carry within themselves the strongest protest against the outdated world of the father, once they have thrown away the aftermath of its macho-leftism.

Because those who call for the alliance of the child and the woman are right: the Feminine is that part of the world separated, put aside, put between parentheses, which suffers the economic and cultural dictatorship of the Father; the first and most insurmountable taboo, the kind of incest most severely cursed, is the amorous rapport between mother and son; the other bans are secondary. Why? Because it is the wall that phallocracy felt it necessary to raise against women's power. Above and beyond the stories of sexual jealousy born of the Freudian triangle, anthropology has been able to discern, in obligatory exogamy, the dictatorship of the patriarchal society and its fear of women, its misogyny which becomes gynophobia in the face of this terrible menace: that the Mother and the Child would unite against the Father.[39]

A massive end of productive work is not a utopia; it has been shown that 7 to 10 percent of present production would suffice amply to meet food, clothing, and housing needs. Even though *The Year 01* does not emphasize the ecological necessity of this solution, one sequence ties pollution to the demented inflation of industry—the one in which the fisherman, stammering from emotion, cries into the telephone: "I did it, I caught a fish!" It is a real cause for celebration: "We welcome the first fish of the Year 01!" When we learn, thanks to our Cassandra preaching in the futurologist desert, that more marine species have disappeared in thirty years than in *the entire geological period following the Pleistocene epoch*, we can ask the question: Which is worth more, fish or gadgets?[40]

The current consumption-production cycle inevitably linked to industrial expansion, fruit of the mental structure of phallocracy, can be broken down in the following way: 80 percent superfluous products, about 20 percent of which are entirely useless, must be thrown on the market at the price of poisoning and destroying the patrimony in an ascending curve; to do this, work time roughly equivalent to 80 percent of a human life must be available, in other words, equivalent to practically total alienation. That is not all— these superfluous or useless objects must be short-lived and replaceable, which increases the poisoning and destruction. Finally, supreme alienation: since the products must be consumed, desire must be inspired by a techno-promotional circuit; that is, fabricated from start to finish. Since the producer is also a consumer, he will thus be alienated and fooled at every level, inside, outside, and all around, as Gébé would say. A scam aimed at time, which is the web of life, at sensitivity, which is its valuable side, a gigantic, planetary, monstrous frustration; these are the results of the cycle born five thousand years ago, beginning with the caging of the second sex and the appropriation of the earth by males: "progress."

This root of the problem remains totally unnoticed in 1974 by modern revolutionary movements.[41] Even for the anticonformists of Marxism, the ecological problem comes back to the evils due to capitalism (and we can be sure that, given the system of profit, the two are closely connected, but the presence of the problem in socialist systems proves that it is not identified solely with capitalism), in much the same way that the inequity in the rapport between the sexes is defined as a superstructure that would be changed by the substitution of one economic regime for another. In instances in which youths have become involved in the protest only through some political label or other, they are subsumed by the System that they fight without knowing it, since they do not recognize, beyond the economico-social aspect, its profound and primordial motivation of war between the sexes. But in the instances in which youths understand that their sensitivity belongs to this battered Feminine, oppressed and repressed in the world order of the Father, where even their economic position situates them, in general, on the same plane as women—supported by the Father who expects a service, a return on his investment—youth of both sexes come more and more to the realization that their cause is not only that of the Mother but of all women the world over. When the male options of macho-leftists (including for girls, we must repeat) have been entirely eradicated by the consciousness of an emergency, of a burning necessity—to explode the consumption-production cycle instead of giving it a new form doomed to the same failure and leading to the same death—feminism will have won because the Feminine will have triumphed.

The case is clear—we are in no way pleading for an illusory superiority of women over men, or even for the "values" of the Feminine, which exist only on a cultural level and not at all on a metaphysical one. We are saying: Do you want to live or die? If you refuse planetary death, you will have to accept the revenge of women, for their personal interests join those of the human community, whereas the males' interests, on an individual basis, are separate from those of the general community, and this holds true even at the level of the current male System. For proof, we have merely to consider the contradiction between the supreme instances of its power, pushing women into production (and they have announced: "1975; the Year of Women"), and the private interests of the males living under this same power, who furiously resist the idea of depriving themselves of their personal maids. We have merely to consider the contradiction between the effort of this same power to diffuse and aid contraception in the goal of using, for its production, the feminine time taken away from the nourishing function, and the same indignant resistance from individual males against the fact that their females could control their procreation!

This brings me back to the beginning of this work: awareness of femininitude, of the misfortune of being a woman, is taking place today in a contradiction and an ambiguity which announce the end of the same misfortune. Starting at real-life events, at radical subjectivity, her experience as a species treated as a minority, separated, reified, looked at, a woman of my generation discovers that her "little problem," her "secondary question," this so-minute detail of the subversive front, indeed, her "fragmentary struggle," is no longer content to link with but identifies directly with the number one question, with the original problem; the basis, even, of the indispensable need to change the world, not just to improve it, but *so that there can still be a world.*

What revenge for the sole human majority to be treated as a minority! Until now it was difficult for women to comprehend the source of the misfortune of femininitude; they limited themselves to demands for "fragments" of world management before getting to the roots, free sexual disposition, which suddenly revealed a sense of totality. She who until now had been not man's "companion" but at once the alchemical crucible of his reproduction, his beast of burden, his scapegoat, his spittoon; with whom he sometimes amused himself by covering her in stones and by proclaiming her to be his Holy Grail; the one who always brought about in him, by virtue of a constant threat of victory, a hostile distrust which has gone as far in certain cultures as the hatred that engenders ritual mutilation (Africa) and death (sexocide of witches in the Middle Ages, of the "debauched" of the Mediterranean basin or the Orient) and that has evolved into a true "gynophobia," as W. Lederer puts it, she has now become in law and in possibility, if not in fact, what she was in prephallocratic times—the sole controller of procreation. As soon as this right can be freely lived through massive contraception and uninhibited access to abortion, the disappearance of half the human nightmare will depend on her as she implements the "stress of the white female rat."

This immense power which she is going to acquire, which is already coming within sight, has nothing in common with the one that organizes, decides, represents, and oppresses, and which still remains in male's hands; it is exactly in this area that she can most efficiently bring about its defeat and sound the death knell of ancient oppression. In short, according to a slogan of the Ecology-Feminism Center, we have to tear the planet away from the male today in order to restore it for humanity of tomorrow. That is the only alternative, for if the male society persists, there will be no tomorrow for humanity.

Her very life threatened, as well as the one she passes on (that she *chooses* whether or not to pass on), she who is the keeper of this life source, in whom the forces of the future are realized and through whom they pass, is thus doubly concerned in finding the fastest solution to the ecological problem. And, what is more, she represents, in the purest Marxist fashion in the world, this producing class frustrated from its production by male distribution, since this source of collective wealth (procreation) is possessed by a minority, males, the feminine species being the human majority.

The specialists themselves recognize it; along with Edgar Morin in the ecological conference of the *Nouvel Observateur* (where no women appeared), they admit that "we are starting to comprehend that the abolition of capitalism and the liquidation of the bourgeoisie merely give way to a new oppressive structure." Reimut Reiche had already brought out and explained this fact in *Sexuality and Class Struggle* that the "core" resists all revisions of the order. That core, as I have shown here, is the phallocracy. It is at the base of an order which can only assassinate nature in the name of profit if it is capitalist, and in the name of progress if it is socialist. The problem for women is, first, demography, then nature, and thus the world; her urgent problem, the one in common with youth, is autonomy and control of her destiny. If humanity is to survive, she must resign herself to this fact.

Consequently, with a society finally in the Feminine, which will mean nonpower (and not power to women), it will be proven that no other human category could have accomplished the ecological revolution, for no other was as directly concerned at all levels. And the two sources of wealth diverted for male profit will once again become an expression of life and no longer an elaboration of death; and the human being will finally be treated *first* as a person, and not above all else as a male or female.

And the planet placed in the feminine will flourish for all.

1971–74

Notes

1. Translator's note: The initials indicate the French expression "*Mouvement de Libération des Femmes.*" For background information and updated details concerning the feminist groups mentioned here, see Claire Duchen, *Feminism in France* (Boston: Routledge and Kegan Paul, 1986); Claire Duchen, *French Connections: Voices from the Women's Movement in France* (Amherst: University of Massachusetts Press, 1987); Elaine Marks and Isabelle de Courtivron, eds., *New French Feminisms* (New York: Schocken Books, 1981); and Nicole Ward Jouve, *White Woman Speaks with Forked Tongue* (New York: Routledge, 1991). These authors discuss the period between the 1970s and the 1990s, discussing the beginnings and evolution of the various groups. Marks and de Courtivron succinctly describe the origins of the MLF: "The 'Mouvement de Libération des Femmes'... commonly referred to as the MLF, is not an organization. It is the name invented by the French press during the summer of 1970 to identify the diverse radical women's groups that had been visible in Paris, Lyon, and Toulouse since the fall of 1968" (30). Although the MLF roughly corresponds, therefore, to the women's liberation movement in the United States, the initials MLF will be used throughout this translation due to the specificity to French feminist activity.
2. *Le Grand Soir*, a prositationist paper, copied by *Le Fléau Social*, May 1973. (Translator's note: The work has not appeared in English. Its title translates as *The Social Plague*.)
3. The antipsychiatric movement has engendered an "International Association of Looney Lunatics" which contests "normality" even more radically than do the movements for the liberation of homosexuals. (Translator's note: The original text reads: "Internationale des Fous Furieux." I have chosen to translate "fous furieux" as "looney lunatics" to approximate the strength of the alliteration. In doing so, I may have obscured the fury implicit in "furieux"—a concept which is particularly important in this context.)
4. See *Halte à la croissance* (Paris: Fayard, 1972). (Translator's note: The work appeared first in English: Donella H. Meadows, Dennis L. Meadows, Jorgen

Randers, and William W. Behrens, III, *The Limits to Growth: A Report for the Club of Rome's Project on the Predicament of Mankind* (London: Earth Island, 1972). *Halte à la croissance* is a translation into French of a part of it written by Janine Delaunay.

5. Translator's note: "Underground" is in English and italicized in the original text.

6. Translator's note: On page 10 of the French text, in a section entitled "Femininitude," D'Eaubonne discusses the criticism espoused by "certain extremists of the left" who call women "a category among others of oppressed people," and claim that a struggle targeting women only is "parcellary."

7. Translator's note: The original text gives the expression in English.

8. Translators note: Published in Paris at Union Générale de l'édition, 1973, the book has not appeared in English. The title translates as *Ecology: Vicious Circle or Alleviation.*

9. Actually, the figures claimed to be exaggerations have since been greatly surpassed.

10. In *Anticlichés sur l'Amérique* (Paris: Robert Laffont, 1971). The general proportion of these figures has risen 2 to 6 percent in noxious substances. (Translator's note: The title of this work translates as *Anti-Clichés about America.* It has not appeared in English.)

11. Jean-François Revel, *Ni Marx ni Jésus* (Paris: Robert Laffont, 1971). Translator's note: The work has appeared in English as *Without Marx or Jesus: The New American Revolution Has Begun* (Garden City, N.Y.: Doubleday, 1971).

12. This masterpiece has recently been reedited in the collection *Présence du Futur* (Paris: Denoël, 1971). Well before the question of pollution and of the destruction of the environment, this novel predicted a devastating future in which oak wood has become such a precious material that rings are made from it, in which the privileged class is composed of publicity technicians and the immense and impoverished mass is composed of "consumers," and in which the struggle to protect nature is the clandestine domain of "Conservers" who are treated . . . as communists. Translator's note: Neither the original work nor the collection has appeared in English. The title translates roughly as *Planet out of Control.*

13. *Actuel*, Oct. 1971. The numbers have risen sharply since that time.

14. At this moment, agriculture has lightheartedly joined forces with industrial polluters. During the summer 1973, one of my female friends contracted bycosis and bronchial poisoning while working on an apple plantation near Montpellier. The farmer sprayed the trees with a mixture of *three* violently toxic products, designed . . . to avoid marks on the fruit. On top of that, he painted the apples with a layer of wax the night before taking them to the markets.

15. Translator's note: The title itself is a linguistic pun, operating on several levels. A *voiture survireuse* is an expression for a car which oversteers and, as such, cannot be properly controlled. The root of the adjective, *vireux*, refers to a noxious material. Pravda here refers to the name of the comic strip heroine. The comic strip no longer exists.

16. Elizabeth Draper, *Conscience et contrôle des naissances* (Paris: Robert Laffont, 1971). She predicted six billion inhabitants by the year 2000. Today, these predictions are closer to seven billion and may still be surpassed, as I discussed earlier. Translator's note: The book was published first in English as *Birth Control in the Modern World: The Role of the Individual in Population Control* (London: Allen and Unwin, 1965).

17. Translator's note: The initials HLM designate *habitations à loyer modéré*—literally, moderated-rent housing. The rent is regulated according to income.

18. The Right's answer to the classical argument: overpopulation, Third-World problem.

19. Translator's note: The title, literally, means *The Corrupted Peasant. Le Paysan Perverti*, published in 1775, was a best seller in France. In 1784, its author published *Le Paysan et la Paysanne Pervertis*, derived in part from the earlier work. The latter has been translated in English as *The Corrupted Ones* (London: Neville Spearman, 1967).

20. Translator's note: Sarcelles and Billancourt are Parisian suburbs, made up of HLMs (see note 17 above). These areas are plagued by gang violence, drug abuse, and crime.

21. Simone de Beauvoir, *The Second Sex* (New York: Vintage Books, 1974).

22. Cf. my contribution to the subject: *Le Féminisme: Histoire et actualité*. Translator's note: The title of D'Eaubonne's work on the topic translates as *Feminism: Its History and Present State*. It was published in Paris by Alain Moreus, 1972. Lionel Tiger's book appeared in English as *Men in Groups* (New York: Random House, 1969).

23. Translator's note: The original text gives this last condition in English, hence in quotation marks.

24. See the film with this title. Translator's note: The French expression, given to the film referred to here, is *Petits meurtres sans importance*.

25. *Tout 12*, banned as an affront to morality.

26. *Actuel*, Oct. 1971.

27. Translator's note: The title translates as "to heal."

28. [See August Bebel, *Women in the Past, Present, and Future* (San Francisco: G. B. Benham, 1897).] Wolfgang Lederer, *Gynophobia ou la peur des femmes* (Paris: Payot, 1971). Translator's note: The work appeared a few years earlier in English: Wolfgang Lederer, *The Fear of Women* (New York: Harcourt Brace Jovanovich, 1968).

29. Translator's note: See note 8.

30. Cf. *Libération des Femmes, année zéro* (Paris: Maspero, 1972). Translator's note: The work has not appeared in English. The title translates as *Women's Liberation, the First Year*.

31. See Claire Duchen, *Feminisim in France: From May '68 to Mitterand* (London: Routledge and Kegan Paul, 1986), 12–14. The following passage is particularly relevant to d'Eaubonne's text: "The event that dominated media attention to feminism was the publication in the weekly news magazine *Le Nouvel Observateur* on 5 April 1971 of a Manifesto signed by 343 women, declaring that they had had illegal abortions. . . . The women who signed included the best-known figures of stage, screen, and intellectual life, with Simone de Beauvoir at the head" (12). Duchen's work includes the text of the manifesto and discusses the surrounding events.

32. Pierre Gordon, *Initiation Sexuelle et Morale Religieuse* (Paris: Presse universitaire de France, 1945), and my work *Le Féminisme*. Translator's note: Gordon's work has not appeared in English. The title translates as *Sexual Initiation and Religious Morality*.

33. The Feminist Front, spurred on by an ecofeminist, wrote to all the representatives to the Parliament from Paris to demand an investigation into the financing of this manipulative operation that, many affirm, comes from these diary-product advertisements. Translator's note: The Nestlé-Guigoz company is known in the United States as Nestles.

34. An assertion which will make my Marxist friends indignant. But if women of the proletariat have something to gain from the "victory of the proletariat," it would only be crumbs of the power supposedly acquired by male proletarians. In the meantime, women of the bourgeoisie would lose these same crumbs (or

even large parts) to which their men gave them access. Never will power as such be given to women. At any rate, it is not even desirable, as the only significant revolution is that which abolishes even the notion of power, and the proletarian state at the same time as sexism.

35. Translator's note: The work was published in Paris by Stock, 1962. It has not been translated into English. The title translates as *After* or *Afterwards*.

36. It is significant that at the present time the collective unconscious is accepting more and more the idea of a necessary apocalypse and is consoling itself with theories about "afterward," instead of thinking of avoiding it. As Victor Hugo said, "catastrophes have a somber way of arranging things."

37. The film was released in 1972. It is an early film by Jacques Doillon with a cast of young, now well-known, actors and can be viewed at the Vidéothèque de Paris. The title translates as *The Year 01*. It has not been released with English subtitles.

38. Translator's note: D'Eaubonne conflates Gébé's name with that of Jean-Jacques Rousseau, thus drawing a parallel between his utopian vision and the "perfect, unspoiled world" Rousseau presented in *Emile ou l'éducation* and other works advocating a return to nature in order to escape the corruption of civilization.

39. See Gilles Deleuze and Felix Guattari, *Anti-Oedipus: Capitalism and Schizophrenia*, trans. Robert Hurley, Mark Seem, and Helen R. Lane (New York: Viking Press, 1977), and especially Serge Moscovici, *Society against Nature: The Emergence of Human Societies*, trans. Sacha Rabinovich (Atlantic Highlands, N.J.: Humanities Press, 1976). Both support, at different levels, *Crachons sur Hegel*, which I have discussed at length in an earlier chapter. Translator's note: The last title translates as *Let's Spit on Hegel*. The original Italian title is Carla Lonzi, *Sputiamo su Hegel*, 1970. It has not been translated into French or English, but it was regarded as a manifesto of the Italian feminist movement, Rivolta Femminile.

40. A community established two years ago in the Cevennes has accomplished this same miracle; trout have come back to the streams, butterflies to the meadows; things which had not been seen in that area for fifteen years. "Nature has prepared the seeds and all man has to do is join in," the founder of this community told me. Join in to love it and not to exploit it.

41. Not long ago, Lenin described the incapacity of the proletariat to go beyond the level of "trade-unionist." That was correct during his time, but he made the mistake of not realizing that it was a question of a historical stage of capitalism which did not create conditions favorable to a revolutionary "spontaneity," given the economy of shortage, and above all, he made the mistake of wanting to "give an a-historical and eternal value to this contingent reality" (Cf. *La Bande à Baader, ou, La violence révolutionnaire*, [Paris: Champ Libre, 1972], 25). The author of this text refers to "Lenin's incomprehension of the dialectic between *production* relations and *distribution* relations." It may not be out of place to establish a parallel between this limitation evinced by Lenin and the one he shows regarding the place of the sexual question in the revolutionary analysis. A correct dialectic which differentiates "production" (procreation) from "distribution" (eros) is also missing there. Translator's note: The work, originally written in German, has not appeared in English. The subtitle of the French version is *La violence révolutionnaire (Revolutionary Violence)*.

Feminism and the Revolt of Nature

Ynestra King

ECOLOGY IS A FEMINIST issue. But why? Is it because women are more a part of nature than men? Is it because women are morally superior to men? Is it because ecological feminists are satisfied with the traditional female stereotypes and wish to be limited to the traditional concerns of women? Is it because the domination of women and the domination of nature are connected?

The feminist debate over ecology has gone back and forth and is assuming major proportions in the movement; but there is a talking-past-each-other, not-getting-to-what's-really-going-on quality to it. The differences derive from unresolved questions in our political and theoretical history, so the connection of ecology to feminism has met with radically different responses from the various feminisms.

RADICAL FEMINISM AND ECOLOGY

Radical feminists of one genre deplore the development of connections between ecology and feminism and see it as a regression which is bound to reinforce sex-role stereotyping. Since the ecological issue has universal implications, so the argument goes, it should concern men and women alike. Ellen Willis, for instance, wrote recently:

> From a feminist perspective, the only good reason for women to organize separately from men is to fight sexism. Otherwise women's political organizations simply reinforce female segregation and further the idea that certain activities and interests are inherently feminine. All-female groups

From: *Heresies*, no. 13 (1981), 12–16.

that work against consumer fraud, or for the improvement of schools, implicitly acquiesce in the notion that women have special responsibilities as housewives and mothers, that it is not men's job to worry about what goes on at the supermarket, or the conditions of their children's education. Similarly, groups like Women Strike for Peace, Women's International League for Peace and Freedom, and Another Mother for Peace perpetuate the idea that women have a specifically female interest in preventing war. . . . If feminism means anything, it's that women are capable of the full range of human emotions and behavior; politics based on received definitions of women's nature and role are oppressive, whether promoted by men or by my alleged sisters.[1]

Other radical feminists—most notably Mary Daly and Susan Griffin—have taken the opposite position. Daly believes that women should identify with nature against men, and that *whatever* we do, we should do it separately from men. For her, the oppression of women under patriarchy and the pillage of the natural environment are basically the same phenomenon.[2] Griffin's book is a long prose poem (actually the form defies precise description; it is truly original).[3] It is not intended to spell out a political philosophy, but to let us know and feel how the woman/nature connection has been played out historically in the victimization of women and nature. It suggests a powerful potential for a movement linking feminism and ecology.

So how do women who call themselves radical feminists come to such divergent positions? Radical feminism roots the oppression of women in biological difference. It sees patriarchy (the systematic dominance of men) preceding and laying the foundation for other forms of oppression and exploitation; it sees men hating and fearing women (misogyny) and identifying us with nature; it sees men seeking to enlist both women and nature in the service of male projects designed to protect men from feared nature and mortality. The notion of women being closer to nature is essential to such projects. If patriarchy is the archetypal form of human oppression, then radical feminists argue that getting rid of it will also cause other forms of oppression to crumble. But the essential difference between the two or more types of radical feminists is whether the woman/nature connection is potentially liberating or simply a rationale for the continued subordination of women.

Other questions follow from this theoretical disagreement: (1) Is there a separate female life experience in this society? If so, is there a separate female culture? (2) If there is, is a female culture merely a male-contrived ghetto constructed long ago by forcibly taking advantage of our physical vulnerability born of our child-bearing function and smaller stature? Or does it suggest a way of life providing a critical vantage point on male society? (3) What are the implications of gender difference? Do we want to do away with gender difference? (4) Can we recognize "difference" without shoring up dominance based on difference?

Rationalist radical feminists (Willis's position) and radical cultural feminists (Daly's and Griffin's position) offer opposite answers, just as they come to opposite conclusions on connecting feminism to ecology—and for the same reasons. The problem with both analyses, however, is that gender identity is neither fully natural nor fully cultural. And it is neither inherently oppressive nor inherently liberating. It depends on other historical factors, and how we consciously understand woman-identification and feminism.

SOCIALIST FEMINISM AND ECOLOGY

Socialist feminists[4] have for the most part yet to enter feminist debates on "the ecology question." They tend to be uneasy with ecological feminism, fearing that it is based on an ahistorical, antirational woman/nature identification; or they see the cultural emphasis in ecological feminism as "idealist" rather than "materialist." The Marxist side of their politics implies a primacy of material transformation (economic/structural transformation precedes changes in ideas/culture/consciousness). Cultural and material changes are not completely separate. There is a dialectical interaction between the two, but in the last instance the cultural is part of the superstructure and the material is the base.

Historically socialism and feminism have had a curious courtship[5] and a rather unhappy marriage,[6] characterized by a tug of war over which is the primary contradiction—sex or class. In an uneasy truce, socialist feminists try to overcome the contradictions, to show how the economic structure and the sex-gender system[7] are mutually reinforced in historically specific ways depending on material conditions, and to show their interdependence. They suggest the need for an "autonomous" (as opposed to a "separatist") women's movement, to maintain vigilance over women's concerns within the production and politics of a mixed society. They see patriarchy as different under feudalism, capitalism, and even under socialism without feminism.[8] And they hold that nobody is free until everybody is free.

Socialist feminists see themselves as integrating the best of Marxism and radical feminism. They have been weak on radical cultural critique and strong on helping us to understand how people's material situations condition their consciousnesses and their possibilities for social transformation. But adherence to Marxism with its economic orientation means opting for the rationalist severance of the woman/nature connection and advocating the integration of women into production. It does not challenge the culture-versus-nature formulation itself. Even where the issue of woman's oppression and its identification with that of nature has been taken up, the socialist feminist solution has been to align women with culture in culture's ongoing struggle with nature.[9]

HERE WE GO AGAIN; THIS ARGUMENT IS AT LEAST ONE HUNDRED YEARS OLD!

The radical feminist/socialist feminist debate does sometimes seem to be the romantic feminist/rationalist feminist debate of the late nineteenth century revisited.[10] We can imagine nineteenth-century women watching the development of robber-baron capitalism, "the demise of morality" and the rise of the liberal state which furthered capitalist interests while touting liberty, equality, and fraternity. Small wonder that they saw in the domestic sphere vestiges of a more ethical way of life, and thought its values could be carried into the public sphere. This perspective romanticized women, although it is easy to sympathize with it and share the abhorrence of the pillage and plunder imposed by the masculinist mentality in modern industrial society. But what nineteenth-century women proclaiming the virtues of womanhood did not understand was that they were a repository of organic social values in an increasingly inorganic world. Women placed in male-identified power positions can be as warlike as men. The assimilation or neutralization of enfranchised women into the American political structure has a sad history.

Rationalist feminists in the nineteenth-century, on the other hand, were concerned with acquiring power and representing women's interests. They opposed anything that reinforced the idea that women were "different" and wanted male prerogatives extended to women. They were contemptuous of romantic feminists and were themselves imbued with the modern ethic of progress. They opposed political activity by women over issues not seen as exclusively feminist for the same reasons rationalist radical feminists today oppose the feminism/ecology connection.

THE DIALECTIC OF MODERN FEMINISM

According to the false dichotomy between subjective and objective—one legacy of male Western philosophy to feminist thought—we must root our movement *either* in a rationalist-materialist humanism *or* in a metaphysical-feminist naturalism. This supposed choice is crucial as we approach the ecology issue. *Either* we take the anthropocentric position that nature exists solely to serve the needs of the male bourgeois who has crawled out of the slime to be lord and master of everything, *or* we take the naturalist position that nature has a purpose of its own apart from serving "man." We are *either* concerned with the "environment" because we are dependent on it, *or* we understand ourselves to be *of* it, with human oppression part and parcel of the domination of nature. For some radical feminists, only women are capable of full consciousness.[11] Socialist feminists tend to consider the naturalist position as historically regressive, antirational, and probably fascistic. This is the crux of the anthropocentric/naturalist debate, which is emotionally loaded for both sides, but especially for those who equate progress and rationality.

However, we do not have to make such choices. Feminism is both the product and potentially the negation of the modern rationalist world view and capitalism. There was one benefit for women in the "disenchantment of the world,"[12] the process by which all magical and spiritual beliefs were denigrated as superstitious nonsense and the death of nature was accomplished in the minds of men.[13] This process tore asunder women's traditional sphere of influence, but it also undermined the ideology of "natural" social roles, opening a space for women to question what was "natural" for them to be and to do. In traditional Western societies, social and economic relationships were connected to a land-based way of life. One was assigned a special role based on one's sex, race, class, and place of birth. In the domestic sphere children were socialized, food prepared, and men sheltered from their public cares. But the nineteenth-century "home" also encompassed the production of what people ate, used, and wore. It included much more of human life and filled many more human needs than its modern corollary—the nuclear family—which purchases commodities to meet its needs. The importance of the domestic sphere, and hence women's influence, declined with the advent of market society.

Feminism also negates capitalist social relations by challenging the lopsided male-biased values of our culture. When coupled with an ecological perspective, it insists that we remember our origins in nature, our connections to one another as daughters, sisters, and mothers. It refuses any longer to be the unwitting powerless symbol of all the things men wish to deny in themselves and project onto us—the refusal to be the "other."[14] It can heal the splits in a world divided against itself and built on a fundamental lie: the defining of culture in opposition to nature.

The dialectic moves on. Now it is possible that a conscious visionary feminism could place our technology and productive apparatus in the service of a society based on ecological principles and values, with roots in traditional women's ways of being in the world. This in turn might make possible a total cultural critique. Women can remember what men have denied in themselves (nature), and women can know what men know (culture). Now we must develop a transformative feminism that sparks our utopian imaginations and embodies our deepest knowledge—a feminism that is an affirmation of our vision at the same time it is a negation of patriarchy. The skewed reasoning that opposes matter and spirit and refuses to concern itself with the objects and ends of life, which views internal nature and external nature as one big hunting ground to be quantified and conquered is, in the end, not only irrational but deadly. To fulfill its liberatory potential, feminism needs to pose a *rational reenchantment* that brings together spiritual and material, being and knowing. This is the promise of ecological feminism.

DIALECTICAL FEMINISM: TRANSCENDING THE RADICAL FEMINIST/SOCIALIST FEMINIST DEBATE

The domination of external nature has necessitated the domination of internal nature. Men have denied their own embodied naturalness, repressed memories of infantile pleasure and dependence on the mother and on nature.[15] Much of their denied self has been projected onto women. Objectification is forgetting. The ways in which women have been both included in and excluded from a culture based on gender differences provide a critical ledge from which to view the artificial chasm male culture has placed between itself and nature. Woman has stood with one foot on each side. She has been a bridge for men, back to the parts of themselves they have denied, despite their need of women to attend to the visceral chores they consider beneath them.

An ecological perspective offers the possibility of moving beyond the radical (cultural) feminist/socialist feminist impasse. But it necessitates a feminism that holds out for a separate cultural and political activity so that we can imagine, theorize or envision from the vantage point of *critical otherness*. The ecology question weights the historic feminist debate in the direction of traditional female values over the overly rationalized, combative male way of being in the world. Rationalist feminism is the Trojan horse of the women's movement. Its piece-of-the-action mentality conceals a capitulation to a culture bent on the betrayal of nature. In that sense it is unwittingly both misogynist and anti-ecological. Denying biology, espousing androgyny and valuing what men have done over what we have done are all forms of self-hatred which threaten to derail the teleology of the feminist challenge to this violent civilization.

The liberation of women is to be found neither in severing all connections that root us in nature nor in believing ourselves to be more natural than men. Both of these positions are unwittingly complicit with nature/culture dualism. Women's oppression is neither strictly historical nor strictly biological. It is both. Gender is a meaningful part of a person's identity. The facts of internal and external genitalia and women's ability to bear children will continue to have social meaning. But we needn't think the choices are external sexual warfare or a denatured (and boring) androgyny. It is possible to take up questions of spirituality and meaning without abandoning the important insights of materialism. We can use the insights of socialist feminism, with its debt to Marxism, to understand how the material conditions of our daily lives interact with our bodies and our psychological heritages. Materialist insights warn us not to assume an *innate* moral or biological superiority and not to depend on alternative culture alone to transform society. Yet a separate radical feminist culture within a patriarchal society is necessary so we can learn to speak our own bodies and our own experiences, so the male culture representing itself as the "universal" does not continue to speak for us.[16]

We have always thought our lives and works, our very beings, were trivial next to male accomplishments. Women's silence is deafening only to those who know it's there. The absence is only beginning to be a presence. Writers like Zora Neale Hurston, Tillie Olsen, Grace Paley, and Toni Morrison depict the beauty and dignity of ordinary women's lives and give us back part of ourselves. Women artists begin to suggest the meanings of female bodies and their relationships to nature.[17] Women musicians give us the sounds of loving ourselves.[18] The enormous and growing lesbian feminist community is an especially fertile ground for women's culture. Lesbians are pioneering in every field and building communities with ecological feminist consciousness. Third-World women are speaking the experience of multiple otherness—of race, sex, and (often) class oppression. We are learning how women's lives are the same and different across these divisions, and we are beginning to engage the complexities of racism in our culture, our movement, and our theory.

There is much that is redemptive for humanity as a whole in women's silent experience, and there are voices that have not yet been heard. Cultural feminism's concern with ecology takes the ideology of womanhood which has been a bludgeon of oppression—the woman/nature connection—and transforms it into a positive factor. If we proceed dialectically and recognize the contributions of both socialist feminism and radical cultural feminism, operating at both the structural and cultural levels, we will be neither materialists nor idealists. We will understand our position historically and attempt to realize the human future emerging in the *feminist present*. Once we have placed ourselves in history, we can move on to the interdependent issues of feminist social transformation and planetary survival.

Toward an Ecological Culture

Acting on our own consciousness of our own needs, we act in the interests of all. We stand on the biological dividing line. We are the less-rationalized side of humanity in an overly rationalized world, yet we can think as rationally as men and perhaps transform the idea of reason itself. As women, we are naturalized culture in a culture defined against nature. If nature/culture antagonism is the primary contradiction of our time, it is also what weds feminism and ecology and makes woman the historic subject. Without an ecological perspective which asserts the interdependence of living things, feminism is disembodied. Without a more sophisticated dialectical method which can transcend historic debates and offer a nondualistic theory of history, social transformation and nature/culture interaction, feminism will continue to be mired in the same old impasse. There is more at stake in feminist debates over "the ecology question" than whether feminists should organize against the draft, demonstrate at the Pentagon, or join mixed antinuke organizations. At stake is the range and potential of the feminist social movement.

Ecological feminism is about reconciliation and conscious mediation, about recognition of the underside of history and all the invisible voiceless activities of women over millennia. It is about connectedness and wholeness of theory and practice. It is the return of the repressed—all that has been denigrated and denied to build this hierarchal civilization with its multiple systems of dominance. It is the potential voice of the denied, the ugly, and the speechless—all those things called "feminine." So it is no wonder that the feminist movement rose again in the same decade as the ecological crisis. The implications of feminism extend to issues of the meaning, purpose, and survival of life.

> Never to despise in myself what I have been taught
> to despise. Not to despise the other.
> Not to despise the *it*. To make this relation
> with it: to know that I am it.
> —Muriel Rukeyser, "Despisals"[19]

Notes

1. Ellen Willis, *Village Voice*, 23 June 1980. In the *Village Voice*, July 1980, Willis began with the question: "Is ecology a feminist issue?" and was more ambiguous than in her earlier column, although her theoretical position was the same.
2. Mary Daly, *Gyn/Ecology: The Meta-Ethics of Radical Feminism* (Boston: Beacon Press, 1978).
3. Susan Griffin, *Woman and Nature: The Roaring Inside Her* (New York: Harper and Row, 1979).
4. For an overview of socialist feminist theory, see Zillah Eisenstein, ed., *Capitalist Patriarchy and the Case for Socialist Feminism* (New York: Monthly Review Press, 1979).
5. See Batya Weinbaum, *The Curious Courtship of Women's Liberation and Socialism* (Boston: South End Press, 1978).
6. See Heidi Hartman, "The Unhappy Marriage of Marxism and Feminism: Towards a More Progressive Union," *Capital and Class* 8 (Summer 1979).
7. The notion of a "sex-gender system" was first developed by Gayle Rubin in "The Traffic in Women," in *Toward an Anthropology of Women*, ed. R. R. Reiter (New York: Monthly Review Press, 1975).
8. See Hilda Scott, *Does Socialism Liberate Women?* (Boston: Beacon Press, 1976).
9. See Sherry B. Ortner, "Is Female to Male as Nature Is to Culture?" in *Woman, Culture, and Society*, ed. M. Z. Rosaldo and L. Lamphere (Stanford, Calif.: Stanford University Press, 1974).
10. For a social history of the nineteenth-century romanticist/rationalist debate, see Barbara Ehrenreich and Dierdre English, *For Her Own Good* (New York: Doubleday, Anchor, 1979).
11. Mary Daly comes very close to this position. Other naturalist feminists have a less clear stance on *essential* differences between women and men.
12. The "disenchantment of the world" is another way of talking about the process of rationalization discussed above. The term was coined by Max Weber.

13. See Carolyn Merchant, *The Death of Nature: Women, Ecology and the Scientific Revolution* (San Francisco: Harper and Row, 1980).
14. For a full development of the idea of "woman as other," see Simone de Beauvoir, *The Second Sex* (New York: Modern Library, 1968).
15. Dorothy Dinnerstein in *The Mermaid and the Minotaur* (New York: Harper and Row, 1977) makes an important contribution to feminist understanding by showing that although woman is associated with nature because of her mothering role, this does not in itself explain misogyny and the hatred of nature.
16. See de Beauvoir, *Second Sex*.
17. See Lucy Lippard, "Quite Contrary: Body, Nature and Ritual in Women's Art," *Chrysalis* 1, no. 2 (1977): 31–47.
18. Alive, Sweet Honey in the Rock, Meg Christian, Holly Near, Margie Adam—the list is long and growing.
19. Muriel Rukeyser, *The Complete Poems of Muriel Rukeyser* (New York: McGraw-Hill, 1978).

Ecosocial Feminism as a General Theory of Oppression

VAL PLUMWOOD

THE UNEASY PRESENT RELATIONSHIP between the various social change movements has been reflected in the vigorous and often bitter debate which has
been taking place in the area of green theory and political ecology. Three
of the main articulated positions involved in this dialogue, which has taken
place in a major way in the United States, are each linked to critiques of
domination associated with the respective movements. The positions—social ecology, Deep Ecology, and ecological feminism—are usually treated as
presenting alternative and competitive analyses of the destruction of the
biosphere. Thus Deep Ecology is perhaps the best known branch of what
has been called "deep green theory," a set of positions or critiques treating
anthropocentrism[1] and the human domination of nature as one of the major roots of environmental problems. Social ecology, whose best-known
exponent is Murray Bookchin, draws on radical tradition and focuses on an
analysis of ecological problems in terms of human social hierarchy, while
ecofeminism sees androcentrism or the domination of women as a model
for other kinds of domination and as linked especially to the domination
of nature.

The debate involves real issues of genuine importance, but is also motivated by competitive reductionism and has an unnecessarily dismissive character. An example will illustrate what is meant by "reductionism" here. Back
in the days when Marxism was king of radical discourses, other discourses
and critiques, such as those of the women's movement and the environment movement, were reduced to subject status to be subsumed, incorporated into the kingdom of the sovereign. Their insights and problems were

From: Ronnie Harding, ed., *Ecopolitics V Proceedings*, Centre for Liberal and General Studies, University of New South Wales, Kensington, New South Wales, Australia, 1992, 63–72.

recognized and accorded legitimacy and attention just to the extent that they could be so absorbed (for example, those aspects of the feminist critique which could be reduced to questions of "class").[2] The reducing position claims to have the "fundamental" or "master" critique into which the reduced others can be absorbed as colonies. What cannot be reduced is dismissed.

Thus Bookchin flatly denies the legitimacy and relevance of the new rival critique of anthropocentrism, the human domination of nature (which of course is the hardest thing for the older radical tradition to try to take account of). The domination of nature, he assures us, came after the domination of human by human and is entirely secondary to it. His practice coheres with his theoretical view of the inferior status of this critique, since in *Remaking Society* he rarely mentions nonhuman nature without attaching the word "mere" to it. (Thus deep ecologists are said to want to "equate the human with mere animality," to "dissolve humanity into a mere species within a biospheric democracy," and reduce humanity "to merely one life form among many").[3] The egalitarian approach of Deep Ecology is roundly condemned as debasing to humans and involving a denial of their special qualities of rationality. Many ecological writers (especially Elizabeth Dodson Gray) have made the point that we construe difference in terms of hierarchy, and that it is not a question of denying human difference but of ceasing to treat it as the basis of superiority and domination.[4] Despite this Bookchin persistently interprets the denial of human hierarchy over nature as the denial of human distinctness. His concept of humans as "second nature" (nature rendered self-conscious) makes it difficult to conceptualize conflicts between the interests of humans and the interests of "first" nature. Thus Bookchin writes of second nature in an ecological society as "first nature rendered self-reflexive, a thinking nature that knows itself and can guide its own evolution."[5]

The critique of human domination of nature seems to me to be a new and inestimably important contribution to our understanding of domination. However, the alternative position in the debate presented by leading exponents of Deep Ecology is hardly less unsatisfactory or incomplete than Bookchin's account of social ecology and is equally intent on masculine strategies of colonization or on denying connection. Leading deep ecologist Warwick Fox makes repeated counterclaims to "most fundamental" status for his own critique of the domination of nature, arguing that it accounts for forms of human domination also.[6] At the same time (and inconsistently) Fox attacks critiques of other forms of domination as irrelevant to environmental concern, claiming for example that feminism has nothing to add to the conception of environmental ethics.[7] This exemplifies a common tendency among deep ecologists to dismiss human hierarchy as simply irrelevant to the problem of the destruction of nature and to distance themselves from other radical critiques. The mistreatment of nature is seen as a problem created by the blanket category of humans, among whom no relevant distinction need be made in explanation of environmental problems.

Each of the three major critiques would appear to share an approach via the concept of *domination* which could provide ground for a new synthesis and a common political orientation. However, as it stands, this potential for meeting remains undeveloped mainly because Deep Ecology has chosen to develop in ways which suppress the potential for a truly *political* understanding of its theme of human/nature domination, that is, as involving the operations of power. Thus Deep Ecology under the guidance of Warwick Fox has chosen for its core concept the notion of *identification*, understood as an individual psychic act rather than a political practice, yielding a theory which is both individualist (in the sense that it fails to look beyond the individual) and psychological-reductionist (in the sense that it discounts factors beyond psychology).[8] A similarly apolitical understanding is given to its core concept of ecological selfhood. Part of the motivation for this way of interpreting the central concepts seems to be desire to distance from other radical traditions and movements which are seen as less respectable and dangerously political. Such an account of Deep Ecology lends itself to being served up as a religious or spiritual garnish for a main course that turns out to be an uncritical approach to change via liberal political theory.[9]

AN ALTERNATIVE COOPERATIVE APPROACH

That fact that the debate is necessary and important should not obscure the fact then that much of its divisiveness is unnecessary. For example, the recognition of nature and animals as oppressed is *in no way incompatible* with the recognition of that domination as linked to, and continuous with, human domination. In fact it greatly enlarges and extends our understanding of each to see them as so linked. The point that it is possible to have essentially linked forms of oppression, and correspondingly relations of co-operation between movements which can fruitfully explore connection without attempting reduction, has been implicit in the approach of many ecofeminists. The concern to avoid absorption or colonization, not to lose what is new and distinctive in the environmental critique, is a legitimate one, but does not demand either isolationist or competitive reductionist stances in relation to other movements. The choice between reducing other critiques or being reduced by them is a false one. The barriers to a theoretical synthesis are political, not theoretical.

It is possible to respect the differences which different struggles and forms of oppression give rise to, to recognize their irreducibility, without paying the price of isolationism. As with individuals, so with movements, connection is risky but nevertheless essential for many reasons. First, some of us understand ourselves as oppressed in several different but connected ways, and work for a world in which sexism, the destruction of nature, racism, and militarism, for example, do not exist. For us movement connection is not an intellectual matter alone; its necessity is felt in the bones, in the attempt to understand and change our lives, in which various forms of oppression may be exemplified. Incompatibilities between critiques in such contexts appear as direct and personal discomforts and conflicts which impel us to change them. But to seek connection is not to require uniformity, or the absorption of each critique and movement into a single oceanic theory or movement, as I argue below.

From such a third perspective many of the criticisms Deep Ecology and social ecology have made of each other seem valid, but can be avoided by such a third position. Thus Deep Ecology has, on this view, been right in criticizing social ecology's human chauvinism and continued subscription to traditional doctrines of human supremacy and difference.[10] But social ecology is similary correct in its criticism of the insensitivity of Deep Ecology to differences within the category of humans and disregard for the role of human hierarchy in creating environmental problems.[11]

The strategies of isolation and colonization followed by both critiques are both bad methodology and bad politics. They are bad methodology because they involve a false choice, and bad politics because they pass up important opportunities for connection and strengthening. They are bad politics also because it is essential for critiques which purport to treat hierarchy to be prepared to meet others on a basis of equality, not with an agenda of inferiorizing or absorbing them. Maximizing chances for change must involve broadening the base of those who desire change, who can see how change is relevant to their lives, and this involves maximizing connection with a wide variety of issues and social change movements.

ECOFEMINISM AND CONNECTION

Unless social and Deep Ecology rework their positions, it seems that if we are to obtain such a theoretical base adequate to encompass and link the concerns of the environment movement then it is to ecological feminism that we will have to turn. Ecofeminism is a very diverse position, and there are ecofeminists who are closely associated with Deep Ecology and other ecofeminists who are close to social ecology, as well as others close to radical and other forms of feminism. Some forms of feminism and ecofeminism, principally those emerging from radical feminism, have a reductionist slant of their own, taking patriarchy to be the basis of all hierarchy, the basic form of domination to which other forms (including not only the domination of nature but capitalism and other forms of human social hierarchy) can be reduced. But many other feminists do not find this approach convincing, and see women's oppression as one (although perhaps a key one and certainly an irreducible one), among a number of forms of oppression.[12] Their approach has centered on enriching and networking various different critiques and exploring connections with the critique of the domination of women. Thus unlike the other two positions, ecofeminism as a general position is not committed to reducing or dismissing either the critique of anthropocentrism and the domination of nature with which Deep Ecology is concerned or that of the human hierarchy with which social ecology is concerned.[13] Many try to combine both perspectives. Thus Karen Warren writes that "transformative feminism would expand upon the traditional conception of feminism as a movement to end women's oppression by recognizing and making explicit the interconnections between all systems of oppression. . . . Feminism, properly understood, is a movement to end *all* forms of oppression. . . . A transformative feminism would build on these insights [of socialist and black feminism] to develop a more expansive and complete feminism, one which ties the liberation of women to the liberation of all systems of oppression."[14]

Major forms of ecofeminism have been concerned with cooperative rather than competitive movement strategies. The domination of women is of course central to the ecofeminist understanding of domination, but is also an illuminating model for many other kinds of domination, since the oppressed are often both feminized and naturalized. The ecofeminism of writers such as Rosemary Ruether has always stressed the links between the domination of women, of human groups such as blacks, and of nature. "An ecological ethic" she writes, "must always be an ethic of ecojustice that recognizes the interconnection of social domination and the domination of nature."[15] Thus the work of many ecofeminists foreshadows the development of ecosocial feminism as a general theory of oppression.

THE NETWORK OF OPPRESSION

Ecofeminism has particularly stressed that the treatment of nature and of women as inferior has supported and "naturalized" not only the hierarchy of male to female but the inferiorization of many other groups of humans seen as more closely identified with nature. It has been used to justify for example the supposed inferiority of black races or indigenes (conceived as more animal), the supposed inferiority of "uncivilized" or "primitive" cultures, and the supposed superiority of master to slave, boss to employee, mental to manual worker. For Western society, which has particularly employed a strongly genderized concept of nature as a way of imposing a hierarchical order on the world, feminization and naturalization have been crucial and connected strands supporting pervasive human relations of inequality and domination both within Western society and between Western society and non-Western societies. The interwoven dualisms of Western culture, of human/nature, mind/body, male/female, reason (civilization)/nature, have been involved here to create a logic of interwoven oppression consisting of many strands coming together.[16]

This interweaving occurs not only at the level of ideology, of the dualistic conceptions of the world, but also at the level of material practices, of production and reproduction. Domination must be seen as material and cultural, not as happening just at the level of ideas. Maria Mies has come closest to giving a general account of how this interweaving happens at the material level, and has traced a plausible historical path from the monopoly of means of coercion in the hands of male hunting bands, to the development of proto-military forms of organization and weaponry.[17] This enables a form of parasitism directed first against women and other tribal groups, who are exploited as agricultural laborers or slaves, and later against wider groups as a process of accumulation. In this form of accumulation (a growth-driven process envisaged as much wider than capitalism itself and including most known forms of socialism) domination has been directed in more recent times against a variety of human groups, especially women and the colonized, and also against nature itself. Mies's framework integrates without reduction, anarchist, antimilitarist, class, race, anticolonialist, feminist, and environmental concerns.

The example of the sealing industry, the first form of accumulation witnessed on this continent, will serve to illustrate some of these processes, and especially how the inferiorization of the sphere of nature and the feminine combines with the definition of oppressed groups as part of this sphere to yield both an ideology and a material practice of linked oppression. The history of the convicts, of the Aborigines and of the seals and whales whose deaths fueled these processes is interwoven at all levels. At the level of *production*, the convict system helped maintain the savagely repressive internal order of the class and property structure of Britain, the product of a long-term previous accumulation process. The slaughter of seals and whales provided fuel, oil, and a commercial basis for the convict transportation industry.

It took Australia's first export industry a mere eight years from the first sealing expedition in 1798 to reduce the numbers of seals in Bass Strait and Tasmania to a point below the levels capable of commercial exploitation. When the seals were finished, the sealers moved on to other states and to New Zealand. Sealers typically killed all sizes and ages, clubbing or stabbing seals as they came ashore. From 1806, when William Collins set up a bay whaling station in the Derwent (where whales were reported to be so thick in season that collisions with boats were a problem) the seal story was immediately repeated with whales. Female southern right whales (or bay whales), now some of the world's rarest whales, were killed along with their young when they entered the bays to give birth. Within a few decades the industry could boast the virtual local extinction of bay whales. Again the industry moved on to repeat the same story in New Zealand.[18]

The ships which had delivered their human cargoes of convicts to the disciplinary system went sealing or whaling and returned profitably to the "civilized" world with holds filled with oil or skins. In turn, the runaway convicts and soldiers provided suitably hardened and desperate workers for these industries and helped clear the country of the despised natives, described by the *Hobart Town Gazette* of 1824 as "the most peaceful creatures in the world."[19] The industry involved the abduction and enslavement of large numbers of Aboriginal women, who were subject to cruelty and to rape, and the killing of other Aborigines. Settlement along these lines led to the virtual annihilation of Aboriginal Tasmanians, who survive today as a distinct grouping with mixed ancestry, claiming Aboriginal identity but almost entirely dispossessed of their culture and lands.[20]

At the level of *culture*, the ideology which linked these common oppressions stressed the inferiority of the order of "nature," which was construed as barbaric, alien, and animal, and also as passive and female. It was contrasted with the truly human realm, marked by patriarchal, Eurocentric, and body-hating concepts of reason and "civilization," maximally distanced from "nature." Aborigines were seen as part of this inferior order, supposedly being "in a state of nature," and without culture. Aborigines "lived like the beasts of the forest," writes Cook at Adventure Bay in 1770; they were "strangers to every principle of social order."[21] Early journal reports consistently stressed their nudity and propensity to leave open to public view "those parts which modesty directs us to conceal." Where clothes are construed as the mark of civilization and culture, nudity confirms an animal and cultureless state, a reduction to body. The ideology of nature, reason, and civilization made it possible to deny kinship and see the natural world as an inferior realm open to merciless exploitation. The unclothed bodies of Aborigines and the technological economy of Aboriginal life meant that they could be seen in terms of such an ideology as not fully human, but as part of the realm of nature, to be treated in much the same way as the seals. Such was the philosophical basis not only for the annihilation of Tasmanian Aborigines but for the annexation of Australia under the rubric of *Terra nullius*, a land without occupiers, under which Australians still live.[22]

Such a history bears witness to a particular way of treating the sphere of *reproduction*, the sphere our culture has associated with women and the feminine principle. (I use reproduction here, following Carolyn Merchant, to include the reproduction of all nature, as well as human reproduction.)[23] The systematic wiping out of breeding colonies and killing of animals, whales, and seals in the act of giving birth seems to involve contempt for the very processes of life. It implies ignorance or denial of human dependence on these processes, and a view of humans as apart, outside of nature, which is treated as limitless provider. Such an economy is only possible in a culture which has systematically backgrounded, "disappeared," the sphere of reproduction. The inferiorization and denial of this sphere of reproduction involves the inferiorization and denial of women's labor, including reproductive labor.[24] The real heart of the problem of sustainability lies in this kind of consciousness with respect to the reproductive economy, a consciousness which is the product of an alienated and masculinized account of human identity.

These connections are not just historical curiosities. Aborigines are still killed, not now through open mass annihilation, with guns and poisoned flour, but through police violence. *Terra nullius* is still the fiction under which the Australian commonwealth is constituted. Seals too are now legally "protected," but the slaughter of seals, in large enough numbers to keep the population in danger, continues unabated into the present.[25] The sphere of reproduction remains unacknowledged. The network of oppression stretches from the past into the present.

A Cooperative Movement Strategy: Methodology and Politics

The conception of oppression as a network of multiple, interlocking forms of domination raises a number of new methodological dilemmas and requires a number of adjustments for liberation movements. The associated critiques cannot for example simply be added together, for there are too many discrepancies between them. Should we say for example that opposition struggle involves one movement or many? Each answer to the one/many dilemma has its problems.

One way to deal with the multiplicity of oppressions is to say that each involves all, for example that feminism should be thought of as a movement to end all forms of oppression.[26] But this seems to imply an oceanic view of the movements as submerged in a single great movement, for example that there should not and cannot be an autonomous women's movement concerned primarily with women's oppression (or indeed any other autonomous movement). But this would deny the specificity of, for example, women's oppression and the need for accounts to relate to lived experience, as well as the possibility of difference of direction and conflicts of interest between movements (for example ethnic, race, and sexual oppression). And even if struggles have a common origin point, enemy, or conceptual structure, it does not follow that they then become the same struggle. The women's movement especially has had good historical reason to distrust the submergence of women's struggle in the struggles of other movements, and has wanted to insist on the importance of movement autonomy and separate identity. And if a struggle which is too narrow and aimed at only a small part of an interlocking system will fail, so too will one which is too broad and lacking in a clear focus and a basis in personal experience. On the other hand, treating the women's movement as isolated from other struggles is equally problematic, because there is no neutral, apolitical concept of the human or of society in which women can struggle for equality, and no pure, unqualified form of domination which is simply male and nothing else which oppresses them. And since most women are oppressed in multiple ways, as particular kinds of women,[27] women's struggle is inevitably interlinked with other struggles.

The dilemma is created by setting up a choice between viewing liberation struggles as a shifting multiplicity only fortuitously connected (as in poststructuralism), versus viewing them as a monolithic, undifferentiated, and unified system. But if there are reasons for seeing it both as a multiple structure and as a unified structure, any model which does not recognize both these aspects is distorting. It is possible to bypass this one/many dilemma if these forms of oppression are seen as very closely, perhaps essentially, related, and working together to form a single system without losing a degree of distinctness and differentiation. A good working model which is easily visualized and which enables such an escape from the one/many dilemma is that of oppressions as forming a net or web. In a web there are both one and many, both distinct foci and strands with room for some independent movement of the parts, but a unified overall mode of operation, forming a single system. The objections which some feminists have raised to what has been called "dual systems theory" in the case of capitalism and patriarchy[28] focus on the links and unified operation of the web rather than the differentiated aspects of the structure. The interconnectedness of forms of oppression provides another reason for viewing these oppressions as forming a *single mutually supporting system*. The sorts of consideration which tell against the oceanic view provide a reason for viewing them as forming a *differentiated system*, with distinct parts which can and must be focused upon separately as well as together, as in a web.[29] Bell hooks's conception of feminism as retaining its own identity but as necessarily overlapping with and participating in a wider struggle captures the politics implied by the weblike nature of oppression and enables a balance between the requirements of identity politics and the requirements of connected opposition which arises from the connected nature of oppressions.[30]

If oppressions form a web,[31] it is a web which now encircles the whole globe and begins to stretch out to the stars, and whose strands grow ever tighter and more inimical to life as more and more of the world becomes integrated into the system of the global market and subject to the influence of its global culture. In the methodology and strategy for dealing with such a web it is essential to take account of both its connectedness and the capacity for independent movement among the parts. Rarely can it be said, "Once we have cut this section, solved this problem, all the rest will follow, other forms of oppression will wither away." A web can continue to function and repair itself despite damage to localized parts of its structure. The parts can even be in conflict and perhaps move for a limited time in opposite directions.

The strategies for dealing with such a web require cooperation. A cooperative movement strategy suggests a methodological principle for both theory and action, that whenever there is a choice of strategies or of possibilities for theoretical development, then other things being equal those strategies and theoretical developments which take account of or promote this wider, connected set of objectives are to be preferred to ones which do not. This should be regarded as a minimum principle of cooperative strategy. But it is one which the major green positions of social and Deep Ecology currently fail.

Thus Deep Ecology has chosen the company of American nature mysticism and of religious Eastern traditions such as Buddhism over that of various radical movements, including feminism, it might have kept better company with. Elsewhere I have argued[32] that Deep Ecology gives various accounts of the ecological self, as indistinguishable (holistic), as expanded and as transpersonal, and that all of these are problematic both from the perspective of ecological philosophy, and from that of other movements. Deep Ecology could do virtually everything it needs to do with a different account of the self as relational which does form a relevant connecting base for other movements. Social ecologists have rightly pointed to some of the political implications of this choice, that they lead away from connections with the radical movements and traditions, and lead toward these being seen as only accidentally connected to environmental concerns. That is, Deep Ecology has chosen a theoretical base which leads to a weakening of connection with other radical movements that are then seen as only accidentally connected, which from a different perspective appear essentially connected. Social ecology, however, also follows a noncooperative strategy in choosing to discount the critique of anthropocentrism, since, as we have seen, there is no incompatibility between rejecting both human hierarchy and also the human domination of nature.

If there is some reason for hope in our current situation, I believe it mainly lies in this: that we now have the possibility of obtaining a much more complete and connected understanding of the web of domination than we have ever had before, and hence a much more comprehensive and connected oppositional practice. What may be especially significant about this point in history is not only that the now global power of the web places both human and biological survival itself for the first time in the balance, but that several critically important parts of its fabric have recently become for the first time the subject of widespread conscious, self-reflective opposition. Domination must always be conceived of in terms of an open and not a closed set. We can never afford to become complacent, to feel that ours is the complete tally, the final form of understanding. Nevertheless the picture now seems much more comprehensive. Some forms of socialist feminism have already moved to integrate colonial/racial domination as a third parameter for analysis, but have still to take explicit account of nature.[33] A further merger with ecofeminism to form a social ecofeminism would provide a broader and deeper basis of oppositional theory and practice, and fill out some crucial connections.

Previous major oppositional theories such as Marxism now seem so defective because they had so many blind spots, such an incomplete, reductionistic, and fragmentary understanding of the web.[34] And that very incompleteness, the failure to address a wide enough range of concerns led to their failure in practice as well as in theory, leaving domination ever ready to renew and consolidate itself in a different but related form, as state and bureaucratic tyranny, as sexism, as militarism, as power over nature. But this does not mean that we should abandon the entire set of radical traditions, as Porritt suggests (and certainly not that we should abandon them because they are presently unfashionable), but rather that we must come to an understanding of them as limited and partial. The problems of human inequality and hierarchy which the radical traditions addressed over the centuries have not gone away and are taking new and even more sinister "environmental" forms. Their visions of human equality and the immense creative energies they harnessed over long periods of history have helped to form the green vision of a world where all species matter, and are still highly relevant to our understanding of the web. We must somehow balance recognition of the power and strength of past radical traditions with recognition of the need for major revision and reworking, and so come to build better.

Notes

1. The critique of anthropocentrism or "deep green theory" has developed over a period of twenty years and has included Deep Ecology as a proper subset. See, for example, Arne Naess, "The Shallow and the Deep, Long-Range Ecology Movement: A Summary," *Inquiry* 16 (1973): 95–100; Arne Naess, *Ecology, Community, and Lifestyle* (New York: Cambridge University Press, 1989); Val Plumwood, "Critical Notice: John Passmore's *Man's Responsibility for Nature*," *Australasian Journal of Philosophy* 53 (1975): 171–85; Richard Routley and Val Plumwood, "Against the Inevitability of Human Chauvinism," in *Ethics and Problems of the 21st Century*, ed. K. E. Goodpaster and K. M. Sayre, (Notre Dame, Ind.: University of Notre Dame Press, 1979); Richard Routley and Val Plumwood, "Human Chauvinism and Environmental Ethics," in *Environmental Philosophy*, ed. Don Mannison, Michael McRobbie, and Richard Routley, Monograph Series 2, Philosophy RSSS, Australian National University, Canberra, Australia, 1980, pp. 96–189; Bill Devall and George Sessions, *Deep Ecology: Living as if Nature Mattered* (Salt Lake City: Peregrine Smith Books, 1985).
2. This tendency is discussed by various contributors, especially Sandra Harding, "What is the Real Material Base of Patriarchy and Capital," in *Women and Revolution*, ed. Lydia Sargent (Boston: South End Press, 1981).
3. Murray Bookchin, *Remaking Society* (Montreal: Black Rose Books, 1989), 44, see also 39, 42. See also Murray Bookchin, *Kick It Over*, Special Supplement (1988), 6A; Murray Bookchin, "Social Ecology Versus Deep Ecology, *Green Perspectives* 4/5 (Summer 1987).
4. Elizabeth Dodson Gray, *Green Paradise Lost: Re-mything Genesis* (Wellesley, Mass.: Roundtable Press, 1979), 19.

5. Murray Bookchin, *The Philosophy of Social Ecology* (Montreal: Black Rose Books, 1990), 182.

6. Warwick Fox, "The Deep Ecology–Ecofeminism Debate and Its Parallels," *Environmental Ethics* 11, no. 1 (1989): 5–25.

7. On the irrelevance of feminism, see Fox, "Deep Ecology–Ecofeminism Debate," 14. For a discussion of this point, see Karen Warren, "The Power and Promise of Ecofeminism," *Environmental Ethics* 12 (1988): 123–46, see 144–45.

8. Fox, "Deep Ecology–Ecofeminism Debate"; Warwick Fox, *Towards a Transpersonal Ecology: Developing New Foundations for Environmentalism* (Boston: Shambala, 1990).

9. This treatment was suggested by the remarks of Jonathon Porritt's "Green Politics," address to the Ecopolitics V conference, University of New South Wales, Sydney, Australia, April 1991.

10. For a general critique of Bookchin's position, see Robyn Eckersley, "Divining Evolution: The Ecological Ethics of Murray Bookchin," *Environmental Ethics* 11 (1989): 99–116.

11. George Bradford, "Return of the Son of Deep Ecology," *Fifth Estate* 24, no. 1 (1989): 5–35; Brian Tokar, "Exploring the New Ecologies: Social Ecology, Deep Ecology and the Future of Green Political Thought," *Fifth Estate* 24, no. 1 (1989): 18–19.

12. See the work of ecofeminists such as Rosemary Radford Ruether, *New Woman, New Earth* (Minneapolis: Seabury Press, 1975); Karen Warren, "Feminism and Ecology: Making Connections," *Environmental Ethics* 9 (1987): 3–20; Warren, "Power and Promise of Ecofeminism"; and Ynestra King, "The Ecology of Feminism and the Feminism of Ecology," in *Healing the Wounds*, ed. Judith Plant (Philadelphia: New Society Publishers, 1989); Ynestra King, "Healing the Wounds: Feminism, Ecology, and the Nature/Culture Dualism," in *Reweaving the World*, ed. Irene Diamond and Gloria Feman Orenstein (San Francisco: Sierra Club Books, 1990).

13. This orientation to connection is not grasped by those Deep Ecology critics of ecofeminism who take their own imperialist urges to be universal and understand ecofeminism as the claim that "there is no need to worry about any form of human domination other than that of androcentrism." See Fox, "Deep Ecology–Ecofeminism Debate," 18, and Michael Zimmerman, "Feminism, Deep Ecology, and Environmental Ethics," *Environmental Ethics* 9 (1987): 21–44, see 37.

14. Warren, "Feminism and Ecology," 18.

15. Rosemary Radford Ruether, "Toward an Ecological-Feminist Theology of Nature," in *Healing the Wounds*, ed. Judith Plant (Philadelphia: New Society Publishers, 1989), 145–50, quotation on 149.

16. See also Joan L. Griscom, "On Healing the Nature/History Split in Feminist Thought," *Heresies* 13, no. 4 (1981): 4–9. Of course, other node points on the net might equally well be starting points for generalization. The model does not imply that all strands are equally important.

17. Maria Mies, *Patriarchy and Accumulation on a World Scale* (London: Zed Books, 1986).

18. James Bonwick, *The Last of the Tasmanians* (London, 1870).

19. Bonwick, *Last of the Tasmanians*, quotation on 45.

20. Val Plumwood, "SealsKin," *Meanjin* 51, no. 1 (1992): 45–58.

21. Bonwick, *Last of the Tasmanians*, quotation on 56.

22. A similar ideology has covered other slaughters in other new worlds. The Spanish priest Las Casas, historian of Columbus's conquests, noted that the Christians

despised the natives and held them as fit objects for enslavement "because they are in doubt as to whether they are animals or beings with souls" (Frederick Turner, *Beyond Geography: The Western Spirit Against the Wilderness* (New Brunswick, N.J.: Rutgers University Press, 1986), 142.

23. Carolyn Merchant, *Ecological Revolutions* (Chapel Hill: University of North Carolina Press, 1989), 6–7.

24. Hazel Henderson, *Creating Alternative Futures: The End of Economics* (New York: Berkeley Publishing, 1978); Marilyn Waring, *Counting for Nothing* (London: Allen and Unwin, Port Nicholson Press, 1988).

25. See Andrew Darby, "Seal Kill: The Slaughter in Our Southern Seas," *Good Weekend*, 5 Jan. 1991, pp. 12–15. As many as three thousand fur seals are killed by the Tasmanian fishing industry each year, many in macho shootouts that wipe out whole colonies. Others are killed even more cruelly as their playfulness leads them to become entangled in the plastic fish-bait packaging and discarded nets tossed overboard from boats. The industry dumps plastic garbage of all kinds, which is a killer of marine life and is found on the most remote beaches.

26. Warren, "Power and Promise of Ecofeminism," 133.

27. Elizabeth V. Spelman, *Inessential Woman: Problems of Exclusion in Feminist Thought* (Boston: Beacon Press, 1988).

28. Mies, *Patriarchy and Accumulation;* Iris Young, "Beyond the Unhappy Marriage: A Critique of Dual Systems Theory," in *Women and Revolution,* ed. Lydia Sargent (Boston: South End Press, 1981), 43–70.

29. The net or web analogy is an alternative to the pillar analogy of some ecofeminists, as in "racism, sexism, class exploitation and ecological destruction form interlocking pillars upon which the structure of patriarchy rests" (Sheila Collins quoted in Warren, "Feminism and Ecology," 7). This seems unnecessarily limiting and quantified.

30. bell hooks, *Talking Back* (Boston: South End Press, 1989), 22.

31. The web model was of course suggested by Foucault (Michel Foucault, "Disciplinary Power and Subjection" in *Power/Knowledge: Selected Interviews and Other Writings of Michel Foucault 1972–1977,* ed. Colin Gordon (New York: Pantheon, 1980), 234 and his version has recently been criticized by Hartsock (Nancy Hartsock, "Foucault on Power: A Theory for Women?" in *Feminism/Postmodernism* [New York: Routledge, 1990]). As Hartsock notes, taking individuals to be the agents of such power relations can result in a view of power as so pervasive and reciprocal that domination disappears and power is "everywhere and therefore nowhere" (170). But we do not have to understand the strands of the net, as Hartsock argues Foucault does, as an all-pervading and utterly diffuse set of power relations operating equally between all individuals, with no central focus or organizing principle. Rather, as I argue (in Val Plumwood, *Feminism and the Mastery of Nature* [New York: Routledge, 1993]), we can understand them as the large-scale multiple but linked social structures of oppression.

32. Val Plumwood, "Nature, Self, and Gender: Feminism, Environmental Philosophy, and Critique of Rationalism," *Hypatia* 6, no. 1 (1991): 3–27.

33. Donna Haraway, "A Manifesto for Cyborgs," in *Feminism/Postmodernism,* ed. Linda J. Nicholson (New York: Routledge, 1991), 190–233; Donna Haraway, *Simians, Cyborgs and Women* (London: Free Association Press, 1991); Hartsock, "Foucault on Power," 164.

34. Val Plumwood, "On Karl Marx as an Environmental Hero," *Environmental Ethics* 3 (1981): 237–44.

Essentialism in Ecofeminist Discourse

ELIZABETH CARLASSARE

ON 17 NOVEMBER 1980 in Washington, D.C., two thousand women encircled the Pentagon, blocking entrances, weaving doors closed with brightly colored yarn, and weaving webs of yarn and ribbon that contained artifacts from their daily lives. These women also planted cardboard tombstones on the Pentagon's lawn, inscribed with the names of victims of U.S. militarism, colonialism, toxic contamination, and sexual violence. This nonviolent direct action, known as the Women's Pentagon Action, was the first explicitly "ecofeminist" action in the United States, and in it women protested against military violence, ecological violence, racism, and social, sexual, and economic violence toward women.

Ecofeminism emerged in the 1970s as part of the women's liberation movement and more recently has begun being articulated in the margins of academic discourse. The term "ecofeminism" is used by some activists and academics to refer to a feminism that connects ecological degradation and the oppression of women. Much of ecofeminist direct action seeks to resist and subvert political institutions, economic structures, and daily activities that are against the interests of life on earth. Much of theoretical and academic ecofeminism seeks to identify, critique, and overthrow ideological frameworks and ways of thinking, such as value-hierarchical dualistic thinking,[1] that sanction ecological degradation and the oppression of women. Beyond this, ecofeminism seeks to bring forth different, nondominating forms of social organization and human-nature interaction.

Ecofeminism does not lend itself to easy generalization. It consists of a diversity of positions, and this is reflected in the diversity of voices and modes of expression represented in ecofeminist anthologies. The ecofeminist anthologies *Reclaim the Earth*,[2] *Healing the Wounds*,[3] and *Reweaving the World*,[4] and the issues of *Heresies*[5] and *Hypatia*[6] on feminism and ecology include the work of women from different countries and social situations, and their work does not adhere to a single form or outlook. Poems, art, photographs, fiction, prose, as well as theoretical/philosophical/"academic" works are included. Ecofeminism's diversity is also reflected by its circulation in a variety of arenas, such as academia, grass-roots movements, conferences, books, journals, and art.

Because of this diversity, I agree with the contention of ecofeminist activist and theorist Carolyn D'Cruz that it is more useful to consider ecofeminism as a discourse than as a unified, coherent epistemology. D'Cruz's view of ecofeminism as discourse is useful because it makes room for the voices of a variety of positioned subjects that share political and ethical concerns.[7] I am emphasizing the diversity within ecofeminism and the usefulness of considering ecofeminism as a discourse to illustrate that there is no epistemological position that all ecofeminists can be said to share. Ecofeminism derives its cohesion not from a unified epistemological standpoint, but more from the shared desire of its proponents to foster resistance to formations of domination for the sake of human liberation and planetary survival.

THE ESSENTIALIST/CONSTRUCTIONIST TENSION IN ECOFEMINISM

Critics of ecofeminism reside both inside and outside of the ecofeminist movement. Within the movement, social and socialist ecofeminists employ materialist methods to analyze class and capitalist economic systems, whereas cultural ecofeminists often employ spiritual or associative, poetic modes to explore oppression on a personal as well as on a larger social level. Many critics within ecofeminism align themselves with the constructionist position of social or socialist ecofeminism, and dismiss cultural ecofeminism for being "essentialist." Essentialism usually refers to the assumption that a subject (for example, a "woman") is constituted by presocial, innate, unchanging qualities. Constructionism, on the other hand, usually refers to the assumption that a subject is constituted by social, historical, and cultural contexts that are complex and variable. Essentialist arguments posit that women and men are endowed with innate qualities or essences that are not historically or culturally contingent, but eternal and unchanging, an outcome of their biology, which is understood as fixed.

For social and socialist ecofeminists who wish to transcend traditional stereotypes of women that naturalize their nature in terms of biology, essentialism in cultural ecofeminism poses serious problems. Social ecofeminism, the form of ecofeminism developed at the Institute for Social Ecology in Vermont, primarily by Chaia Heller, emphasizes that the association of women with nature in Western capitalist patriarchy is largely a social (historical and cultural) construction and that the liberation of nature will only come about through revolutionary social change in which systems that feed on human oppression, most notably capitalist patriarchy, are replaced by nonhierarchical, nondominating forms of social organization. Social ecofeminism combines radical feminism's "body politics" with the social ecological perspective developed by eco-anarchist Murray Bookchin to link ecological and social domination, and explore their specific effects on women.

Socialist ecofeminism, the form of ecofeminism proposed and advocated by environmental historian and ecofeminist Carolyn Merchant, shares social ecofeminism's constructionist position and belief that revolutionary social change and the overthrow of capitalist patriarchy are required for human liberation and planetary survival. Unlike social ecofeminism, socialist ecofeminism advocates new forms of socialism (rather than state socialism or a social ecological form of anarchy) as potentially ecologically sustainable economic and political frameworks.[8] In socialist ecofeminism, changes in the spheres of social reproduction, biological reproduction, and production are required to restructure gender relationships and human-nature interactions in order to achieve an egalitarian and ecological transformation of society. The use of essentialism in cultural ecofeminism appears to be at odds with the shared constructionist position of social and socialist ecofeminism.

Because I do not pretend to play the "god trick,"[9] I identify myself as a social ecofeminist who believes that there is something to be learned from cultural ecofeminism and that essentialism within cultural ecofeminism does not necessitate its rejection. Based on these beliefs, I direct the following questions (à la Diana Fuss)[10] toward the social/ist[11] and cultural[12] positions within ecofeminism: If cultural ecofeminist texts are essentialist, what is the motivation behind the use of this concept? Can essentialism ever be used strategically as a form of resistance? To what extent is there slippage between the social/ist and cultural positions within ecofeminism? Do the constructionism and essentialism of these respective positions imply and rest on each other? It is with these questions in mind that I explore the criticism of essentialism levied against cultural ecofeminism by some social/ist ecofeminists.

Some social/ist ecofeminists respond to the presence of essentialism in cultural ecofeminism by criticizing and dismissing cultural ecofeminism because of a fear that implying the existence of innate links between women and nature reinforces the patriarchal stereotypes based upon women's "innate" biological and psychological characteristics. To social/ist ecofeminists, ideas such as woman as nurturer, woman as caretaker, and woman as closer to nature have been used to oppress women, limit their sphere of activity, and squelch their potency as social and cultural agents.[13]

Mistrust and criticism of essentialism is not new to feminism. Most feminists are acutely aware of the many ways in which essentialist arguments have been used to oppress women. One example of the perpetuation of sexist representations of women through essentialist arguments, illustrated by feminist Judith Genova, has to do with the size of their brains:

Around the turn of the century, researchers were convinced that the five ounce difference in weight between male and female brains was the cause of female cognitive inferiority. . . . Happily, in what soon came to be known as the "elephant problem," elephants and whales rescued women from this particular argument. If intelligence were a matter of absolute brain weight, elephants and whales would outscore men on intelligence tests handily. Since this was clearly absurd (species chauvinism remains unchanged today), absolute brain weight was quickly abandoned as a measure of intelligence. In its place various other relative measures were proposed: brain weight as an expression of body weight, body height, thigh bone weight or cranial height. . . . The only alternative that gave men significantly heavier brains and thus preserved the already established prejudice was to express brain weight as a measure of body height.[14]

This example illustrates why many feminists have a knee-jerk reaction of dismay and dismissal to biological or psychological essentializing of women—essentialist arguments have been a common ploy used to mark "woman" as incapable of fully acting as an agent of culture, society, and history. That biological essentializing of women still informs common working definitions of woman today is evidenced by media headlines such as "Brain Dead Mother Has Her Baby."[15] As feminist and cultural critic Valerie Hartouni points out, such headlines make sense only if motherhood is understood solely as a "biologically rooted, passive—indeed, in this case, literally mindless—state of being. Within this understanding, motherhood is cast as 'natural' or 'instinctual,' a synonym for female, the central aspect of women's social and biological selves, the expression and completion of 'female nature.'"[16]

It is distaste for essentialism's "regressive potential" that causes some social/ist ecofeminists to dismiss cultural ecofeminism. The tension between social/ist ecofeminism's constructionist position and cultural ecofeminism's use of essentialism in its efforts to forge a specifically woman-based culture and spirituality echoes a similar tension within feminism and all politics based upon a shared identity. Feminist critic of science Evelyn Fox Keller has articulated the same essentialist/constructionist tension within feminism as follows: "Discussions of gender tend to lean towards one of two poles—either toward biological determinism, or toward infinite plasticity, a kind of generic anarchy."[17] To social/ist ecofeminists, essence has been socially inscribed on women for the purpose of legitimating their domination by men. Social/ist ecofeminists make it clear that although women and nature have been related in Western culture *historically* to oppress women, neither of the two entities have either innate essences or an essential connection. Essence, they argue, is not in itself pansocial or presocial or a determinant of social structure. In the words of Carolyn Merchant, "There are no unchanging 'essential' characteristics of sex, gender, or nature."[18]

CONSTRUCTING ESSENCES

Two cultural ecofeminist texts that have been repeatedly dismissed as essentialist by some social/ist ecofeminists are Susan Griffin's *Woman and Nature*[19] and Mary Daly's *Gyn/Ecology*.[20] Griffin's *Woman and Nature* has been accused of essentialism by former social ecofeminist Janet Biehl in her book-long critique of ecofeminism, *Rethinking Ecofeminist Politics*:

> Psycho-biological ecofeminists believe that women, owing to their biological makeup, have an innately more "caring" and "nurturing" way of being than men. . . .
>
> That associations were created by patriarchal and patricentric cultures to debase women does not deter psycho-biologistic ecofeminists from consciously identifying themselves with nonhuman nature itself. They are often inspired by Griffin, an early ecofeminist writer, who proclaimed, "We know ourselves to be made from this earth. We know this earth is made from our bodies. For we see ourselves. And we *are* nature. We are *nature seeing nature*. We are nature with a concept of nature. Nature weeping. Nature speaking nature to nature."[21]

Although Biehl did at one time consider herself a social ecofeminist and put effort into developing a theory of social ecofeminism, in this text she explains that she was impelled to revoke her affiliation with ecofeminism altogether because of its infusion with cultural[22] ecofeminists. It is her belief that the presence of cultural ecofeminists within ecofeminism renders the movement contradictory, incoherent, and ineffective.

Janet Biehl interprets Griffin's text as regressively perpetuating the patriarchal essentialist associations of women with nature and men with culture that have been used to oppress women. I argue, however, that alternative interpretations of Griffin's text are possible. It seems unlikely that Griffin as a feminist would be interested in perpetuating the oppression of women. If she is not using essentialism for this purpose, how else might she be using it? Can Griffin's text be interpreted as an instance of deploying essentialism strategically as a form of resistance? Furthermore, is it possible to read constructionism within Griffin's use of an "essential" association between women and nature?

Griffin's *Woman and Nature* is a dialogue between two voices, the voice of patriarchy and the voice of women and nature. In this text, the patriarchal voice repeats from a disembodied, universal subject position the reasons the domination of women and nature is justified. While the separation of women and nature from men and culture is cemented with the use of these two voices, the essentialist association of women with nature and men with culture can be interpreted as a tactic of resistance just as easily as it can be interpreted as compliant with hegemonic uses of dualism, and the content of the text can be read to substantiate this interpretation.

Through the use of the repetitive patriarchal voice, which espouses essentialist arguments about "woman's nature" to explain the reasons women are necessarily subordinate to men, the text illustrates the extent to which women's essence has been *historically constructed* by patriarchal scientific discourse as inferior to that of men to perpetuate masculine privilege. *Woman and Nature* illuminates one view of how Western woman's essence has been created and transformed over time by specific historical acts, and the text functions, in part, as a critique of objectivity and a challenge to the hegemony of scientific discourse. The maintenance of a voice that is simultaneously the voice of both women and nature can be interpreted, however, as leaving in place essentialist notions of women as closer to nature than men, at the same time that the content of this text shows these essentialist notions to be historically constructed. Because *Woman and Nature* explores the historical construction of essences of women and nature, which implies that essences do change, this text cannot be unambiguously classified as essentialist or constructionist; constructionism is implicated in Griffin's use of essentialism.

Mary Daly's *Gyn/Ecology* has been criticized along with *Woman and Nature* for being essentialist, for putting forth the idea that women and nature are innately linked. Socialist feminist Alison Jaggar has written:

> Mary Daly, for instance, appears to endorse the "native talent and superiority of women." None of these authors attempts to provide a systematic account of just what are women's special powers, other than their capacity to give birth, nor of the relation of these powers to female biology. Moreover, the authors' style of writing is invariably poetic and allusive rather than literal and exact. But there is a repeated suggestion that women's special powers lie in women's special closeness to non-human nature. The radical feminist author Susan Griffin, for instance, has written a very popular book that draws parallels between men's attitudes toward women and their attitudes toward non-human nature. Of course, as we shall see later, such parallels are capable of a number of interpretations, but Griffin herself suggests that women and non-human nature are inseparable from each other.[23]

Gyn/Ecology, like *Woman and Nature*, however, can be interpreted as simultaneously essentialist and constructionist, asserting women's essentialized gender characteristics while acknowledging the construction of woman's essence within a particular social, cultural, and historical context. Daly's *Gyn/Ecology* is concerned with uncovering the ways in which patriarchal Western religion, myth, and language have constructed an essential notion of woman that has been used to legitimate the subordination and oppression of women, as well as with seizing and reconstructing the gender category "woman" for the sake of empowering women.

The essentialism that Griffin's and Daly's texts contain can be interpreted as a conscious oppositional strategy rather than as unconsciously regressive. None of the criticisms of these cultural ecofeminist texts that I have come across acknowledge that all essences are contextual, arising out of specific historical, cultural, and social situations. Feminist Donna Haraway makes the related point that ecofeminism in general, and the work of Susan Griffin in particular, can only be understood "as oppositional ideologies fitting the late twentieth century. They would simply bewilder anyone not preoccupied with the machines and consciousness of late capitalism."[24] This statement points to the strategic deployment of essentialism within particular historical and political contexts and, in this sense too, essentialism rests on constructionism.

Many of the critics of cultural ecofeminism could be called guilty of resorting to an essentialist notion of essentialism—dismissing it as unconditionally "bad" without examining the specific ways in which it is used in specific situations. Many of them could also be called guilty of resorting to an essentialist notion of cultural ecofeminism or ecofeminism, dismissing all their multifaceted projects based on the interpretation of certain texts as essentialist.[25]

MARGINALIZING ECOFEMINIST DISCOURSE WITH THE LABEL "ESSENTIALIST"

If, as is the case with *Woman and Nature* and *Gyn/Ecology*, it is possible to read constructionism within essentialism and to interpret the use of essentialism by cultural ecofeminists as a conscious strategy working for, not against, the liberation of women, what is going on when texts such as these are labeled "essentialist" and dismissed as regressive? It is true that the texts of Griffin and Daly do not employ the same stylistic devices as the works of most social/ist ecofeminists. Griffin writes in a way that she calls "associative"; *Woman and Nature* is poetry as much as it is prose. In *Gyn/Ecology*, Daly writes in a style that she calls "gynocentric." Neither of these texts ascribe to the stylistic devices of traditional academic writing or traditional philosophical discourse. To what extent does the criticism of cultural ecofeminism as "essentialist" by social/ist ecofeminists serve to privilege certain positions and discursive practices within ecofeminism over others, and police what counts most in ecofeminist discourse? How does the criticism serve to privilege some ways of knowing over others? To what extent do these criticisms serve to maintain the very same dominant discursive practices and ways of knowing that voices within ecofeminism and feminism call into question?

The move toward privileging traditional discursive practices that takes place with the accusation of essentialism comes through in the complaint of Alison Jaggar from the previous section. She argues that both Daly and Griffin put forth essentializing notions of woman's nature and write in styles that are "invariably poetic and allusive rather than literal and exact." By implying that Griffin's and Daly's texts are essentialist and then describing them as "poetic and allusive," Jaggar draws an association between essentialism and associative poetic writing. This association implies that there is something better about discursive practices that employ "literal and exact" language, and the criticism works to marginalize the work of ecofeminists who do not write in this way.

Whereas Jaggar marginalizes the discursive practices of some (cultural eco)feminists on the basis that they are essentialist, Janet Biehl marginalizes their epistemological practices. The epistemological privileging that often accompanies the accusation of essentialism is made transparent when Biehl moves from calling (cultural) ecofeminism "irrational" to criticizing it for being essentialist.[26] The epithet of "irrational" is interesting, because it is one that has historically been used so often in the service of essentializing "the other" and dismissing their subject positions and knowledges as "false." In calling (cultural) ecofeminism as "irrational," Biehl implies that her epistemological position is superior, a common use of the word "irrational." What is curious, though, is that she uses this word to dismiss works such as *Woman and Nature*, which by means of both their style and content, contest notions of rationality, objectivity, totalizing epistemologies, and discourses of truth as masculinist and oppressive and situate these notions as historical and cultural events. Biehl's criticism also implies that one of the goals of ecofeminism is the articulation of a coherent, totalizing epistemology. Although this may be the desire of a few ecofeminists (and one former ecofeminist, Biehl herself), such criticisms work to obscure the many ecofeminist voices that do not strive for a unified epistemological position.

The way of knowing that Janet Biehl assumes to be rational, in contrast to cultural ecofeminism's "irrationality," is social ecofeminism. Although both cultural and social/ist ecofeminists resist patriarchy, social/ist ecofeminists in general believe that changes in social, political, and economic systems are required for the liberation of both women and nature. Cultural ecofeminists, on the other hand, generally believe that changes in human consciousness and spirituality are inseparable from the changes in institutions that are required for the liberation of women and nature. To them, oppression is a sign of spiritual crisis—political and cultural transformation will not occur without a concurrent shift in human consciousness.

It would seem that in practice the criticism of cultural ecofeminism as "essentialist" circulates within ecofeminism to privilege social/ist ecofeminism and its materialist ways of knowing (which are also more "academic") and to marginalize cultural ecofeminism and its spiritual and intuitive ways of knowing, and their respective voices. The criticism of cultural ecofeminism as "irrational" implies that the goal of ecofeminism is the development of a unified theory of oppression (using historical materialism). In so doing, the partial perspective that materialist methods offer is not acknowledged and the plurality of perspectives that do fall under the umbrella of ecofeminism is not encouraged.

That the essentialist criticism functions, in part, to privilege social/ist ecofeminism and materialist methods became clear when I realized in the course of reading ecofeminist texts that cultural ecofeminists had been named by social/ist ecofeminists. Although former social ecofeminist Janet Biehl, social ecofeminist Ynestra King, and socialist ecofeminist Carolyn Merchant appropriate the term "cultural" feminism, or ecofeminism, in most of the texts referred to in this way, the author did not refer to herself as a cultural feminist or ecofeminist. Cultural ecofeminism appeared to be a label social/ist ecofeminists used to designate what they did not want to be affiliated with, rather than a self-appropriated term used by ecofeminists interested in creating a woman-based culture and spirituality.[27] Furthermore, after reading cultural ecofeminist texts, it became clear that the discomfort between cultural and social/ist ecofeminists was one-sided. Some social ecofeminists spoke of cultural ecofeminism disparagingly, but not the other way around. It seemed that so-called cultural ecofeminists did not perceive a rift between themselves and social/ist ecofeminists. In this case it would seem that the power of naming has been deployed, like the label "essentialist," partially in the interest of privileging the position of social/ist ecofeminists.

Cultural ecofeminists such as Griffin and Daly have been criticized by social ecofeminists for being apolitical. Biehl has written, "By asking women to value their essential natures above men's, ecofeminism becomes an exercise in personal transformation rather than a concerted political effort."[28] In relying on more spiritual, psychological, and intuitive explorations of the oppression of women, instead of on materialist analyses of institutions, some social ecofeminists believe that the work of cultural ecofeminists does not spell out a path for political or social action, only a path for personal transformation. Some cultural ecofeminists have addressed the concern of social ecofeminists by suggesting that if social ecofeminists believe that cultural ecofeminism is apolitical, perhaps they are using a definition of politics that is too narrow. The essentialist language of Daly and Griffin indicates that for them the politics of women's liberation includes revaluing that which has been associated with women and devalued. Politics takes many forms, including discursive and linguistic resistance, as well as direct action. It is possible to read the works of Griffin and Daly as strategically deploying and reconstructing the characteristics that have been traditionally associated with women for the sake of fostering social and psychological transformations and empowering women while working to liberate women from their gender construct. Texts that assert in content and style the gender characteristics associated with women can be simultaneously essentialist and politically strategic. Feminist Elizabeth Weed describes the dilemma involved with writing essentially in feminism as the "ever difficult issue of what is at stake in the deployment of the very terms that one might be trying to displace."[29]

The criticism of cultural ecofeminism as essentialist, and hence apolitical, works to marginalize both ecofeminist writers that employ nontraditional discursive styles and those that press the necessity of a spiritual transformation of consciousness for the liberation of women and nature. The marginalization of these voices within ecofeminism unconsciously reinscribes traditional discursive practices and ways of knowing. It works to privilege the work of social/ist ecofeminists with their emphasis on transforming the material conditions of life by overthrowing oppressive institutions, and it works to discount the work of cultural ecofeminists with their emphasis on transforming consciousness, reclaiming women's history, and fostering a woman-based culture and spirituality. Although the ultimate goals of both positions are the same, namely, women's liberation and an end to ecological degradation, social/ist ecofeminist criticisms of cultural ecofeminism privilege one means of achieving these goals over another, a transformation of social structures over a psychic transformation.

Feminists of color have been particularly critical of the essentializing move involved with asserting the claims of the gender category "woman." Like social/ist ecofeminists, some feminists of color have been concerned about a tendency within feminism to erase differences between women and universalize women's experience by employing an essentialist notion of women. Audre Lorde has criticized Daly's *Gyn/Ecology* for just this reason in her essay titled "An Open Letter to Mary Daly":

> To imply . . . that all women suffer the same oppression simply because we are women is to lose sight of the many varied tools of patriarchy. It is to ignore how those tools are used by women without awareness against each other.
> . . . I ask that you be aware of the effect that this dismissal has upon the community of Black women and other women of Color, and how it devalues your own words. . . . When radical lesbian feminist theory dismisses us, it encourages its own demise.[30]

Lorde, like some social/ist ecofeminists, criticizes employing an essential notion of women because it is invariably accompanied by an erasure of difference between women and it allows women with race, class, or national privileges to sidestep their obligation to take responsibility for their own power and participation in structures of domination. Essentializing women further assumes that certain experiences represent the experience of all women. In celebrating the commonalities of women and asserting a unified essentialized gender category "woman," the diversity of women's lives and histories across the boundaries of race, class, nationality, age, and sexuality is ignored.

Lorde's criticism is an extremely important one because it points to the potential of universalized notions of women or women's experience to incorporate some women and their experiences while marginalizing others. The criticism also points to the persistent problem in feminism and all politics based on common or collective identity of obscuring difference regardless of whether identity is taken to be essential or constructed.

ESSENTIAL CONSTRUCTIONS

I have explored how constructionism is implied within essentialism and how the criticism of some ecofeminist texts as "essentialist" works to privilege materialist voices and the position of social/ist ecofeminists, and to marginalize spiritual voices and the position of cultural ecofeminists within the movement. Now I shift my focus to a brief exploration of the following correlative question: To what extent are essentialist notions implicated in the social/ist ecofeminist position, which is typically considered to be more of a constructionist position than cultural ecofeminism? Having located the dependency of cultural ecofeminism's use of essentialism on variable historical and political contexts, here I will briefly explore the possibility that social/ist ecofeminists' constructions can remain fixed.

The social/ist position within ecofeminism focuses on economic, social, and political institutions such as capitalism, patriarchy, and the nation-state for the purpose of showing the ways in which they have fostered the domination of nature and human oppression along the lines of gender, race, class, sexuality, age, etc. Social/ist ecofeminism unites ecological degradation with the personal concerns of women by rebelling against the ways in which women's bodies and minds are "poisoned" along with nature by capitalist patriarchy. Social/ist ecofeminism draws on the radical feminist insights of body politics and the idea that the "personal is political" to recognize and rebel against the many ways in which the social and ecological "poison" of capitalist patriarchy is inscribed in women's bodies and minds.[31] Social/ist ecofeminism also makes women's roles in biological and social reproduction central in theory construction.[32]

Social/ist ecofeminists' interest in women's bodies as a site of power struggle is one point at which essentialism steps into their constructionist position. The body and biological sex for some social/ist ecofeminists are part of material nature; they are "natural," not socially constructed. The constructionist position of social/ist ecofeminism, then, depends on a bit of essentialism: women are made, not born, but they are socially constructed out of naturally sexed anatomical raw material. Basing gender on material nature or biological sex necessarily naturalizes and essentializes gender.[33]

Although social/ist ecofeminists might, in general, be more straightforward in their recognition of differences between women than are cultural ecofeminists, their politics retain the category "woman" as basic. Social/ist ecofeminists recognize that "woman" is a mobile construction that rests on spatially and temporally variable social relations, and they deny that there is any immutable eternal essence that defines women. Social/ist ecofeminists often work, however, with a historically continuous, simple, essentialized notion of "woman" despite their recognition that "woman" is a construction, a mutable representation with a history.

Social/ist ecofeminists, as social constructionists, also run into another contradiction: If "woman" rests solely on variable social relations, how can the category be asserted without essentializing women? In the words of Diana Fuss, "Is it possible to generate a theory of feminine specificity that is not essentialist?"[34] It is not my goal to answer these questions. I raise them to illustrate the point that just as it is possible to read constructionism within cultural ecofeminism's use of essentialism, it is possible to detect essentialism within the constructionist position of social/ist ecofeminism.

Interrogating the use of the label "essentialist" within ecofeminism shows that it is a site of power struggle within ecofeminism's politics of liberation over the production of ecofeminist knowledges, circulating within ecofeminism in part to maintain hegemonic discursive practices by privileging materialist ways of knowing over spiritual and intuitive ways of knowing. The application of the label "essentialist" by some social/ist ecofeminists is performing one of the same acts they fear from essentialism—erasure of difference. Discussion and debate over the efficacy of different political strategies is important in ecofeminism, but cultural ecofeminism has often been too quickly labeled "essentialist" and dismissed. Criticism of ecofeminism's essentializing tendencies is important to ensure critical self-reflexivity and for examining the ways in which essentializing may sometimes work against the goals of women's liberation by homogenizing the diversity of women's experiences. Dismissing cultural ecofeminism on this basis, however, precludes the possibility of learning from this position and obscures the legitimacy of the variety of positions and discursive forms under ecofeminism's umbrella.

Rather than despairing about ecofeminism's "incoherence" and employing tactics to marginalize certain voices within the movement, ecofeminists and feminists can take the variety of positions within ecofeminism as a sign of the movement's vitality. The uses of essentialism by ecofeminists can be explored for their political intentions and their points of effectiveness and ineffectiveness in liberatory politics, instead of dismissed unconditionally as "regressive." Instead of throwing out ecofeminism altogether because of a presumed incompatibility of social/ist and cultural ecofeminism, feminists and ecofeminists can recognize that both are resistive strategies that share the goals of ending oppression and fostering planetary survival. The extent to which there is slippage between the social/ist and cultural positions within ecofeminism can be recognized, as can the extent to which the essentialism and constructionism of these respective positions imply and rest upon each other.

It could very well be, as I have argued, that the membrane that separates cultural ecofeminism's use of essentialism from social/ist ecofeminism's use of constructionism is not as impermeable as it would at first seem. By showing that constructionism can be read within cultural ecofeminism's use of essentialism and that essentialism is not far beneath the surface of social/ist ecofeminism's use of constructionism, it is possible to destabilize the social/ist ecofeminist criticism that cultural ecofeminism is essentialist. Feminists

and ecofeminists can recognize that the different orientations of the social/
ist and cultural ecofeminist positions may not be so different after all and
that the essentialist/constructionist tension in the production of ecofeminist
knowledges does not forestall ecofeminism's political effectiveness. Ecofeminism
demonstrates that it is possible to unite politically without assuming a unified
position or a totalizing epistemology. One of the recurring themes in
ecofeminism is "unity in diversity"—the idea that difference does not have
to mean domination. Just as recognizing differences among women does not
preclude the possibility of connection and sharing a common identity for
the sake of struggling together for liberation, recognizing differences among
ecofeminists does not preclude the possibility of uniting on the basis of
shared political and ethical concerns. The vitality of ecofeminist actions,
such as the Women's Pentagon Action described in the opening paragraph
of this essay, lends substance to this claim.

Notes

I am grateful to Carolyn Merchant and Richard Norgaard for their insightful suggestions and support. I also wish to thank Chaia Heller for her helpful
comments.

1. For a discussion of value-hierarchical thinking as a patriarchal conceptual
   framework, see Karen Warren, "Feminism and Ecology: Making Connections,"
   Environmental Ethics 9 (1987): 6.
2. Leonie Caldecott and Stephanie Leland, eds., *Reclaim the Earth: Women Speak
   Out for Life on Earth* (London: Women's Press, 1983).
3. Judith Plant, ed., *Healing the Wounds: The Promise of Ecofeminism* (Santa Cruz,
   Calif.: New Society Publishers, 1989).
4. Irene Diamond and Gloria Feman Orenstein, eds., *Reweaving the World: The
   Emergence of Ecofeminism* (San Francisco: Sierra Club Books, 1990).
5. *Heresies: A Feminist Journal of Art and Politics*, no. 13 (1981). Special issue on
   feminism and ecology.
6. *Hypatia* 6, no. 1 (Spring 1991). Special issue on ecological feminism.
7. Carolyn D'Cruz, *Ecofeminism as Practice, Theory, Discourse: An Archaelogical
   and Genealogical Study* (Thesis, Communication Studies, Murdoch University,
   Murdoch, Western Australia, 1990), 41–42.
8. Mary Mellor promotes a similar position under the name "feminist green socialism"
   in *Breaking the Boundaries: Toward a Feminist Green Socialism* (London: Virago
   Press, 1992).
9. Donna Haraway, *Simians, Cyborgs, and Women: The Reinvention of Nature* (New
   York: Routledge, 1991), 189. Donna Haraway uses this phrase to refer to the
   masculinist viewpoint that sees "everything from nowhere."
10. This line of questioning is an ecofeminist-specific version of that used by feminist
    Diana Fuss in *Essentially Speaking: Feminism, Nature, and Difference* (New York:
    Routledge, 1989). In this text, Fuss locates essentialist/constructionist debates
    within the identity politics of feminism, cultural studies, and gay and lesbian
    studies, and explores the two positions, exposing their mutual reliance and
    destabilizing the prevasive assumption that essentialism is unconditionally "bad."

I owe much to this work—Fuss's arguments provide the theoretical grounding of this essay in many places.

11. I use the term "social/ist" to refer to the shared constructionist position of social and socialist ecofeminism.

12. Some ecofeminist writers use other terms (such as "radical," "radical cultural," and "psycho-biologistic") to refer to this position in ecofeminism and feminism.

13. Such criticism abounds in the work of social/ist ecofeminists and has been applied to both cultural ecofeminism and cultural feminism. A few examples follow: "Ecofeminist writers influenced by cultural feminism ... tend to see women as 'closer to nature' than men, and women's essential nature is inherently more ecological than men's" (Janet Biehl, "What Is Social Ecofeminism?" *Green Perspectives* 11 [October 1988]: 2). "Some radical feminists (e.g., so-called 'nature feminists,' Mary Daly, Susan Griffin, Starhawk) ... applaud the close connections between women and nature and urge women to celebrate our bodies, rejoice in our place in the community of inanimate and animate beings, and seek symbols that can transform our spiritual consciousness so as to be more in tune with nature.... [Radical feminism] mystifies women's experiences to locate women closer to nature than men" (Warren, "Feminism and Ecology: Making Connections," 14–15). "If women overtly identify with nature and both are devalued in modern Western culture, don't such efforts work against women's prospects for their own liberation? Is not the conflation of woman and nature a form of essentialism? Are not women admitting that by virtue of their own reproductive biology they are in fact closer to nature than men and that indeed their social role is that of caretaker? Such actions seem to cement existing forms of oppression against both women and nature, rather than liberating either" (Carolyn Merchant, Preface to *The Death of Nature* [San Francisco: Harper and Row, 1990], xvi). "Yet in emphasizing the female, body, and nature components of the dualities male/female, mind/body, and culture/nature, radical ecofeminism runs the risk of perpetuating the very hierarchies it seeks to overthrow ... any analysis that makes women's essence and qualities special ties them to a biological destiny that thwarts the possibility of liberation. A politics grounded in women's culture, experience, and values can be seen as reactionary" (Carolyn Merchant, "Ecofeminist and Feminist Theory," in Diamond and Orenstein, *Reweaving the World*, 102). "The cultural feminist, in effect, believes that women may create a new, improved culture based on female natural law. The implicit nature philosophy within cultural feminism suggests that there exist certain inextricable female principles which women can know and incorporate in the creation of a radical women's culture. Women's 'innate' ability to cooperate, our increased ecological sensibility, and our peace-loving nature are simply a few of the female principles by which female nature abides" (Chaia Heller, "Toward a Radical Eco-Feminism," in *Renewing the Earth: The Promise of Social Ecology*, ed. John Clark [London: Green Print, 1990], 158).

14. Judith Genova, "Women and the Mismeasure of Thought," in *Feminism and Science*, ed. Nancy Tuana (Bloomington: Indiana University Press, 1989), 211.

15. For example see, Yumi L. Wilson, "Brain-Dead Woman Gives Birth, Then Dies," *San Francisco Chronicle*, 4 Aug. 1993, pp. A1, A13.

16. Valerie Hartouni, "Containing Women: Reproductive Discourse in the 1980s," in *Technoculture*, ed. Constance Penley and Andrew Ross (Minneapolis: University of Minnesota Press, 1991), 30.

17. Evelyn Fox Keller, "The Gender/Science System; Or, Is Sex to Gender as Nature Is to Science?" in Tuana, *Feminism and Science*, 34.

18. Merchant, Preface to *The Death of Nature*, xvi.
19. Susan Griffin, *Woman and Nature: The Roaring Inside Her* (New York: Harper and Row, 1978).
20. Mary Daly, *Gyn/Ecology: The Meta-Ethics of Radical Feminism* (Boston: Beacon Press, 1978).
21. Janet Biehl, *Rethinking Ecofeminist Politics* (Boston: South End Press, 1991), 12–13.
22. Biehl uses the term "psychobiologistic" rather than "cultural" in *Rethinking Ecofeminist Politics*; see 12–13.
23. Alison M. Jaggar, *Feminist Politics and Human Nature* (Totowa, N.J.: Rowman and Allanheld, 1983), 94.
24. Haraway, *Simians, Cyborgs, and Women*, 174.
25. Carolyn D'Cruz makes a similar point in *Ecofeminism as Practice, Theory, Discourse*, 51.
26. Biehl, *Rethinking Ecofeminist Politics*, 2, 5, 11–17, 99–100.
27. One exception is ecofeminist Charlene Spretnak, who associates herself with cultural feminism in *States of Grace: The Recovery of Meaning in the Postmodern Age* (San Francisco: Harper and Row, 1991), 127–33.
28. Biehl, "What Is Social Ecofeminism?" 2.
29. Elizabeth Weed, ed., *Coming to Terms: Feminism, Theory, Politics* (New York: Routledge, 1989), xxviii.
30. Audre Lorde, *Sister Outsider* (Freedom, Calif.: Crossing Press, 1984), 67–69.
31. The idea that capitalist patriarchy produces social and ecological "poison" in women's bodies, referred to more succinctly as "ecocide in the body," has been developed by Chaia Heller and is central to social ecofeminism.
32. This idea has been put forward by Carolyn Merchant. See Carolyn Merchant, *Ecological Revolutions: Nature, Gender, and Science in New England* (Chapel Hill: University of North Carolina Press, 1989), 269–70.
33. For a feminist, green socialist account of the contradictions between essentialism and materialism, see Mary Mellor, "Eco-Feminism and Eco-Socialism: Dilemmas of Essentialism and Materialism," *Capitalism, Nature, Socialism* 3, no. 2 (June 1992): 43–62.
34. Fuss, *Essentially Speaking*, 69–70.

Ecofeminism and Deep Ecology

FREYA MATHEWS

TWO OF THE MOST important philosophies of nature that have been developing over the past decade or more, namely, Deep Ecology and ecofeminism, have offered different accounts of the way in which we should relate to nature. These are different accounts not so much of how we should *treat* the natural world, in light of its entitlement or otherwise to our moral consideration, but of how we should both construe and experience our *relationship* with it. Deep Ecology, as we shall see, takes a basically holistic view of nature; its image of the natural world is that of a fieldlike whole of which we and other "individuals" are parts.[1] It encourages us to seek our true identity by identifying with wider and wider circles of nature, presenting the natural world as an extension of ourselves, our self-writ-large, as it were. Since on this view our interests are convergent with those of nature, it becomes incumbent on us to respect and serve these common interests.

Ecofeminists, in contrast, tend to portray the natural world as a community of beings, related, in the manner of a family, but nevertheless distinct.[2] We are urged to respect the individuality of these beings, rather than seeking to merge with them, and our mode of relating to them should be via open-minded and attentive encounter, rather than through abstract metaphysical preconceptualization.[3] It is envisaged that the understanding born of such encounters will result in an attitude of care or compassion which can provide the ground for an ecological ethic.[4]

In this debate between Deep Ecology and ecofeminism, I find myself—if I might introduce a personal note here—somewhat theoretically conflicted, since I feel an affinity and a loyalty toward both these positions—toward the grand metaphysical vision of Deep Ecology and toward the ecofeminist

From: "Relating to Nature," in *Ecopolitics V Proceedings*, ed. Ronnie Harding, Centre for Liberal and General Studies, University of New South Wales, Kensington, New South Wales, Australia, 1992, 489–96.

ethic of kinship and care. Nor is the loyalty at issue here of a purely intellectual character—the battle lines of this debate are emphatically gendered.[5]
Moreover, since ecofeminism has developed partly as a critique of Deep
Ecology, the solution cannot be a matter of merely cutting and pasting the
theories together. In spite of this, however, I do believe that a kind of
dialectical reconciliation of these two views of nature can be achieved, though
at the cost perhaps of an irreducible *ambivalence* in the resulting ecological
ethic. But since such ambivalence may in fact be precisely what an understanding of the ecological structure of reality requires of us, I do not think
this counts against the theory I shall present.

In this paper then, I begin with an examination of the metaphysical axioms of Deep Ecology. I argue that these axioms generate a fundamental
dilemma for deep ecologists. In attempting to resolve this dilemma, I find
I have to give up the ethical conclusions to which Deep Ecology is normally assumed to lead, and draw instead on an ethical perspective more
akin to that found in ecofeminist literature.

THE TWO METAPHYSICAL AXIOMS OF DEEP ECOLOGY

The primary metaphysical axiom of Deep Ecology is the thesis of metaphysical interconnectedness. Naess images the natural world as a field of
relations. He advocates:

> Rejection of the man-in-environment image in favor of *the relational, to-
> tal-field image*. Organisms as knots in the biospherical net or field of
> intrinsic relations. An intrinsic relation between two things A and B is
> such that the relation belongs to the definitions or basic constitutions of
> A and B, so that without the relation, A and B are no longer the same
> things. The total field model dissolves not only the man-in-environment
> concept, but every compact thing-in-milieu concept—except when talk-
> ing at a superficial or preliminary level of communication.[6]

In an early paper Warwick Fox identifies as the "central intuition" of Deep
Ecology the idea that "there is no firm ontological divide in the field of
existence. . . . [t]o the extent that we perceive boundaries, we fall short of
deep-ecological consciousness."[7] Deep ecologically minded followers of James
Lovelock (1979) favor the model of an organism.[8] All exponents agree that
individuals, to the extent they can be identified at all, are constituted out
of their relations with other individuals: they are not discrete substances
capable of existing independently of other individuals. The whole is understood to be more than the sum of its parts, and the parts are defined through
their relations to one another and to the whole.

The second metaphysical presupposition of Deep Ecology functions more
as a hidden premise—it is not listed as an axiom, as the interconnectedness
thesis is, but is nevertheless taken for granted in all versions of the theory.

The presupposition in question is that nature can best look after its own interests, that it is only *our* interventions in the natural course of events that give rise to terminal ecological disasters. This assumption is implicit in the injunction to let nature take the lead in ecological matters, to minimize our interference in it, and to try to shape our own interests to those of nature. It is neatly summed up in Barry Commoner's third law of ecology: nature knows best.[9]

Now let us look at the implications of these two metaphysical assumptions for our relation to the natural world. According to deep ecologists, the fact of our interconnectedness with the rest of nature implies that we are ultimately *identifiable* with nature; the fact of the indivisibility of reality implicates us in wider and wider circles of being. We should accordingly shed our confining ego identity, and gradually open up to nature at large. The process of achieving the widest possible identification with nature is equated, in Deep Ecology, with self-realization; self-realization is a matter, according to Warwick Fox, of enlarging one's sphere of identification.[10]

Normative implications are taken to follow hard on the heels of this identification thesis, together with the assumption that nature can and should look after its own interests. For if we are in this sense one with nature, and our interests are convergent with those of nature, then we shall be called upon to *defend* nature from human interference, just as we are called on to defend ourselves against attack. As activist and deep ecologist John Seed puts it, "'I am protecting the rainforest' develops to 'I am part of the rainforest protecting myself.'" Recognition of our identifiability with nature is taken to entail a commitment to *ecological resistance*.[11]

THE IDENTIFICATION DILEMMA

At this point in the argument however, an intractable dilemma raises its head. I shall call it the "identification dilemma." If we are identifiable with nature, as the interconnectedness thesis implies, then whatever we do, where this will include our exploitation of the environment, will qualify as natural. Since nature knows best how to look after itself, it follows that whatever qualifies as natural must be ecologically for the best, at least in the long run. In short, if we are truly part of, or one with, nature, and nature knows best, then our depredations of the natural world must be ecologically, and hence morally, unobjectionable.[12]

To this objection a deep ecologist might reply that although we are ontologically one with nature, we may not consciously recognize this to be the case. In consciousness we may construct our identity in opposition to nature. Our actions vis-à-vis the environment will then reflect this false consciousness, rather than the underlying ontological fact: we shall be acting *as if* we were ontologically detached even though this is not in fact the case. Such action may then be regarded as unnatural, in the sense that it does not testify to our actual interconnectedness with the rest of the world.

This reply however would appear to conflate the natural with the true. It may be perfectly natural for consciousness to belie the ontological facts, for there may be adaptive value in its doing so in certain circumstances. After all, there are many species which, though ontologically interconnected with the rest of life (according to the interconnectedness thesis), nevertheless appear to act out of narrow self-interest and exploit the environment to the best of their ability for their own ends. ("Plagues" of locusts and mice spring to mind in this connection; but many species, even in normal circumstances, tread anything but lightly on their lands, relying on the regenerative powers of nature rather than on their own restraint to ensure the continuing health of their environments. The noble elephant is a case in point.) Such a gap between consciousness and the ontological underpinnings of a species' identity may well serve nature's own purposes—it may be part of the long-term ecological scheme of things. If this is the case, then such a gap would be ecologically and hence ethically unobjectionable. If we consider it desirable that our consciousness reflect our true ontological estate, then we cannot claim that this is because such fidelity to ontology is natural; we must rather admit that it is because we value truth. But then there is no reason to suppose that the present self-interested, exploitative behavior of humanity is unnatural; and if it is natural—if it is in accordance with the ways of nature—it cannot, from a deep ecological viewpoint, count as wrong.

In sum, it is plausible to argue, in the light of the interconnectedness thesis, that whatever we do to the environment is natural, and that, since nature knows best, our present despoliation of the environment must in fact be in nature's long-term interests. We might wish to change our ways on our own behalf, recognizing that we are at present orchestrating our own extinction.[13] But we have no grounds for changing our ways on behalf of nature, which is to say, on grounds of ecological morality. To suppose otherwise is in fact to perpetuate the old division between humanity and nature, and with it the old assumption of human supremacism. For to suppose that we can destroy nature is to deny that nature knows best, where this is to admit that we had really better take the rudder after all, and steer nature through this crisis that we have created for it. In other words, to allow that what we are doing to the environment is natural, and yet to insist that it needs to be changed by us, is to deny that nature knows what it is doing; it is subtly to re-usurp control. If we are true to the metaphysical premises of Deep Ecology, if we accept both our oneness with nature and nature's fitness to conduct its own ecological affairs without our assistance, then we should allow our own evolution to run its "natural" course, whatever that turns out to be, on the understanding that by doing so we shall be advancing the cause of life on earth. It may well be that our massive impact on the planetary ecosystem is paving the way for an epoch-making transition in evolution—perhaps analogous to the transition from anaerobic to aerobic life in the early stages of the history of life on earth.[14]

The insistence of deep ecologists that we are one with a nature which best knows how to look after itself then, does seem directly to imply that we have no ecological, nor hence moral, grounds for intervening in the spontaneous course of human affairs as these affect the environment. This poses a dilemma for Deep Ecology, since deep ecologists have no desire so to acquiesce in the present regime of environmental degradation and destruction. If they persist—as I have no doubt they will—in exhorting us to engage in active "ecological resistance," then we have to conclude that there is an inconsistency at the heart of Deep Ecology.

HOLISTIC AND INDIVIDUALISTIC READINGS OF THE TWO AXIOMS

If, as environmentalists, we are already committed to ecological resistance, the conclusion of the previous section forces us to reexamine the two metaphysical premises of Deep Ecology. One or both of them will have to be modified, in some way, if Deep Ecology is to retain its activist appeal. Let us then review each of these axioms in turn.

THE INTERCONNECTED THESIS

Is there anything logically amiss with the idea of interconnectedness that is so central to Deep Ecology, anything that would account for the counterintuitive conclusion to which, when conjoined with the thesis that nature knows best, it was found to lead? I think the problem with this thesis, in the present connection, is not that its interpretation within Deep Ecology is in any way logically flawed, but merely that it is *partial*. Deep ecologists have, in the main, given the idea of interconnectedness a holistic reading; they have taken it to mean that nature, as a metaphysical whole, is logically prior to its parts, and that the identity of each part is functionally determined by way of its relation to the whole. They concede a degree of autonomy to individuals, but ultimately they view that autonomy as apparent only, without fundamental ontological significance.

It is arguable however that this reading of the interconnectedness thesis captures only one side of its meaning. If a systems-theoretic approach is adopted, it is possible to see interconnectedness as entailing the identities of both wholes and individuals. From a systems-theoretic viewpoint, the world (particularly the biological world) appears as a field of relations, a web of interconnections, which does indeed cohere as a whole, but within which a genuine form of individuation is nevertheless possible. An individual is, from this viewpoint, an energy configuration or system which maintains itself by way of its continuous interactions with its environment. Since it is only able to maintain its integrity by way of this continuous give and take with the environment, its existence is a function of its relations, its interconnections. But since these interactions do indeed enable it actively to maintain its integrity, it does enjoy a genuine, though relative, individuality. In this way the world may be seen as both a seamless whole and a manifold of individuals.

On this reading then, metaphysical interconnectedness implies an irreducible ontological ambivalence at the level of individuals: individuals are, in this scheme of things, analogous to the "wavicles" of quantum mechanics. In quantum mechanics, light is analyzed in terms of these wavicles: looked at from one point of view, a ray of light manifests as a stream of particles (photons), while from another point of view it manifests as a wave phenomenon (a pattern in a field). Light cannot be reduced to either photons or field. Ontological ambivalence is thus intrinsic to its nature.

Under the sway of the interconnectedness thesis, Deep Ecology tends to view the natural world from the holistic perspective exclusively, and therefore considers individuals as fieldlike rather than as particulate. This one-sided reading of the interconnectedness thesis inevitably also affects its reading of the principle that nature knows best. The principle that nature knows best will be understood to mean that nature knows best for itself as a whole; but it is not taken to imply that nature knows best for the individuals that are its elements. Reading the principle in this latter sense raises obvious questions about its validity. Let us look at the principle in the light of this double reading, and consider whether it can be retained.

THE THESIS THAT NATURE KNOWS BEST

The principle that nature knows best implies that nature is the best servant of its own interests, and therefore that, from the viewpoint of environmental ethics, whatever nature does is right. It follows from this that the natural order is a moral order, that within this natural order everything ultimately turns out for the best, so far as nature is concerned. Can this assumption be defended? In order to answer this we need, as I have pointed out, to look at the principle under both its holistic and its individualistic interpretations. I shall argue that under the holistic interpretation, the natural order is indeed a moral order, but that under the individualistic interpretation it is not.

The answer to the question whether nature knows best, when nature is viewed under its holistic aspect, depends to some extent on the empirical question of whether or not we or any other particular life form have the capacity to extinguish life altogether on the planet. On current evidence this appears to be extremely unlikely: it is widely believed that even full-scale nuclear holocaust would fail to eliminate *microbial* life forms, and that the adaptations of these life forms to the new conditions would usher in a new evolutionary epoch.[15] In light of this assumption that the demise of one order of life creates an opportunity for another, I think we can say that, from the viewpoint of the whole, nature inevitably works toward its own good.

Nature—understood under its holistic aspect—knows best not only in the sense that it is capable of looking after its own interests; it appears to know best in a wider moral sense as well, since the ecological order not only secures its own self-perpetuation, but also appears to exemplify both justice and generosity. I have argued elsewhere for the justice of the ecological order. Such ecological justice consists, in the first place, in the fact that ecological "transgressors" pay for their ecological "transgressions" by being selected out of existence; and it consists, in the second place, in the fact that such self-elimination of actual individuals provides possible individuals with their opportunity to gain entry into the actual world. Such perfect impartiality between the actual and the possible must surely represent the acme of justice! If it is objected that it is scarcely just to condemn an entire ecosystem to extinction on account of the ecological "transgressions" of one of its elements, it must be remembered that from the holistic point of view there is no absolute distinction between an element and its ecosystem. The various elements of an ecosystem are merely different expressions of its own intrinsic logic or theme. It makes no sense, from this holistic perspective, to say that we, as ecological deviants, are endangering our otherwise ecologically viable ecosystems, or the ecologically innocent elements of those ecosystems. For if we are deviant, so are the ecosystems with which we are holistically or internally related, and so too are all the elements of those ecosystems. If we deserve to be selected out for our mistakes, so too does the ecosystem, or even the entire order of life, which defines us.

From the holistic point of view then, the natural order is arguably an order of justice, and as such qualifies as a moral order in a richer sense than that implied in the original maxim that nature knows best. Lest such a moral order seem too stern for us to countenance however, there is, as I remarked earlier, a second way in which the natural—still viewed from an holistic perspective—is equivalent to the right. The moral significance of nature, understood in this second sense, resides in its boundless *generosity*. Etymologically, "nature," as Holmes Rolston III points out, is derived from the Latin *natus*, meaning birth.[16] Nature is the source, the wellspring, of life, and life is, after all, an entirely gratuitous gift, owed to no one. "When nature slays," says Rolston, "she takes only the life she gave . . . and she gathers even that life back to herself by reproduction and reenfolding organic resources and genetic materials, and produces new life out of it." Because nature does not favor those who have life over those who do not, life is dealt out lavishly; the dispensability of the actual is a necessary condition for this lavishness. Nature is not only just, but infinitely generous. The natural order then, viewed from the holistic perspective, is moral not only in that it secures the long-term good of nature, but also in its justice and its generosity.

When nature is examined from the individualistic rather than the holistic viewpoint however, does it still qualify as a moral order? Is the natural still the right? We have seen that, from the point of view of the whole, individuals are generously given life and justly sacrificed that the gift of life might be passed on. As long as we are (quite properly) identifying with the whole, we can appreciate both the effectiveness and the justice of this arrangement, and concur in the price that is paid for it. When we (equally properly) identify ourselves as individuals however, we are likely to see things differently. Nature no longer appears to know best, if by its "knowing best" we mean that it is capable of looking after the interests of individuals. Nor does it appear as just: the situation of actual individuals is importantly different from that of possible individuals. As actual individuals we have actual interests, urgent needs, and desires; we can suffer, and suffer terribly. There is neither justice nor generosity in trading in actual individuals for possible ones, from this perspective. The stern though admittedly life-giving "plan" of nature-as-a-whole then has less to commend it from down here. Nor is it only *our* fate which assumes a larger moral significance from this perspective: that of other actual individuals does likewise. Fellow-feeling for them, familiarity with the imperative which drives them, identification with the shivering vulnerability that their actuality implies, gives rise to concern, to a moral interest in their plight.

Ironically then the impulse to resist the progressive destruction of the present order of life springs not, as Deep Ecology claims, from our identification with nature as a whole—though that identification is perfectly proper, in light of the holistic interpretation of interconnectedness—but rather from our commitment to our individuality. It is as individuals that we feel concern for other individuals. In defending nonhuman beings against human depredations we may even in a sense be *resisting* the greater moral order, the grand order of ecological justice. The compassion which forms the basis of our environmental ethic, from this individualistic point of view, is a function of our finitude rather than of our cosmic self-realization. In securing the conditions for the ongoing unfolding of life, nature (in its holistic aspect) is morally more far-sighted than we; in the name of compassion we seek to block that unfolding by clinging to those individuals which already exist, out of a sense of solidarity with them. As individuals we give our allegiance to individuals, if necessary even against the moral requirements of nature-as-a-whole.

DEEP ECOLOGY AND ECOFEMINISM: COMPLEMENTARY PERSPECTIVES?

This view of the basis of environmental ethics is much closer to ecofeminism than to Deep Ecology. Although ecofeminism is by no means a position or a theory, but simply a fairly open field of inquiry, it could nevertheless be taken to subscribe, either implicitly or explicitly, to the interconnectedness thesis.[17] It tends to interpret interconnectedness in the individualistic rather than in the holistic sense however: nature, from the ecofeminist perspective, is a community of beings, related, in the manner of a family, but nevertheless distinct. We are urged to respect the otherness, the distinct individuality of these beings, rather than seeking to merge with them, in pursuit of an undifferentiated oneness.[18]

Since ecofeminism does not identify us directly with nature-as-a-whole, it does not fall foul of the identification dilemma. In other words, since it does not define us as identifiable with a monolithic nature, it does not have to see our destruction of the environment as a case of nature "destroying" itself, where seeing our action in this way renders it morally unobjectionable. On the contrary, since it sees us as related to nature as to the members of a community or family, to whom the proper attitude is one of familial consideration and care, born of an empathetic understanding made possible by our common origins, or our mutually defining relations, ecofeminism is able to condemn our abuse of the environment outright: this is no way to treat one's family! So for ecofeminism concern for nature is the product of a re-awakening to our kinship with our individual nonhuman relatives; it is grounded in our individuality, rather than in any kind of cosmic identification, and it springs out of a sense of solidarity with our fellow beings.

It seems to me, as I indicated at the outset, that ecofeminism and Deep Ecology, with their complementary interpretations of the interconnectedness thesis, each capture an important aspect of our metaphysical and ethical relationship with nature. For if reality is indeed internally interconnected, if it does consist in a web of relations, then, as I explained earlier, it may be seen as both a whole and as a manifold of individuals. From the viewpoint of the whole it does appear to qualify as a moral order, though from the viewpoint of the individual, it does not. Since I claim both these viewpoints need to be taken into account in our attempt to determine how we should relate to nature, we find ourselves committed in the end to an irreducible moral ambivalence consisting of compassionate intervention on behalf of nature on the one hand, and enlightened acquiescence in the natural tide of destruction on the other. In accepting this ambivalence, we discover on the one hand that it is our humanity—our very finitude and limitation—rather than any grand plan in the stars, that impels us to act on behalf of our embattled fellow creatures. In this way the moral loftiness of Deep Ecology is brought down to the ground, rendered human. But on the other hand we discover that our compassion—the value taken for granted by ecofeminism—is not beyond moral question either. In light of the grand plan that is in the stars, compassion is seen to come down to our love of the familiar, our solidarity with the things that remind us of ourselves. Ecofeminism is, in its turn, divested of its moral smugness, and forced to

ask questions it had not hitherto even thought of. While ecofeminism human-
izes Deep Ecology, you might say, Deep Ecology does indeed deepen ecofeminism.

The recognition that our grounds for ecological resistance lie in our humanity, rather than in our self-writ-large, or in the stars, is particularly
important for environmentalists, I think. For many environmentalists, face
to face with the heart-breaking consequences of human rapaciousness, become embittered toward humankind, and come to see our species as a curse
upon the earth. Out of such a relapse into dualistic thinking, no true healing or affirmation of life can come. To recognize that our humanity is the
wellspring not only of a consuming destructiveness but also of the precious
compassion which counters it, may be a redeeming thought, which will
help to lead us out of the moral impasse created by the divorce between
humanity and nature.

Notes

For an expanded version of this article, see "Relating to Nature," *The Trumpeter*,
11, no. 2 (Spring 1994).

1. This is typical of earlier expositions of Deep Ecology. See Bill Devall and
   George Sessions, *Deep Ecology: Living as if Nature Mattered* (Salt Lake City:
   Peregrine Smith Books, 1985); Warwick Fox, "Deep Ecology: A New Philosophy of Our Time?" *Ecologist* 14 (1984): 194–200; Arne Naess, "The Shallow
   and the Deep, Long-Range Ecology Movement: A Summary" *Inquiry* 16 (1973):
   95–100. It is with the "central intuition articulated in these earlier versions of
   Deep Ecology that I am concerned in this paper.
2. Jim Cheney brought this point our very clearly ("Ecofeminism and Deep Ecology," *Environmental Ethics* 9, no. 2 [1987]: 115–45). It is implicit or explicit in
   many ecofeminist writings; for instance, Judith Plant, "The Circle is Gathering," in *Healing the Wounds: the Promise of Ecofeminism*, ed. J. Plant (Philadelphia: New Society Publishers, 1989), 242–53. However, as ecofeminism is not
   typically expounded systematically as a philosophy, other views of nature are
   also represented in ecofeminist works. Conversely, the view of nature that I
   have here identified as ecofeminist is also espoused by writers who make no
   reference to feminist theory at all. See, for instance, J. Baird Callicott's account of American Indian views of nature (*In Defence of the Land Ethic* [Al-
   bany: SUNY Press, 1989], 177–202). Both Callicott and Aldo Leopold, the
   architect of the land ethic Callicott is concerned to defend, tend to view
   nature as a community of natural elements and beings, but both also seem to
   adopt a holistic interpretation of community for ethical purposes, whereas this
   would run counter to the ecofeminist tendency. I am not really concerned to
   discuss either Deep Ecology or ecofeminism *per se* here, but rather a certain
   complex of issues which are central, but not exclusive, to these two positions.
   The issues in question concern the relative merits of the individualistic and
   holistic views of our relationship with nature. An author who has recently
   addressed certain aspects of these issues without referring to either Deep Ecology or ecofeminism is Robert W. Gardiner ("Between Two Worlds: Humans
   in Nature and Culture," *Environmental Ethics* 12, no. 4 [1990]: 339–52).

3. Cheney, again, has tellingly developed this point—this time along postmodernist, rather than feminist, lines (Jim Cheney, "The Neo-stoicism of Radical Environmentalism," *Environmental Ethics* 11, no. 4 [1989]: 293–324).

4. Evelyn Fox Keller develops a sophisticated argument along these lines. (*Reflections on Gender and Science* [New Haven, Conn.: Yale University Press, 1985]).

5. Indeed, in light of Cheney's arguments, women who subscribe to the view of Deep Ecology invite the charge of exemplifying a masculine psychology (Cheney, "Ecofeminism and Deep Ecology," 115–45).

6. Naess, "Shallow and the Deep," 95–100, quotation on 95.

7. Fox, "Deep Ecology," 194–200, see 196. In his later work, Fox has introduced a relative form of individuality in his ecological metaphysic. See Warwick Fox, *Toward a Transpersonal Ecology* (Boston: Shambala Press, 1990).

8. James Lovelock, *Gaia: A New Look at Life on Earth* (Oxford: Oxford University Press, 1979).

9. Barry Commoner, *The Closing Circle* (London: Jonathan Cape, 1973).

10. Fox, "Deep Ecology," 194–200.

11. John Seed, "Anthropocentrism," appendix E, in Devall and Sessions, *Deep Ecology*, 243–46.

12. Richard Watson, "A critique of anti-anthropocentric biocentrism," *Environmental Ethics* 5, no. 3 (1983): 245–56.

13. Ibid.

14. The details of this transition—and its implications for the future prospects of life on earth—are fascinatingly described in Lynn Margulis and Dorion Sagan, *Microcosmos* (New York: Summit Books, 1986).

15. Lovelock, *Gaia*; Margulis and Sagan, *Microcosmos*.

16. Holmes Rolston III, "Can We and Ought We to Follow Nature?" *Environmental Ethics* 1, no. 1 (Spring 1979): 7–30, quotation on 28–29.

17. This is evident in the web imagery which is so central to ecofeminism, and which appears in a number of ecofeminist titles, for example, Judith Plaskow and Carol Christ, *Weaving the Visions* (New York: Harper and Row, 1989) and Irene Diamond and Gloria Feman Orenstein, eds., *Reweaving the World: The Emergence of Ecofeminism* (San Francisco: Sierra Club Books, 1990).

18. Cheney, "Ecofeminism and Deep Ecology," provides a good account of why ecofeminists adopt this position.

PART V

Environmental Justice

The Importance of
Environmental Justice

PETER WENZ

PEOPLE PLAY MULTIPLE ROLES in environmental systems. We are all constituents, as well as observers of the environment. So when we discuss the environment, we are always to some extent discussing ourselves. . . . *Environmental Justice* is primarily about theories of distributive justice. . . . There are three reasons for concentrating on contexts of environmental concern. . . . [First], we are involved in the environment more than we sometimes realize, and greater self-awareness can facilitate prudent behavior. Second, theories of distributive justice have not been related as often to the environmental area as to some other areas of concern. Third, and most important, environmental concerns involve relationships not only among people who live in the same society at the same time; but also among people who live in different societies at the same time, between people of the present and those of the future, between human and nonhuman animals, and between people and the biosphere in general. Because environmental concerns are uniquely global, . . . theories of distributive justice are tested most thoroughly for their comprehensiveness when they are applied to environmental matters. . . .

Justice usually becomes an issue in contexts in which people's wants or needs exceed the means of their satisfaction. . . . The difference that an ample supply can make is illustrated well in the case of water. People need water wherever they live; but water is scarce in some places, ample in others. Where it is scarce, societies have devised elaborate methods of apportioning the water among those who need and desire it. Infringing upon the water rights of another is considered a serious injustice. In societies with

From: *Environmental Justice* (Albany: State University of New York Press, 1988), xi–xii, 6–21, excerpts.

advanced legal systems, such as in the United States, the apportionment of water in areas of scarcity is governed by intricate legal rules. But where water is plentiful, the situation is entirely different. In some parts of England, residents are charged a quarterly fee for the maintenance of the water works and sewage installation. The fee does not vary according to the amount of water used. In fact, there is no water meter to determine the amount used. Because there is enough water to serve everyone's wants and needs, people do not care how much they or their neighbors consume.

In sum, questions about justice arise concerning those things that are, or are perceived to be, in short supply relative to the demand for them. In these situations, people are concerned about getting their fair share, and arrangements are made, or institutions are generated, to allocate the scarce things among those who want or need them.

These generalizations are subject to two qualifications. First, the people sharing the scarce good must care enough about what they receive to desire their fair share. . . . In a drought, I am likely to want my house to be allocated its fair share of water by those at our local utility. Water is very important to me. I am not personally acquainted with most people in my town. I am not so benevolent as to want those among them who are no worse off than I to be given more water than I am given. So limited benevolence is, along with scarcity, an element of situations in which issues of justice arise.

The second qualification is this: Arrangements or institutions designed to allocate scarce things make sense only for those things that people are able to distribute. . . . The people at City Water, Light, and Power are able to distribute water from the reservoir. But the ability people have to distribute rain, good fortune, and perfect pitch is very limited. These things may be scarce, and people may want their fair shares, but there are no arrangements or institutions designed to allocate them, because people lack the power of distribution. . . .

THE NEED FOR COORDINATED ENVIRONMENTAL RESTRAINT

Allowing people to make, take, or receive whatever they can get—so long as they do not directly attack, brutalize, or steal from others—will sometimes make scarcity worse, hurting everyone in the long run. Garrett Hardin calls this "The Tragedy of the Commons." Imagine a pasture that can be used in common by many herdsmen. It is a limited resource, but one capable of supplying enough food for the animals that the herdsmen depend upon for their livelihood. Suppose that one of the herdsmen wants to increase his income. He can do this by doubling his herd. He will have to work harder to care for the larger herd, but he feels that the increased income is worth the effort. He is not directly attacking, brutalizing, or stealing from anyone else, so his extra income is earned within the rules that prohibit these things. The pasture is not appreciably damaged by the grazing of these extra animals because, though they double the individual's herd, they do not significantly increase the total number of animals grazing on the common pasture.

Since it is a common pasture, other herdsmen who wish to increase their income are also free to double or triple their flocks. However, as more and more animals graze in the common pasture, the flora is ruined owing to overgrazing. The result is the destruction of the common resource, the pasture, on which all had depended for a livelihood. No one committed barbarous acts against anyone else; yet, while literally minding their own business, they ruined the basis of that business. Their efforts to increase their incomes were self-defeating. In situations like this one, allowing people to take whatever they want of a scarce resource results in its destruction. The appropriation made by each person must be coordinated with those of others to ensure that the collective appropriation is not excessive and ruinous. So in these situations it is practical to determine each person's fair share of the collective good in order to avoid the tragedy of the commons. Such a determination can be made only by reference to an agreed standard of justice. Philosophical investigations into the nature and principles of justice are required.

Many environmental resources are like the pasture in Hardin's story. The oceans, the air, and the ozone layer are as important to our lives as was the pasture to the herdsmen. Yet no one owns them. If, in the pursuit of her own pleasure, gain, or preferred lifestyle, each person is free to use or despoil these natural resources in any way that does not directly brutalize other human beings, everyone will suffer in the long run. The ozone layer, for example, protects us all from solar radiation that can cause cancer. Suppose that the use of aerosol sprays diminishes the ozone layer. No single individual's use of aerosol sprays has an appreciable effect on the layer, so anyone can use such sprays without harming anyone else. But if millions use these sprays over long periods of time, the protection provided by the layer against harmful solar radiation could be significantly diminished. This could harm everyone, including those who make and use aerosol sprays. So if millions of people would like to use these sprays, some restraint must be exercised. One way of effecting such restraint would be to devise and enforce a system that permits people to use only limited amounts of aerosol spray. But what should the limit be? As in the case of the common pasture, it makes sense for each to be given her fair share, whatever that may be. Practicality again suggests that mutually agreeable principles of justice be discovered and employed in order to determine everyone's fair share.

This reasoning applies to pollution, generally. Whether it is a banana peel tossed from a car window on the highway or a gram of carbon monoxide emitted from the car's exhaust, every litter bit does *not* hurt. It is the concentration of litter bits that hurts, and this concentration usually hurts those who do the polluting as well as others. Our waste products are integral to our life processes. In limited concentrations, they aid the life process as a whole. But our preferred life-styles and activities tend to generate kinds and levels of waste that pollute our environment and can eventually make it uninhabitable. So restraint is necessary in a great many matters, ranging from our use of the national parks to our consumption of fossil fuels. And again, when restraint is necessary to preserve the environment, it seems that everyone should receive a fair share, and be restrained to a fair degree, in accordance with reasonable principles of justice. This is environmental justice. . . .

THE VULNERABILITY OF MODERN SOCIETIES

It is an essential condition of a nation-state, especially of one that seeks to provide some liberty for its people, that the vast majority of people perceive the divisions of benefits and burdens to be reasonably just. The importance of this condition, like the importance of many essentials, is most clearly evident when the condition is not met. Consider, for example, the situation in Northern Ireland. In the mid-1960s, the condition of Catholics in Northern Ireland resembled that of blacks in the South of the United States. They went to separate schools, received poorer educational opportunities, had lower incomes, were discriminated against in employment, and held fewer government jobs. A movement of peaceful protest against these conditions was begun in the late 1960s. It was modeled on the protest of Martin Luther King, Jr., in the American South. Unlike the government of the United States, which compromised with the civil rights movement, the British government made few concessions. . . .

If a modern, vulnerable social order is perceived by a significant percentage of its members to be irremediably and grossly unjust, it can be crippled or destroyed by relatively few people, with the active and passive support of others. This has happened in Northern Ireland. The application of force is inadequate to prevent this. The modern social order requires the voluntary cooperation of the vast majority, especially in a relatively free society. In order to receive this cooperation, the social order must be perceived as tolerably just. Discussions of the nature and principles of justice are therefore a practical necessity. We live in a society that many believe to be characterized by significant injustices. The society is vulnerable to radical disruption from within. But beliefs concerning justice are sufficient to mollify the vast majority of people so as to dissuade them from engaging in or supporting terrorism. . . . The perception that society's institutions and policies are reasonably just is necessary for such (relatively) willing submission to restraint. . . .

THE NEED FOR ENVIRONMENTAL JUSTICE

The conditions of justice recur frequently with respect to the environment. Arrangements must often be made to allocate access to activities and commodities so as to insure that the uses people make of the environment are compatible with one another, and with the environment's continued habitability. For example, it is now widely believed that burning large amounts of coal that contain significant quantities of sulfur results in acid rain. This rain is blamed for the defoliation of forests in the northeastern United States, southeastern Canada, and southern Germany. Many people use these forests as sources of recreation. They are also used by the lumber industry, and serve to retard soil erosion. So there is pressure to decrease significantly the amount of high sulfur coal that is burned, or to require that smokestacks of furnaces using this coal be fitted with scrubbers that prevent a very high percentage of the sulfur from escaping into the atmosphere. The coal is burned hundreds of miles away from the affected forests. It is used to power factories and to provide people with electricity. Significant reductions in the use of this coal would adversely affect the owners of the mines from which the coal is extracted, as well as the workers in those mines. The factories that currently use this coal would have either to use alternate, usually more expensive fuels, or else install expensive scrubbers. Either course could make their products more expensive. Products that become too expensive will no longer be marketable, and the factories that produce them will have to shut down, putting many people out of work. Electricity that is produced for household consumption would have to become more expensive for the same reasons.

Because the areas where the mines are located and where the coal is used are so far from the areas where the forests are damaged, the people who benefit from the current use of high sulfur coal, and who would be adversely affected by proposed changes, are for the most part different from the direct beneficiaries of forest preservation. Their interests are opposed. The more that one group gets of what it wants, the less the other group can get. . . .

Decisions of public policy are required concerning the use of high-sulfur coal. People will not feel well served by their government if this policy, which could put them out of work, or result in a mudslide covering their houses, is not clearly defensible. People will want to know why they should have to make the sacrifices that are required of them, and how these sacrifices compare to those that are required of others. The government will have to employ defensible principles of justice in fashioning its environmental policy if those affected by it are to believe that the sacrifices required of them are justified.

The same is true of most other environmental policies. They require people to make important sacrifices. Many policies are designed to deal with situations resembling the tragedy of the commons. As already noted, the air we breathe is like the commons. Uncontrolled automobile emissions of carbon monoxide would jeopardize the health of many people, especially children, the aged, and those with emphysema. Emissions could be reduced by improving mass transit systems at public expense, on the assumption that the use of mass transit would then replace some automobile use. The improvement in air quality would in this case be paid for by the taxpaying public. This may seem equitable, since clean air is a public good. But members of the public would not benefit equally from this policy. Those living in urban areas would benefit most for two reasons: The air quality would be improved most in those areas, and urban residents would have available inexpensive public transportation. Should all taxpayers have to pay for benefits that accrue disproportionately to urban residents?

Alternatively, automobile emissions could be reduced by placing heavy taxes on gasoline, thereby discouraging its use. The improvement in air quality would be more widespread, because automobile use would be reduced generally, not just in urban areas. But do rural areas require such improvement? And are the burdens allocated equitably? On this plan, the user pays. This would hit hardest lower-income people and those in rural areas who must use their cars to get to work. Rich people would be virtually unaffected.

Many other plans could be devised to reduce air pollution caused by automobile emissions, and various plans can be combined with one another. Although all are aimed at reducing air pollution, each plan, and each combination of plans, benefits different people and/or places different burdens on different groups. Because these benefits and burdens can be significant, it will be necessary to assure people that they are receiving their fair share of benefits and are not being required unfairly to shoulder great burdens. The social fabric will not be destroyed by any one environmental policy that is perceived to be unjust. But the number and extent of environmental policies has increased and will continue to increase considerably. The perception that these policies are consistently biased in favor of some groups and against others could undermine the voluntary cooperation that is necessary for the maintenance of social order. Voluntary cooperation is especially necessary if the social order is to be maintained in a relatively open society where authoritarian measures are the exception rather than the rule. Thus, because social solidarity and the maintenance of order in a relatively free society require that people consider their sacrifices to be justified in relation to the sacrifices of others, environmental public policies will have to embody principles of environmental justice that the vast majority of people consider reasonable.

Environmental Racism and the Environmental Justice Movement

ROBERT BULLARD

Communities are not all created equal. In the United States, for example, some communities are routinely poisoned while the government looks the other way. Environmental regulations have not uniformly benefited all segments of society. People of color (African Americans, Latinos, Asians, Pacific Islanders, and Native Americans) are disproportionately harmed by industrial toxins on their jobs and in their neighborhoods. These groups must contend with dirty air and drinking water—the byproducts of municipal landfills, incinerators, polluting industries, and hazardous waste treatment, storage, and disposal facilities.

Why do some communities get "dumped on" while others escape? Why are environmental regulations vigorously enforced in some communities and not in others? Why are some workers protected from environmental threats to their health while others (such as migrant farmworkers) are still being poisoned? How can environmental justice be incorporated into the campaign for environmental protection? What institutional changes would enable the United States to become a just and sustainable society? What community organizing strategies are effective against environmental racism? ... The pervasive reality of racism is placed at the very center of the analysis.

From: Robert Bullard, ed., *Confronting Environmental Racism: Voices from the Grassroots* (Boston: South End Press, 1993), 15–24, 38–39.

INTERNAL COLONIALISM AND WHITE RACISM

The history of the United States has long been grounded in white racism. The nation was founded on the principles of "free land" (stolen from Native Americans and Mexicans), "free labor" (cruelly extracted from African slaves), and "free men" (white men with property). From the outset, institutional racism shaped the economic, political, and ecological landscape, and buttressed the exploitation of both land and people. Indeed, it has allowed communities of color to exist as internal colonies characterized by dependent (and unequal) relationships with the dominant white society or "mother country." In their 1967 book, *Black Power*, Stokely Carmichael and Charles Hamilton were among the first to explore the "internal" colonial model as a way to explain the racial inequality, political exploitation, and social isolation of African Americans. As Carmichael and Hamilton write:

> The economic relationship of America's black communities [to white society] . . . reflects their colonial status. The political power exercised over those communities goes hand in glove with the economic deprivation experienced by the black citizens.
>
> Historically, colonies have existed for the sole purpose of enriching, in one form or another, the "colonizer"; the consequence is to maintain the economic dependency of the "colonized."[1]

Generally, people of color in the United States—like their counterparts in formerly colonized lands of Africa, Asia, and Latin America—have not had the same opportunities as whites. The social forces that have organized oppressed colonies internationally still operate in the "heart of the colonizer's mother country."[2] For Robert Blauner, people of color are subjected to five principal colonizing processes: they enter the "host" society and economy involuntarily; their native culture is destroyed; white-dominated bureaucracies impose restrictions from which whites are exempt; the dominant group uses institutionalized racism to justify its actions; and a dual or "split labor market" emerges based on ethnicity and race. Such domination is also buttressed by state institutions. Social scientists Michael Omi and Howard Winant go so far as to insist that "every state institution is a racial institution."[3] Clearly, whites receive benefits from racism, while people of color bear most of the cost.

ENVIRONMENTAL RACISM

Racism plays a key factor in environmental planning and decision making. Indeed, environmental racism is reinforced by government, legal, economic, political, and military institutions. It is a fact of life in the United States that the mainstream environmental movement is only beginning to wake up to. Yet, without a doubt, racism influences the likelihood of exposure to environmental and health risks and the accessibility to health care. Racism provides whites of all class levels with an "edge" in gaining access to a healthy physical environment. This has been documented again and again.

Whether by conscious design or institutional neglect, communities of color in urban ghettos, in rural "poverty pockets," or on economically impoverished Native American reservations face some of the worst environmental devastation in the nation. Clearly, racial discrimination was not legislated out of existence in the 1960s. While some significant progress was made during this decade, people of color continue to struggle for equal treatment in many areas, including environmental justice. Agencies at all levels of government, including the federal EPA, have done a poor job protecting people of color from the ravages of pollution and industrial encroachment. It has thus been an uphill battle convincing white judges, juries, government officials, and policy makers that racism exists in environmental protection, enforcement, and policy formulation.

The most polluted urban communities are those with crumbling infrastructure, ongoing economic disinvestment, deteriorating housing, inadequate schools, chronic unemployment, a high poverty rate, and an overloaded health-care system. Riot-torn south-central Los Angeles typifies this urban neglect. It is not surprising that the "dirtiest" zip code in California belongs to the mostly African-American and Latino neighborhood in that part of the city.[4] In the Los Angeles basin, over 71 percent of the African Americans and 50 percent of the Latinos live in areas with the most polluted air, while only 34 percent of the white population does.[5] This pattern exists nationally as well. As researchers D. R. Wernette and L. A. Nieves note:

> In 1990, 437 of the 3,109 counties and independent cities failed to meet at least one of the EPA ambient air quality standards . . . 57 percent of whites, 65 percent of African Americans, and 80 percent of Hispanics live in 437 counties with substandard air quality. Out of the whole population, a total of 33 percent of whites, 50 percent of African Americans, and 60 percent of Hispanics live in the 136 counties in which two or more air pollutants exceed standards. The percentage living in the 29 counties designated as nonattainment areas for three or more pollutants are 12 percent of whites, 20 percent of African Americans, and 31 percent of Hispanics.[6]

Income alone does not account for these above-average percentages. Housing segregation and development patterns play a key role in determining where people live. Moreover, urban development and the "spatial configuration" of communities flow from the forces and relationships of industrial production which, in turn, are influenced and subsidized by government policy.[7] There is widespread agreement that vestiges of race-based decision making still influence housing, education, employment, and criminal justice. The same is true for municipal services such as garbage pickup and disposal, neighborhood sanitation, fire and police protection, and library services.

Institutional racism influences decisions on local land use, enforcement of environmental regulations, industrial facility siting, management of economic vulnerability, and the paths of freeways and highways.

People skeptical of the assertion that poor people and people of color are targeted for waste-disposal sites should consider the report the Cerrell Associates provided the California Waste Management Board. In their 1984 report, *Political Difficulties Facing Waste-to-Energy Conversion Plant Siting,* they offered a detailed profile of those neighborhoods most likely to organize effective resistance against incinerators. The policy conclusion based on this analysis is clear. As the report states: "All socioeconomic groupings tend to resent the nearby siting of major facilities, but middle and upper socioeconomic strata possess better resources to effectuate their opposition. Middle and higher socioeconomic strata neighborhoods should not fall within the one-mile and five-mile radius of the proposed site."[8]

Where then will incinerators or other polluting facilities be sited? For Cerrell Associates, the answer is low-income, disempowered neighborhoods with a high concentration of nonvoters. The ideal site, according to their report, has nothing to do with environmental soundness but everything to do with lack of social power. Communities of color in California are far more likely to fit this profile than are their white counterparts.

Those still skeptical of the existence of environmental racism should also consider the fact that zoning boards and planning commissions are typically stacked with white developers. Generally, the decisions of these bodies reflect the special interests of the individuals who sit on these boards. People of color have been systematically excluded from these decisionmaking boards, commissions, and governmental agencies (or allowed only token representation). Grass-roots leaders are now demanding a shared role in all the decisions that shape their communities. They are challenging the intended or unintended racist assumptions underlying environmental and industrial policies.

TOXIC COLONIALISM ABROAD

To understand the global ecological crisis, it is important to understand that the poisoning of African Americans in south-central Los Angeles and of Mexicans in border *maquiladoras* have their roots in the same system of economic exploitation, racial oppression, and devaluation of human life. The quest for solutions to environmental problems and for ways to achieve sustainable development in the United States has considerable implications for the global environmental movement.

Today, more than nineteen hundred *maquiladoras,* assembly plants operated by American, Japanese, and other foreign countries, are located along the two thousand-mile U.S.-Mexican border.[9] These plants use cheap Mexican labor to assemble products from imported components and raw materials, and then ship them back to the United States.[10] Nearly half a million Mexicans work in the *maquiladoras*. They earn an average of $3.75 a day. While these plants bring jobs, albeit low-paying ones, they exacerbate local pollution by overcrowding the border towns, straining sewage and water systems, and reducing air quality. All this compromises the health of workers and nearby community residents. The Mexican environmental regulatory agency is under-staffed and ill-equipped to adequately enforce the country's laws.[11]

The practice of targeting poor communities of color in the Third World for waste disposal and the introduction of risky technologies from industrialized countries are forms of "toxic colonialism," what some activists have dubbed the "subjugation of people to an ecologically-destructive economic order by entities over which the people have no control."[12] The industrialized world's controversial Third-World dumping policy was made public by the release of an internal, 12 December 1991, memorandum authored by Lawrence Summers, chief economist of the World Bank. It shocked the world and touched off a global scandal. Here are the highlights:

"Dirty" Industries: Just between you and me, shouldn't the World Bank be encouraging MORE migration of the dirty industries to the LDCs [Less Developed Countries]? I can think of three reasons:

1) The measurement of the costs of health impairing pollution depends on the foregone earnings from increased morbidity and mortality. From this point of view a given amount of health impairing pollution should be done in the country with the lowest cost, which will be the country with the lowest wages. I think the economic logic behind dumping a load of toxic waste in the lowest wage country is impeccable and we should face up to that.

2) The costs of pollution are likely to be non-linear as the initial increments of pollution probably have very low cost. I've always thought that under-polluted areas in Africa are vastly UNDER-polluted; their air quality is probably vastly inefficiently low compared to Los Angeles or Mexico City. Only the lamentable facts that so much pollution is generated by non-tradable industries (transport, electrical generation) and that the unit transport costs of solid waste are so high prevent world welfare-enhancing trade in air pollution and waste.

3) The demand for a clean environment for aesthetic and health reasons is likely to have very high income elasticity. The concern over an agent that causes a one in a million change in the odds of prostate cancer is obviously going to be much higher in a country where people survive to get prostate cancer than in a country where under 5 [year-old] mortality is 200 per thousand. Also, much of the concern over industrial atmosphere discharge is about visibility impairing particulates. These discharges may have very little direct health impact. Clearly trade in goods that embody aesthetic pollution concerns could be welfare enhancing. While production is mobile the consumption of pretty air is a non-tradable.

The problem with the arguments against all of these proposals for more pollution in LDCs (intrinsic rights to certain goods, moral reasons, social concerns, lack of adequate markets, etc.) could be turned around and used more or less effectively against every Bank proposal.

BEYOND THE RACE-VERSUS-CLASS TRAP

Whether at home or abroad, the question of who *pays* and who *benefits* from current industrial and development policies is central to any analysis of environmental racism. In the United States, race interacts with class to create special environmental and health vulnerabilities. People of color, however, face elevated toxic exposure levels even when social class variables (income, education, and occupational status) are held constant.[13] Race has been found to be an independent factor, not reducible to class, in predicting the distribution of (1) air pollution in our society,[14] (2) contaminated fish consumption,[15] (3) the location of municipal landfills and incinerators,[16] (4) the location of abandoned toxic waste dumps,[17] and (5) lead poisoning in children.[18]

Lead poisoning is a classic case in which race, not just class, determines exposure. It affects between three and four million children in the United States—most of whom are African Americans and Latinos living in urban areas. Among children five years old and younger, the percentage of African Americans who have excessive levels of lead in their blood far exceeds the percentage of whites at all income levels.[19]

The federal Agency for Toxic Substances and Disease Registry found that for families earning less than six thousand dollars annually an estimated 68 percent of African-American children had lead poisoning, compared with 36 percent for white children. For families with incomes exceeding fifteen thousand dollars, more than 38 percent of African-American children have been poisoned, compared with 12 percent of white children. African-American children are two to three times more likely than their white counterparts to suffer from lead poisoning independent of class factors.

One reason for this is that African Americans and whites do not have the same opportunities to "vote with their feet" by leaving unhealthy physical environments. The ability of an individual to escape a health-threatening environment is usually correlated with income. However, racial barriers make it even harder for millions of African Americans, Latinos, Asians, Pacific Islanders, and Native Americans to relocate. Housing discrimination, redlining, and other market forces make it difficult for millions of households to buy their way out of polluted environments. For example, an affluent African-American family (with an income of fifty thousand dollars or more) is as segregated as an African-American family with an annual income of five thousand dollars.[20] Thus, lead poisoning of African-American children is not just a "poverty thing."

White racism helped create our current separate and unequal communities. It defines the boundaries of the urban ghetto, *barrio*, and reservation, and influences the provision of environmental protection and other public services. Apartheid-type housing and development policies reduce neighborhood options, limit mobility, diminish job opportunities, and decrease environmental choices for millions of Americans. It is unlikely that this nation will ever achieve lasting solutions to its environmental problems unless it also addresses the system of racial injustice that helps sustain the existence of powerless communities forced to bear disproportionate environmental costs.

The Limits of Mainstream Environmentalism

Historically, the mainstream environmental movement in the United States has developed agendas that focus on such goals as wilderness and wildlife preservation, wise resource management, pollution abatement, and population control. It has been primarily supported by middle- and upper-middle-class whites. Although concern for the environment cuts across class and racial lines, ecology activists have traditionally been individuals with above-average education, greater access to economic resources, and a greater sense of personal power.[21]

Not surprisingly, mainstream groups were slow in broadening their base to include poor and working-class whites, let alone African Americans and other people of color. Moreover, they were ill equipped to deal with the environmental, economic, and social concerns of these communities. During the 1960s and 1970s, while the "Big Ten" environmental groups focused on wilderness preservation and conservation through litigation, political lobbying, and technical evaluation, activists of color were engaged in mass direct action mobilizations for basic civil rights in the areas of employment, housing, education, and health care. Thus, two parallel and sometimes conflicting movements emerged, and it has taken nearly two decades for any significant convergence to occur between these two efforts. In fact, conflicts still remain over how the two groups should balance economic development, social justice, and environmental protection.

In their desperate attempt to improve the economic conditions of their constituents, many African-American civil rights and political leaders have directed their energies toward bringing jobs to their communities. In many instances, this has been achieved at great risk to the health of workers and the surrounding communities. The promise of jobs (even low-paying and hazardous ones) and of a broadened tax base has enticed several economically impoverished, politically powerless communities of color both in the United States and around the world.[22] Environmental job blackmail is a fact of life. You can get a job, but only if you are willing to do work that will harm you, your families, and your neighbors.

Workers of color are especially vulnerable to job blackmail because of the greater threat of unemployment they face compared to whites and because of their concentration in low-paying, unskilled, nonunionized occupations. For example, they make up a large share of the nonunion contract workers in the oil, chemical, and nuclear industries. Similarly, over 95 percent of migrant farmworkers in the United States are Latino, African American, Afro-Caribbean, or Asian, and African Americans are overrepresented in high-risk, blue-collar, and service occupations for which a large pool of replacement labor exists. Thus, they are twice as likely to be unemployed as their white counterparts. Fear of unemployment acts as a potent incentive for many African-American workers to accept and keep jobs they know are health threatening. Workers will tell you that "unemployment and poverty are also hazardous to one's health." An inherent conflict exists between the interests of capital and that of labor. Employers have the power to move jobs (and industrial hazards) from the Northeast and Midwest to the South and Sunbelt, or they may move the jobs offshore to Third-World countries where labor is even cheaper and where there are even fewer health and safety regulations. Yet, unless an environmental movement emerges that is capable of addressing these economic concerns, people of color and poor white workers are likely to end up siding with corporate managers in key conflicts concerning the environment.

Indeed, many labor unions already moderate their demands for improved work-safety and pollution control whenever the economy is depressed. They are afraid of layoffs, plant closings, and the relocation of industries. These fears and anxieties of labor are usually built on the false but understandable assumption that environmental regulations inevitably lead to job loss.[23]

The crux of the problem is that the mainstream environmental movement has not sufficiently addressed the fact that social inequality and imbalances of social power are at the heart of environmental degradation, resource depletion, pollution, and even overpopulation. The environmental crisis can simply not be solved effectively without social justice. As one academic human ecologist notes, "Whenever [an] in-group directly and exclusively benefits from its own overuse of a shared resource but the costs of that overuse are 'shared' by out-groups, then in-group motivation toward a policy of resource conservation (or sustained yields of harvesting) is undermined."[24]

THE MOVEMENT FOR ENVIRONMENTAL JUSTICE

Activists of color have begun to challenge both the industrial polluters and the often indifferent mainstream environmental movement by actively fighting environmental threats in their communities and raising the call for environmental justice. This groundswell of environmental activism in African-American, Latino, Asian, Pacific Islander, and Native-American communities is emerging all across the country. While rarely listed in the standard environmental and conservation directories, grass-roots environmental justice groups have sprung up from Maine to Louisiana and Alaska.

These grass-roots groups have organized themselves around waste-facility siting, lead contamination, pesticides, water and air pollution, native self-government, nuclear testing, and workplace safety.[25] People of color have invented and, in other cases, adapted existing organizations to meet the disproportionate environmental challenges they face. A growing number of grass-roots groups and their leaders have adopted confrontational direct action strategies similar to those used in earlier civil rights conflicts. Moreover, the increasing documentation of environmental racism has strengthened the demand for a safe and healthy environment as a basic right of all individuals and communities.[26]

Drawing together the insights of *both* the civil rights and the environmental movements, these grass-roots groups are fighting hard to improve the quality of life for their residents. As a result of their efforts, the environmental justice movement is increasingly influencing and winning support from more conventional environmental and civil rights organizations. For example, the National Urban League's *1992 State of Black America* included—for the first time in the seventeen years the report has been published—a chapter on the environmental threats to the African-American community.[27] In addition, the NAACP, ACLU, and NRDC led the fight to have poor children tested for lead poisoning under Medicaid provisions in California. The class-action lawsuit *Matthews v. Coye*, settled in 1991, called for the state of California to screen an estimated 500,000 poor children for lead poisoning at a cost of $15 to $20 million.[28] The screening represents a big step forward in efforts to identify children suffering from what federal authorities admit is the number one environmental health problem of children in the United States. For their part, mainstream environmental organizations are also beginning to understand the need for environmental justice and are increasingly supporting grassroots groups in the form of technical advice, expert testimony, direct financial assistance, fundraising, research, and legal assistance. Even the Los Angeles chapter of the wilderness-focused Earth First! movement worked with community groups to help block the incinerator project in south-central Los Angeles. . . .

CONCLUSION

The mainstream environmental movement has proven that it can help enhance the quality of life in this country. The national membership organizations that make up the mainstream movement have clearly played an important role in shaping the nation's environmental policy. Yet, few of these groups have actively involved themselves in environmental conflicts involving communities of color. Because of this, it's unlikely that we will see a mass influx of people of color into the national environmental groups any time soon. A continuing growth in their own grassroots organizations is more likely. Indeed, the fastest growing segment of the environmental movement is made up by the grass-roots groups in communities of color which are increasingly linking up with one another and with other community-based groups. As long as U.S. society remains divided into separate and unequal communities, such groups will continue to serve a positive function.

It is not surprising that indigenous leaders are organizing the most effective resistance within communities of color. They have the advantage of being close to the population immediately affected by the disputes they are attempting to resolve. They are also completely wedded to social and economic justice agendas and familiar with the tactics of the civil rights movement. This makes effective community organizing possible. People of color have a long track record in challenging government and corporations that discriminate. Groups that emphasize civil rights and social justice can be found in almost every major city in the country.

Cooperation between the two major wings of the environmental movement is both possible and beneficial, however. Many environmental activists of color are now getting support from mainstream organizations in the form of technical advice, expert testimony, direct financial assistance, fundraising, research, and legal assistance. In return, increasing numbers of people of color are assisting mainstream organizations to redefine their limited environmental agendas and expand their outreach by serving on boards, staffs, and advisory councils. Grass-roots activists have thus been the most influential activists in placing equity and social justice issues onto the larger environmental agenda and democratizing and diversifying the movement as a whole. Such changes are necessary if the environmental movement is to successfully help spearhead a truly global movement for a just, sustainable, and healthy society and effectively resolve pressing environmental disputes. Environmentalists and civil rights activists of all stripes should welcome the growing movement of African Americans, Latinos, Asians, Pacific Islanders, and Native Americans who are taking up the struggle for environmental justice.

Notes

1. Stokely Carmichael and Charles V. Hamilton, *Black Power: The Politics of Liberation in America* (New York: Vintage, 1967), 16–17.
2. Robert Blauner, *Racial Oppression in America* (New York: Harper and Row, 1972), 26.
3. Michael Omi and Howard Winant, *Racial Formation in the United States: From the 1960's to the 1980's* (New York: Routledge and Kegan Paul, 1986), 76–77.
4. Jane Kay, "Fighting Toxic Racism: L.A.'s Minority Neighborhood is the 'Dirtiest' in the State." *San Francisco Examiner*, 7 Apr. 1991, p. A1.
5. Paul Ong and Evelyn Blumenberg, "Race and Environmentalism" (Paper read at Graduate School of Architecture and Urban Planning, 14 Mar. 1990, UCLA).
6. D. R. Wernette and L. A. Nieves, "Breathing Polluted Air," *EPA Journal* 18 (Mar./Apr. 1992): 16–17.

7. Joe R. Feagin and Clairece B. Feagin, *Discrimination American Style: Institutional Racism and Sexism* (Malabar, Fla.: Robert E. Krieger, 1986); Mark Gottdiener and Joe R. Feagin, "The Paradigm Shift in Urban Sociology," *Urban Affairs Quarterly* 24, no. 2 (Dec. 1988): 163–87.
8. Cerrell Associates, Inc., *Political Difficulties Facing Waste-to-Energy Conversion Plant Siting*, California Waste Management Board, Technical Information Series (Los Angeles: Cerrell Associates, 1984), 43.
9. Center for Investigative Reporting and Bill Moyers, *Global Dumping Grounds: The International Trade in Hazardous Waste* (Washington, D.C.: Seven Locks Press, 1990); Roberto Sanchez, "Health and Environmental Risks of the Maquiladora in Mexicali," *Natural Resources Journal* 30 (Winter 1990): 163–86; Jo Ann Zuniga, "Watchdog Keeps Tabs on Politics of Environment Along Border," *Houston Chronicle*, 24 May 1992, p. 22A.
10. Matthew Witt, "An Injury to One Is a Gravio a Todo: The Need for a Mexico-U.S. Health and Safety Movement," *New Solutions: A Journal of Environmental and Occupational Health Policy* 1 (Mar. 1991): 28–33.
11. Working Group on Canada-Mexico Free Trade, "Que Pasa? A Canada-Mexico 'Free' Trade Deal," *New Solutions: A Journal of Environmental and Occupational Health Policy* 2 (January 1991): 10–25.
12. Greenpeace, *The International Trade in Wastes: A Greenpeace Inventory* (Washington, D.C.: Greenpeace, USA, 1990), 3.
13. Bunyan Bryant and Paul Mohai, *Race and the Incidence of Environmental Hazards* (Boulder, Colo.: Westview Press, 1992).
14. Myrick A. Freeman, "The Distribution of Environmental Quality," in *Environmental Quality Analysis*, ed. Allen V. Kneese and Blair T. Bower (Baltimore: Johns Hopkins University Press for Resources for the Future, 1971); Michael Gelobter, "The Distribution of Air Pollution by Income and Race" (Paper presented at the Second Symposium on Social Science in Resource Management, Urbana Ill., June 1988); Leonard Gianessi, H. M. Peskin, and E. Wolff, "The Distributional Effects of Uniform Air Pollution Policy in the U.S.," *Quarterly Journal of Economics* (May 1979): 281–301; Wernette and Nieves, "Breathing Polluted Air," 16–17.
15. Pat C. West, M. Fly, and R. Marans, "Minority Anglers and Toxic Fish Consumption: Evidence From a State-Wide Survey of Michigan," *Proceedings of the Michigan Conference on Race and the Incidence of Environmental Hazards*, ed. Bunyan Bryant and Paul Mohai (Ann Arbor: University of Michigan School of Natural Resources, 1989), 108–22.
16. Robert D. Bullard, "Solid Waste Sites and the Black Houston Community," *Sociological Inquiry* 53 (Spring 1983): 273–88; Robert D. Bullard, *Invisible Houston: The Black Experience in Boom and Bust* (College Station: Texas A&M University Press, 1987); Robert D. Bullard, *Dumping in Dixie: Race, Class and Environmental Quality* (Boulder, Colo.: Westview Press, 1990); Robert D. Bullard, "Environmental Justice for All," *EnviroAction* 9 (Nov. 1991).
17. United Church of Christ Commission for Racial Justice, *The First National People of Color Environmental Leadership Summit: Program Guide* (New York: United Church of Christ, 1992).
18. Agency for Toxic Substances and Disease Registry, *The Nature and Extent of Lead Poisoning in Children in the United States: A Reprint to Congress* (Atlanta: U.S. Department of Health and Human Services, 1988).
19. Ibid., 1–12.
20. Nancy A. Denton and Douglas S. Massey, "Residential Segregation of Blacks, Hispanics, and Asians by Socioeconomic Class and Generation," *Social Science*

Quarterly 69 (1988): 797–817; Gerald D. Jaynes and Robin M. Williams, Jr., *A Common Destiny: Blacks and American Society* (Washington, D.C.: National Academy Press, 1989).

21. Kenneth M. Bachrach and Alex J. Zautra, "Coping with Community Stress: The Threat of a Hazardous Waste Landfill," *Journal of Health and Social Behavior* 26 (June 1985): 127–41; Robert D. Bullard, *Dumping in Dixie*; Robert D. Bullard and Beverly H. Wright, "Blacks and the Environment," *Humboldt Journal of Social Relations* 14 (1987): 165–84; Frederick Buttel and William L. Flinn, "The Structure and Support for the Environmental Movement 1968–70," *Rural Sociology* 39 (1974): 56–69; Riley E. Dunlap, "Public Opinion on the Environment in the Reagan Era: Polls, Pollution, and Politics," *Environment* 29 (1987): 6–11, 31–37; Paul Mohai, "Public Concern and Elite Involvement in Environmental Conservation," *Social Science Quarterly* 66 (Dec. 1985): 820–38; Paul Mohai, "Black Environmentalism," *Social Science Quarterly* 71 (Apr. 1990): 744–65; Denton E. Morrison, "The Soft Cutting Edge of Environmentalism: Why and How the Appropriate Technology Notion is Changing the Movement," *Natural Resources Journal* 20 (Apr. 1980): 275–98; Denton E. Morrison, "How and Why Environmental Consciousness has Trickled Down," in *Distributional Conflict in Environmental Resource Policy*, ed. Allan Schnaiberg, Nicholas Watts, and Klaus Zimmermann, 187–220 (New York: St. Martin's Press, 1986).

22. Bryant and Mohai, *Race and the Incidence of Environmental Hazards*; Bullard, *Dumping in Dixie*; Center for Investigative Reporting and Moyers, *Global Dumping Grounds*.

23. Michael H. Brown, *Laying Waste: The Poisoning of America by Toxic Chemicals* (New York: Pantheon Books, 1980); Michael H. Brown, *The Toxic Cloud: The Poisoning of America's Air* (New York: Harper and Row, 1987).

24. William Catton, *Overshoot: The Ecological Basis of Revolutionary Change* (Chicago: University of Illinois Press, 1982).

25. Dana Alston, *We Speak for Ourselves: Social Justice, Race, and Environment* (Washington, D.C.: Panos Institute, 1990); Bryant and Mohai, *Race and the Incidence of Environmental Hazards*; Bullard, *Dumping in Dixie*; Robert D. Bullard, *Directory of People of Color Environmental Groups 1992* (Riverside: University of California, Department of Sociology, 1992); Catton, *Overshoot*.

26. Bullard and Wright, "Blacks and the Environment," 165–84; Robert D. Bullard and Beverly H. Wright, "The Quest for Environmental Equity: Mobilizing the African American Community for Social Change," *Society and Natural Resources* 3 (1991): 301–11; United Church of Christ Commission for Racial Justice, *People of Color Environmental Leadership Summit*.

27. Robert D. Bullard, "Urban Infrastructure: Social, Environmental, and Health Risks to African Americans," in *The State of Black America 1992*, ed. Billy J. Tidwell (New York: National Urban League, 1992), 183–96.

28. Bill Lann Lee, "Environmental Litigation on Behalf of Poor, Minority Children: *Matthews v. Coye*: A Case Study" (Paper presented at the annual meeting of the American Association for the Advancement of Science, Chicago, Apr. 1992).

From Resistance to Regeneration

WINONA LaDUKE

NATURAL LAW IS THE preeminent law, the one we are all accountable to. It is our belief that all societies and individuals are accountable to natural law. We have a code of ethics and a way of living on this land which is based on being accountable to that law. That is the understanding of most indigenous peoples.

Our way of living within natural law has to do with concepts which are indigenous values. In my community we have a term, *minobimaatisiiwin*, which means "the good life." It's alternately translated as "continuous rebirth" and is an essential tenet in the value system which we strive to as individuals and collectively as communities. In order to do that we have a few concepts that keep us in synch with *minobimaatisiiwin* and with natural law.

We have an understanding that time is cyclical. The seasons, the moons, the tides, lives, life, women—all of these things are in cycles. Another concept is called "reciprocity." This means that when I harvest wild rice on my reservation in northern Minnesota or when we hunt deer, I always pray, offer tobacco, offer *saymah*, because I understand in my language all of those things are animate. The deer, the plants, even *asin*, a stone, is animate. All things have standing on their own. I always give thanks for them giving me their life, because I understand that I am totally reliant upon them to continue in my *minobimaatisiiwin*, my way of life on this land. In order to do that you must always give when you take, and when you take you must only take what you need and leave the rest. If you take more than you need, you are out of order with the natural law. That is our way of understanding it.

Indigenous values are values of social societies and peoples who have

From: *The Nonviolent Activist* (September–October 1992): 3–6.

continuously inhabited this land in what I would suggest are sustainable communities, a sustainable way of life, totally integrated into ecosystems. I want to contrast that with industrial thinking.

I went to school, all the way through school, in this country. Industrial thinking is very different from indigenous thinking. For example, when I was taught time in this country, I was taught time with a timeline that began in 1492. Things advanced along this timeline and "progressed." It is my perception that there is a whole set of values associated with the timeline, the idea, for example, of progress, which is defined by indices like economic growth and technological advancement. Progress is something that "you want to have" or is desirable on this timeline. There are other values such as a perception that people are primitive and become civilized or that the wild becomes tamed or cultivated. This is an industrial way of thinking.

The second concept is the idea of capitalism, where labor and capital and resources are put together for the purpose of accumulation. That's the most basic formula for capitalism. The less labor and capital resources you put together and the more you accumulated, the better capitalist you were. I would suggest that in its essence this idea of capitalism means that you inherently and structurally always take more than you need and do not leave the rest. Because of that, capitalism is chronically out of order with natural law. However, the problem is not just with capitalism; it has to do with industrialism, and it's broader.

RETHINKING NORTH AMERICAN GEO-POLITICS

There are over seven hundred indigenous peoples in North America. There are over eighty indigenous nations in California alone, over two hundred in Alaska. Over fifty million indigenous peoples live in the world's rain forests. There are not just toucans and other birds. There are people living there.

In Canada 80 percent of the population lives within one hundred miles of the U.S.-Canadian border. That's basically white folks. When you go north of the fiftieth parallel, around Edmonton, the majority population is native. In the upper two-thirds of Canada the majority population is native. When we look to a new vision of where we need to go for the next five hundred years, indigenous people are essential to what is going on in terms of the environmental crisis in the world. If we are going to change things, to reclaim our lives and save our mother, it has to be a partnership between indigenous peoples and other peoples.

The experience of colonialism in terms of energy development has been devastating in our community. In the United States today we have about 4 percent of our land left as reservation. Today two-thirds of the uranium resources and one-third of all western low-sulphur coal are on our lands. In the late 1970s four of the ten largest coal strip mines in the country were on Indian lands, and 100 percent of all federally controlled uranium production came from Indian reservations. We are the people who have the single largest hydroelectric project on the continent slated for our land at James Bay in Canada.

We are central and essential to the North American industrial development plan—and we are at the end of it. We have the dubious honor of having had over one hundred separate proposals for toxic waste dumps forwarded to our reservations, a result of people in urban working-class communities organizing successfully to keep those dumps out of their communities. The proposed sites moved to Third-World countries and Indian country. Presently fifteen of the eighteen grants for nuclear waste research are on Indian reservations. In order to challenge that, to change that, there has to be a meaningful relationship between indigenous people and people of conscience in this country.

The economic and social devastation caused by energy development in our communities is widespread. Working in the Navajo nation in the Southwest brought a lot of those issues home to me. The Navajo way of life is based on the land. Huge mining and development projects bring a war on that way of life. People are forced from land-based economies into cash economies. They become marginal parts of the dominant culture, not only wage slaves to somebody else's society, but producing things for someone else's society with no benefit to themselves. In the Southwest in the late 1970s and early 1980s, on land on or adjacent to the Navajo reservation, there were forty-two operating uranium mines, ten uranium mills, five coal fired power plants, and four coal strip mines. In the meantime, 85 percent of Navajo households had no electricity. In one year the Navajo nation produced enough electricity to fuel the needs of New Mexico for thirty-two years. The Navajo nation has been and continues to be a colony of the southwestern United States, providing power for the entire region.

The impact on the Navajo community has been not only social transformation, but also widespread radioactive contamination. Miners drank water from the bottom of uranium mines and the government knew full well the miners would contract lung cancer. Ventilation shafts did not operate correctly in the mines and workers breathed air contaminated by radon. Kerr-McGee sent native miners down there under contract with the Atomic Energy Commission and the federal government did studies on them. By the mid-1970s most of the Navajo uranium miners had died from lung cancer. Women who washed miners' clothes developed skin cancer. It is no coincidence that the Shiprock Indian Health Service records a birth defect rate *eight times* the national average.

In the late 1970s and early 1980s the Oil Chemical and Atomic Workers Union went down there and organized to clean up the mines. You know what they do when they clean up a mine? They run a fan and blow the air from the mine out into the community. That's all that happens. There is no safe way to mine uranium.

TAKING STOCK OF NORTH AMERICAN ENERGY POLICY

The antinuclear movement in this country had a number of successes. The nuclear industry wanted to have one thousand nuclear power plants in the United States by the year 2000. There are 120 today. We beat 880 nuclear power plants. For instance, if you go to the visitor's center at Seabrook (New Hampshire) there is a picture of two shiny white domes of a nuclear power plant. But if you look at the plant, one dome is all white and one dome is kind of reddish. It's rusted. They never finished it.

We've had some successes and some victories. We stopped most uranium mining in the United States and a host of other energy projects and dumps. The problem is that we didn't change the consumption of electricity, so the industry shifted from nukes to coal and to hydroelectric power. That's why the James Bay dams in northern Canada were built and new ones proposed. Forty days after the nuclear power plant at Shoreham (Long Island, New York) was canceled, Governor Cuomo signed the contract for James Bay 1. James Bay is at the bottom of Hudson Bay, in Quebec, Ontario, and Manitoba. James Bay 1 flooded out and devastated an area of about seventy-five thousand square kilometers destroying four major rivers. Cree and Inuit peoples live there. For years the only white people they saw were perhaps a missionary, a teacher, a doctor every once in a while. In the 1970s all of a sudden Hydro Quebec appeared and said, "These rivers are ours." The company began construction of a huge set of dams to produce ten thousand megawatts of power to ship down to the States. Now they want to do James Bay 2, the sequel. If they go ahead with all the dams, 356,000 square kilometers of land, an area the size of New England, will be devastated.

James Bay is perhaps the most destructive hydroelectric project in North America, but similar dams are in place or underway spanning the entire north of Canada. In Manitoba a set of dams by Manitoba Hydro on the Hudson Bay is virtually as destructive. Those dams and those in Alberta, British Columbia, and elsewhere are constructed mostly for American export contracts. In British Columbia, for instance, the power is contracted to utilities as far south as San Diego.

RESISTANCE

The James Bay campaign did have a great victory. The single largest piece of financing for James Bay was the New York contract, and Governor Cuomo cancelled that contract in March 1992 because people organized and waged a campaign against the energy plan. But Hydro Quebec has dug in. They are recalcitrant. Do not confuse the Francophone right to sovereignty and cultural integrity with Quebec's right to destroy the north. Their position is that the Crees are making them look bad in the United States. The head of the major labor union called the Crees "social parasites." It's an Indian-hating government. The majority of the financing is still coming from the States. Shearson Lehman, the American Express card people, is the single largest underwriter, and American pension plans pick up a good portion of the remaining financing. To beat that project we've got to cut the money because it's way too capital intensive for them to be able to finance it long term. We also need to look at a new North American energy plan.

The Innu in Labrador beat a NATO base a couple of years ago. The Innu live in an area about the size of France where NATO wanted to site a huge base. The Innu did everything possible including civil disobedience on the runway. Authorities kept arresting them, hauling them into court. The court finally agreed that they couldn't be trespassing on the runway, because they actually believe it's their territory. After ten years, NATO decided not to site the base in Labrador. For indigenous people their survival is at stake. The Innu had no choice but to try to stop the NATO base.

There's another story, that of Milton Born with a Tooth. He's a Peiqan (or Blackfeet) from the Blood Reserve in Alberta, Canada, just north of the Blackfeet Reservation in the States. In the 1920s a river called Old Man River was diverted off the reserve (which is what the government usually does to Indians, takes rivers and moves them off the reserves and leaves the Indians high and dry). It was diverted for some southern Alberta farmers. In the early 1980s a new project was proposed, the Old Man Dam project. The environmentalists and the Peiqan fought it. They were in court for many years. Finally the court ruled that an environmental impact assessment on the dam was required. But Canada is very draconian in their environmental policies: they didn't require the dam construction to stop during the environmental review.

The environmentalists and the Indians litigated, used everything they could in the court system, and in the end they had no recourse. So Milton Born with a Tooth and the Lonefighter Society did something that basically no one else would have done. They borrowed some gas, cars, and bulldozers and went to the site of the first diversion of the river. They broke down the dam from the 1920s to send Old Man River back to its original riverbed through the reserve. About a year and a half ago two-thirds of the river was running through the old riverbed. It left the new dam site high and dry. In response the Royal Canadian Mounted Police and their SWAT team engaged in a full assault on the Indians and arrested them. Milton became a big folk hero. It has prolonged the whole struggle, but the Indians haven't prevailed yet.

VISIONS

It is time to change from a society based on conquest to a society based on survival. There are no more frontiers and there is nothing left to conquer.

We must make the change if we are to survive, if we are to rebuild community and heal ourselves. And we must articulate a vision. It's hard to get people to change if they don't have an idea of what we want to change to. The fight over the spotted owl in the Northwest has pitted environmentalists against working people. Workers are saying they'll be left starving in the dark. We have to articulate a vision of a new society based on decentralized energy production, on deindustrializing the society, on using things like the peace dividend, transforming the society from a war economy into a peace economy.

I understand how power is distributed: on race, class, sex, etc. If you look at a lot of the social movement thinkers in the country, they are white men. It is possible that they have a perception of a vested interest in the society, so it is more difficult for them to make the deeper change that needs to be made. People of color and indigenous people who have been marginalized historically and to this day from their piece of the pie (or indigenous people who say that we don't even want the pie) have some views which may offer leadership for the kind of structural change that we need to make. We need to expand the vision of the social movements in this country to allow for that voice to come forth.

Indigenous peoples need industrial society and white folks to change society, and those groups need indigenous peoples. American society has no concept of how to survive. Ours is the only experience of continuous habitation of this land. The only people who have a concept of how to survive are indigenous peoples, and we cannot be replicated or produced.

People who are environmentalists in this country have gone through a major transformation. How people look at the environment is very much how people look at indigenous or Indian people. In thinking about the environment, for the most part people feel guilty, but guilt does nobody any good. Environmentalists have begun the process of overcoming that guilt and empowering themselves to make change. Those folks have the capacity to do the same thing with indigenous peoples, because all people do is to feel guilty when they think of Indians. . . .

I'm positive that we can make that change, but it requires a lot of work and organizing. I don't have anything better to do with the rest of my life than to figure out how to change that. One of the reasons Americans can't change things sometimes is because they have an idea of Armageddon, a perception that in the end we're going to fry anyway, so why bother changing? I'm sure it has to do with linear thinking. In my experience most indigenous peoples don't have a concept of Armageddon but of earth renewal, a transformation, a constant process of change. *Minobimaatisiiwin*. Continuous rebirth. I'm sure we can make that change, and I want to be part of it.

Development, Ecology, and Women

VANDANA SHIVA

DEVELOPMENT AS A NEW PROJECT OF WESTERN PATRIARCHY

"DEVELOPMENT" WAS TO HAVE been a postcolonial project, a choice for accepting a model of progress in which the entire world remade itself on the model of the colonizing modern West, without having to undergo the subjugation and exploitation that colonialism entailed. The assumption was that Western-style progress was possible for all. Development, as the improved well-being of all, was thus equated with the Westernization of economic categories—of needs, of productivity, of growth. Concepts and categories about economic development and natural resource utilization that had emerged in the specific context of industrialization and capitalist growth in a center of colonial power, were raised to the level of universal assumptions and applicability in the entirely different context of basic needs satisfaction for the people of the newly independent Third-World countries. Yet, as Rosa Luxemburg has pointed out, early industrial development in Western Europe necessitated the permanent occupation of the colonies by the colonial powers and the destruction of the local "natural economy."[1] According to her, colonialism is a constantly necessary condition for capitalist growth: without colonies, capital accumulation would grind to a halt. "Development" as capital accumulation and the commercialization of the economy for the generation of "surplus" and profits thus involved the reproduction not merely of a particular form of creation of wealth, but also of the associated creation of poverty and dispossession. A replication of economic development based on commercialization of resource use for com-

---

From: *Staying Alive: Women, Ecology, and Development* (London: Zed Books, 1988), 1–9, 13.

modity production in the newly independent countries created the internal colonies.[2] Development was thus reduced to a continuation of the process of colonization; it became an extension of the project of wealth creation in modern Western patriarchy's economic vision, which was based on the exploitation or exclusion of women (of the West and non-West), on the exploitation and degradation of nature, and on the exploitation and erosion of other cultures. "Development" could not but entail destruction for women, nature, and subjugated cultures, which is why, throughout the Third World, women, peasants, and tribals are struggling for liberation from development just as they earlier struggled for liberation from colonialism.

The U.N. Decade for Women was based on the assumption that the improvement of women's economic position would automatically flow from an expansion and diffusion of the development process. Yet, by the end of the decade, it was becoming clear that development itself was the problem. Insufficient and inadequate "participation" in "development" was not the cause for women's increasing underdevelopment; it was rather, their enforced but asymmetric participation in it, by which they bore the costs but were excluded from the benefits, that was responsible. Development exclusivity and dispossession aggravated and deepened the colonial processes of ecological degradation and the loss of political control over nature's sustenance base. Economic growth was a new colonialism, draining resources away from those who needed them most. The discontinuity lay in the fact that it was now new national elites, not colonial powers, that masterminded the exploitation on grounds of "national interest" and growing GNPs, and it was accomplished with more powerful technologies of appropriation and destruction.

Ester Boserup[3] has documented how women's impoverishment increased during colonial rule; those rulers who had spent a few centuries in subjugating and crippling their own women into de-skilled, de-intellectualized appendages, disfavored the women of the colonies on matters of access to land, technology, and employment. The economic and political processes of colonial underdevelopment bore the clear mark of modern Western patriarchy, and while large numbers of women and men were impoverished by these processes, women tended to lose more. The privatization of land for revenue generation displaced women more critically, eroding their traditional land-use rights. The expansion of cash crops undermined food production, and women were often left with meager resources to feed and care for children, the aged and the infirm, when men migrated or were conscripted into forced labor by the colonizers. As a collective document by women activists, organizers and researchers stated at the end of the U.N. Decade for Women, "The almost uniform conclusion of the Decade's research is that with a few exceptions, women's relative access to economic resources, incomes and employment has worsened, their burden of work has increased, and their relative and even absolute health, nutritional and educational status has declined.[4]

The displacement of women from productive activity by the expansion of development was rooted largely in the manner in which development projects appropriated or destroyed the natural resource base for the production of sustenance and survival. It destroyed women's productivity both by removing land, water, and forests from their management and control, as well as through the ecological destruction of soil, water, and vegetation systems so that nature's productivity and renewability were impaired. While gender subordination and patriarchy are the oldest of oppressions, they have taken on new and more violent forms through the project of development. Patriarchal categories which understand destruction as "production" and regeneration of life as "passivity" have generated a crisis of survival. Passivity, as an assumed category of the "nature" of nature and of women, denies the activity of nature and life. Fragmentation and uniformity as assumed categories of progress and development destroy the living forces which arise from relationships within the "web of life" and the diversity in the elements and patterns of these relationships.

The economic biases and values against nature, women, and indigenous peoples are captured in this typical analysis of the "unproductiveness" of traditional natural societies: "Production is achieved through human and animal, rather than mechanical, power. Most agriculture is unproductive; human or animal manure may be used but chemical fertilisers and pesticides are unknown. . . . For the masses, these conditions mean poverty."[5]

The assumptions are evident: nature is unproductive; organic agriculture based on nature's cycles of renewability spells poverty; women and tribal and peasant societies embedded in nature are similarly unproductive, not because it has been demonstrated that in cooperation they produce *less* goods and services for needs, but because it is assumed that "production" takes place only when mediated by technologies for commodity production, even when such technologies destroy life. A stable and clean river is not a productive resource in this view: it needs to be "developed" with dams in order to become so. Women, sharing the river as a commons to satisfy the water needs of their families and society, are not involved in productive labor: when replaced by the engineering man, water management and water use become productive activities. Natural forests remain unproductive till they are developed into monoculture plantations of commercial species. Development, thus, is equivalent to maldevelopment, a development bereft of the feminine, the conservation, the ecological principle. The neglect of nature's work in renewing herself and women's work in producing sustenance in the form of basic, vital needs is an essential part of the paradigm of maldevelopment, which sees all work that does not produce profits and capital as non- or unproductive work. As Maria Mies[6] has pointed out, this concept of surplus has a patriarchal bias because, from the point of view of nature and women, it is not based on material surplus produced *over and above* the requirements of the community: it is stolen and appropriated through violent modes from nature (who needs a share of her produce to reproduce herself) and from women (who need a share of nature's produce to produce sustenance and ensure survival).

From the perspective of Third-World women, productivity is a measure of producing life and sustenance; that this kind of productivity has been rendered invisible does not reduce its centrality to survival—it merely reflects the domination of modern patriarchal economic categories which see only profits, not life.

MALDEVELOPMENT AS THE DEATH OF THE FEMININE PRINCIPLE

In this analysis, maldevelopment becomes a new source of male-female inequality. "Modernization" has been associated with the introduction of new forms of dominance. Alice Schlegel[7] has shown that under conditions of subsistence, the interdependence and complementarity of the separate male and female domains of work is the characteristic mode, based on diversity, not inequality. Maldevelopment militates against this equality in diversity, and superimposes the ideologically constructed category of Western technological man as a uniform measure of the worth of classes, cultures, and genders. Dominant modes of perception based on reductionism, duality, and linearity are unable to cope with equality in diversity, with forms and activities that are significant and valid, even though different. The reductionist mind superimposes the roles and forms of power of Western male-oriented concepts on women, all non-Western peoples, and even on nature, rendering all three "deficient," and in need of "development." Diversity, and unity and harmony in diversity, become epistemologically unattainable in the context of maldevelopment, which then becomes synonymous with women's underdevelopment (increasing sexist domination), and nature's depletion (deepening ecological crises). Commodities have grown, but nature has shrunk. The poverty crisis of the South arises from the growing scarcity of water, food, fodder, and fuel, associated with increasing maldevelopment and ecological destruction. This poverty crisis touches women most severely, first because they are the poorest among the poor, and then because, with nature, they are the primary sustainers of society.

Maldevelopment is the violation of the integrity of organic, interconnected, and interdependent systems, that sets in motion a process of exploitation, inequality, injustice, and violence. It is blind to the fact that a recognition of nature's harmony and action to maintain it are preconditions for distributive justice. This is why Mahatma Gandhi said, "There is enough in the world for everyone's need, but not for some people's greed."

Maldevelopment is maldevelopment in thought and action. In practice, this fragmented, reductionist, dualist perspective violates the integrity and harmony of man in nature, and the harmony between men and women. It ruptures the cooperative unity of masculine and feminine, and places man, shorn of the feminine principle, above nature and women, and separated from both. The violence to nature as symptomatized by the ecological crisis, and the violence to women, as symptomatized by their subjugation and exploitation arise from this subjugation of the feminine principle. I want to argue that what is currently called development is essentially maldevelopment, based on the introduction or accentuation of the domination of man over nature and women. In it, both are viewed as the "other," the passive non-self. Activity, productivity, creativity which were associated with the feminine principle are expropriated as qualities of nature and women, and transformed into the exclusive qualities of man. Nature and women are turned into passive objects, to be used and exploited for the uncontrolled and uncontrollable desires of alienated man. From being the creators and sustainers of life, nature and women are reduced to being "resources" in the fragmented, antilife model of maldevelopment.

TWO KINDS OF GROWTH, TWO KINDS OF PRODUCTIVITY

Maldevelopment is usually called "economic growth," measured by the Gross National Product. Jonathon Porritt, a leading ecologist, has this to say of GNP:

> Gross National Product—for once a word is being used correctly. Even conventional economists admit that the hey-day of GNP is over, for the simple reason that as a measure of progress, it's more or less useless. GNP measures the lot, all the goods and services produced in the money economy. Many of these goods and services are not beneficial to people, but rather a measure of just how much is going wrong; increased spending on crime, on pollution, on the many human casualties of our society, increased spending because of waste or planned obsolescence, increased spending because of growing bureaucracies. It's all counted. . . .[8]

The problem with GNP is that it measures some costs as benefits (e.g., pollution control) and fails to measure other costs completely. Among these hidden costs are the new burdens created by ecological devastation, costs that are invariably heavier for women, both in the North and South. It is hardly surprising, therefore, that as GNP rises, it does not necessarily mean that either wealth or welfare increase proportionately. I would argue that GNP is becoming, increasingly, a measure of how real wealth—the wealth of nature and that produced by women for sustaining life—is rapidly decreasing. When commodity production as the prime economic activity is introduced as development, it destroys the potential of nature and women to produce life and goods and services for basic needs. More commodities and more cash mean less life—in nature (through ecological destruction) and in society (through denial of basic needs). Women are devalued first, because their work cooperates with nature's processes, and second, because work which satisfies needs and ensures sustenance is devalued in general. Precisely because more growth in maldevelopment has meant less sustenance of life and life-support systems, it is now imperative to recover the feminine principle as the basis for development which conserves and is ecological. Feminism as ecology, and ecology as the revival of *Prakriti*, the source of all life, become the decentered powers of political and economic transformation and restructuring.

This involves, first, a recognition that categories of "productivity" and growth which have been taken to be positive, progressive, and universal are, in reality, restricted patriarchal categories. When viewed from the point of view of nature's productivity and growth, and women's production of sustenance, they are found to be ecologically destructive and a source of gender inequality. It is no accident that the modern, efficient, and productive technologies created within the context of growth in market economic terms are associated with heavy ecological costs, borne largely by women. The resource and energy intensive production processes they give rise to demand ever increasing resource withdrawals from the ecosystem. These withdrawals disrupt essential ecological processes and convert renewable resources into nonrenewable ones. A forest, for example, provides inexhaustible supplies of diverse biomass over time if its capital stock is maintained and it is harvested on a sustained yield basis. The heavy and uncontrolled demand for industrial and commercial wood, however, requires the continuous overfelling of trees which exceeds the regenerative capacity of the forest ecosystem, and eventually converts the forests into nonrenewable resources. Women's work in the collection of water, fodder, and fuel is thus rendered more energy- and time-consuming. (In Garhwal, for example, I have seen women who originally collected fodder and fuel in a few hours now traveling long distances by truck to collect grass and leaves in a task that might take up to two days.) Sometimes the damage to nature's intrinsic regenerative capacity is impaired not by overexploitation of a particular resource but, indirectly, by damage caused to other related natural resources through ecological processes. Thus the excessive overfelling of trees in the catchment areas of streams and rivers destroys not only forest resources, but also renewable supplies of water, through hydrological destabilization.

Resource intensive industries disrupt essential ecological processes not only by their excessive demands for raw material, but by their pollution of air and water and soil. Often such destruction is caused by the resource demands of nonvital industrial products. In spite of severe ecological crises, this paradigm continues to operate because for the North and for the elites of the South, resources continue to be available, even now. The lack of recognition of nature's processes for survival *as factors in the process of economic development* shrouds the political issues arising from resource transfer and resource destruction, and creates an ideological weapon for increased control over natural resources in the conventionally employed notion of productivity. All other costs of the economic process consequently become invisible. The forces which contribute to the increased "productivity" of a modern farmer or factory worker, for instance, come from the increased use of natural resources. Amory Lovins has described this increase as the amount of "slave" labor presently at work in the world. According to him, each person on earth, on an average, possesses the equivalent of about fifty slaves, each working a forty-hour week. Man's global energy conversion from all sources (wood, fossil fuel, hydroelectric power, nuclear) is currently approximately 8×10^{12} watts. This is more than twenty times the energy content of the food necessary to feed the present world population at the FAO standard diet of 3,600 cal/day. The "productivity" of the western male compared to women or Third-World peasants is not intrinsically superior; it is based on inequalities in the distribution of this "slave" labor. The average inhabitant of the United States, for example, has 250 times more "slaves" than the average Nigerian. "If Americans were short of 249 of those 250 'slaves,' one wonders how efficient they would prove themselves to be?"[9]

It is these resource and energy intensive processes of production which divert resources away from survival, and hence from women. What patriarchy sees as productive work, is, in ecological terms, highly destructive production. The second law of thermodynamics predicts that resource intensive and resource wasteful economic development must become a threat to the survival of the human species in the long run. Political struggles based on ecology in industrially advanced countries are rooted in this conflict between *long-term survival options* and *short-term overproduction and overconsumption*. Political struggles of women, peasants, and tribals based on ecology in countries like India are far more acute and urgent since they are rooted in the *immediate threat to the options for survival* for the vast majority of the people, *posed by resource-intensive and resource-wasteful economic growth* for the benefit of a minority.

In the market economy, the organizing principle for natural resource use is the maximization of profits and capital accumulation. Nature and human needs are managed through market mechanisms. Demands for natural resources are restricted to those demands registering on the market; the ideology of development is in large part based on a vision of bringing all natural resources into the market economy for commodity production. When these resources are already being used by nature to maintain her production of renewable resources and by women for sustenance and livelihood, their diversion to the market economy generates a scarcity condition for ecological stability and creates new forms of poverty for women. . . .

The paradox and crisis of development arises from the mistaken identification of culturally perceived poverty with real material poverty, and the mistaken identification of the growth of commodity production as better satisfaction of basic needs. In actual fact, there is less water, less fertile soil, less genetic wealth as a result of the development process. Since these natural resources are the basis of nature's economy and women's survival economy, their scarcity is impoverishing women and marginalized peoples in an unprecedented manner. Their new impoverishment lies in the fact that resources which supported their survival were absorbed into the market economy while they themselves were excluded and displaced by it.

The old assumption that with the development process the availability of goods and services will automatically be increased and poverty will be removed, is now under serious challenge from women's ecology movements in the Third World, even while it continues to guide development thinking in centers of patriarchal power. Survival is based on the assumption of the sanctity of life; maldevelopment is based on the assumption of the sacredness of "development." Gustavo Esteva asserts that the sacredness of development has to be refuted because it threatens survival itself. "My people are tired of development," he says, "they just want to live."[10]

The recovery of the feminine principle allows a transcendence and transformation of these patriarchal foundations of maldevelopment. It allows a redefinition of growth and productivity as categories linked to the production, not the destruction, of life. It is thus simultaneously an ecological and a feminist political project which legitimizes the way of knowing and being that create wealth by enhancing life and diversity, and which delegitimizes the knowledge and practice of a culture of death as the basis for capital accumulation.

Notes

1. Rosa Luxemburg, *The Accumulation of Capital* (London: Routledge and Kegan Paul, 1951).
2. An elaboration of how "development" transfers resources from the poor to the well endowed is contained in Jayanta Bandyopadhyay and Vandana Shiva, "Political Economy of Technological Polarisations," *Economic and Political Weekly* 7, no. 45 (6 Nov. 1982): 1827–32; and Jayanta Bandyopadhyay and Vandana Shiva, "Political Economy of Ecology Movements," *Economic and Political Weekly* 23, no. 24 (11 Jun. 1988): 1223–32.
3. Ester Boserup, *Women's Role in Economic Development* (London: Allen and Unwin, 1970).
4. DAWN, *Development Crisis and Alternative Visions: Third World Women's Perspectives* (Bergen: Christian Michelsen Institute, 1985), 21.
5. M. George Foster, *Traditional Societies and Technological Change* (Delhi: Allied Publishers, 1973).
6. Maria Mies, *Patriarchy and Accumulation on a World Scale* (London: Zed Books, 1986).
7. Alice Schlegel, ed., *Sexual Stratification: A Cross-Cultural Study* (New York: Columbia University Press, 1977).

8. Jonathan Porritt, *Seeing Green* (Oxford: Blackwell, 1984), 121.

9. Amory Lovins, quoted in S. R. Eyre, *The Real Wealth of Nations* (London: Edward Arnold, 1978), 133.

10. Gustavo Esteva, remarks made at a Conference of the Society for International Development, Rome, 1985.

Radical Environmentalism:
A Third-World Critique

RAMACHANDRA GUHA

Even God dare not appear to the poor man except in the form of bread.
—Mahatma Gandhi

THE RESPECTED RADICAL JOURNALIST Kirkpatrick Sale recently celebrated "the passion of a new and growing movement that has become disenchanted with the environmental establishment and has in recent years mounted a serious and sweeping attack on it—style, substance, systems, sensibilities and all."[1] The vision of those whom Sale calls the "New Ecologists"—and what I refer to in this article as Deep Ecology—is a compelling one. Decrying the narrowly economic goals of mainstream environmentalism, this new movement aims at nothing less than a philosophical and cultural revolution in human attitudes toward nature. In contrast to the conventional lobbying efforts of environmental professionals based in Washington, it proposes a militant defense of "Mother Earth," an unflinching opposition to human attacks on undisturbed wilderness. With their goals ranging from the spiritual to the political, the adherents of Deep Ecology span a wide spectrum of the American environmental movement. As Sale correctly notes, this emerging strand has in a matter of a few years made its presence felt in a number of fields: from academic philosophy (as in the journal *Environmental Ethics*) to popular environmentalism (for example, the group Earth First!).

From: "Radical Environmentalism and Wilderness Preservation: A Third-World Critique," *Environmental Ethics* 11 (Spring 1989): 71–80; original title: "Radical Environmentalism and Wilderness Preservation: A Third World Critique."

In this essay I develop a critique of Deep Ecology from the perspective of a sympathetic outsider. I critique Deep Ecology not as a general (or even a foot soldier) in the continuing struggle between the ghosts of Gifford Pinchot and John Muir over control of the U.S. environmental movement, but as an outsider to these battles. I speak admittedly as a partisan, but of the environmental movement in India, a country with an ecological diversity comparable to the United States, but with a radically dissimilar cultural and social history.

My treatment of Deep Ecology is primarily historical and sociological, rather than philosophical, in nature. Specifically, I examine the cultural rootedness of a philosophy that likes to present itself in universalistic terms. I make two main arguments: first, that Deep Ecology is uniquely American, and despite superficial similarities in rhetorical style, the social and political goals of radical environmentalism in other cultural contexts (e.g., West Germany and India) are quite different; second, that the social consequences of putting Deep Ecology into practice on a worldwide basis (what its practitioners are aiming for) are very grave indeed.

THE TENETS OF DEEP ECOLOGY

While I am aware that the term "Deep Ecology" was coined by the Norwegian philosopher Arne Naess, this article refers specifically to the American variant.[2] Adherents of the deep ecological perspective in this country, while arguing intensely among themselves over its political and philosophical implications, share some fundamental premises about human-nature interactions. As I see it, the defining characteristics of Deep Ecology are fourfold:

First, Deep Ecology argues that the environmental movement must shift from an "anthropocentric" to a "biocentric" perspective. In many respects, an acceptance of the primacy of this distinction constitutes the litmus test of Deep Ecology. A considerable effort is expended by deep ecologists in showing that the dominant motif in Western philosophy has been anthropocentric—that is, the belief that man and his works are the center of the universe—and conversely, in identifying those lonely thinkers (Leopold, Thoreau, Muir, Aldous Huxley, Santayana, etc.) who, in assigning man a more humble place in the natural order, anticipated deep ecological thinking. In the political realm, meanwhile, establishment environmentalism (shallow ecology) is chided for casting its arguments in human-centered terms. Preserving nature, the deep ecologists say, has an intrinsic worth quite apart from any benefits preservation may convey to future human generations. The anthropocentric-biocentric distinction is accepted as axiomatic by deep ecologists, it structures their discourse, and much of the present discussion remains mired within it.

The second characteristic of Deep Ecology is its focus on the preservation of unspoiled wilderness—and the restoration of degraded areas to a more pristine condition—to the relative (and sometimes absolute) neglect of other issues on the environmental agenda. I later identify the cultural roots and portentous consequences of this obsession with wilderness. For the moment, let me indicate three distinct sources from which it springs. Historically, it represents a playing out of the preservationist (read *radical*) and utilitarian (read *reformist*) dichotomy that has plagued American environmentalism since the turn of the century. Morally, it is an imperative that follows from the biocentric perspective; other species of plants and animals, and nature itself, have an intrinsic right to exist. And finally, the preservation of wilderness also turns on a scientific argument—namely, the value of biological diversity in stabilizing ecological regimes and in retaining a gene pool for future generations. Truly radical policy proposals have been put forward by deep ecologists on the basis of these arguments. The influential poet Gary Snyder, for example, would like to see a 90 percent reduction in human populations to allow a restoration of pristine environments, while others have argued forcefully that a large portion of the globe must be immediately cordoned off from human beings.[3]

Third, there is a widespread invocation of Eastern spiritual traditions as forerunners of Deep Ecology. Deep Ecology, it is suggested, was practiced both by major religious traditions and at a more popular level by "primal" peoples in non-Western settings. This complements the search for an authentic lineage in Western thought. At one level, the task is to recover those dissenting voices within the Judeo-Christian tradition; at another, to suggest that religious traditions in other cultures are, in contrast, dominantly if not exclusively "biocentric" in their orientation. This coupling of (ancient) Eastern and (modern) ecological wisdom seemingly helps consolidate the claim that Deep Ecology is a philosophy of universal significance.

Fourth, deep ecologists, whatever their internal differences, share the belief that they are the "leading edge" of the environmental movement. As the polarity of the shallow/deep and anthropocentric/biocentric distinctions makes clear, they see themselves as the spiritual, philosophical, and political vanguard of American and world environmentalism.

TOWARD A CRITIQUE

Although I analyze each of these tenets independently, it is important to recognize, as deep ecologists are fond of remarking in reference to nature, the interconnectedness and unity of these individual themes.

1. Insofar as it has begun to act as a check on man's arrogance and ecological hubris, the transition from an anthropocentric (human-centered) to a biocentric (humans as only one element in the ecosystem) view in both religious and scientific traditions is only to be welcomed.[4] What is unacceptable are the radical conclusions drawn by Deep Ecology, in particular, that intervention in nature should be guided primarily by the need to preserve

biotic integrity rather than by the needs of humans. The latter for deep ecologists is anthropocentric, the former biocentric. This dichotomy is, however, of very little use in understanding the dynamics of environmental degradation. The two fundamental ecological problems facing the globe are (i) overconsumption by the industrialized world and by urban elites in the Third World and (ii) growing militarization, both in a short-term sense (i.e., ongoing regional wars) and in a long-term sense (i.e., the arms race and the prospect of nuclear annihilation). Neither of these problems has any tangible connection to the anthropocentric-biocentric distinction. Indeed, the agents of these processes would barely comprehend this philosophical dichotomy. The proximate causes of the ecologically wasteful characteristics of industrial society and of militarization are far more mundane: at an aggregate level, the dialectic of economic and political structures, and at a microlevel, the life-style choices of individuals. These causes cannot be reduced, whatever the level of analysis, to a deeper anthropocentric attitude toward nature; on the contrary, by constituting a grave threat to human survival, the ecological degradation they cause does not even serve the best interests of human beings! If my identification of the major dangers to the integrity of the natural world is correct, invoking the bogy of anthropocentrism is at best irrelevant and at worst a dangerous obfuscation.

2. If the above dichotomy is irrelevant, the emphasis on wilderness is positively harmful when applied to the Third World. If in the United States the preservationist/utilitarian division is seen as mirroring the conflict between "people" and "interests," in countries such as India the situation is very nearly the reverse. Because India is a long-settled and densely populated country in which agrarian populations have a finely balanced relationship with nature, the setting aside of wilderness areas has resulted in a direct transfer of resources from the poor to the rich. Thus, Project Tiger, a network of parks hailed by the international conservation community as an outstanding success, sharply posits the interests of the tiger against those of poor peasants living in and around the reserve. The designation of tiger reserves was made possible only by the physical displacement of existing villages and their inhabitants; their management requires the continuing exclusion of peasants and livestock. The initial impetus for setting up parks for the tiger and other large mammals such as the rhinoceros and elephant came from two social groups: first, a class of ex-hunters turned conservationists belonging mostly to the declining Indian feudal elite, and second, representatives of international agencies, such as the World Wildlife Fund (WWF) and the International Union for the Conservation of Nature and Natural Resources (IUCN), seeking to transplant the American system of national parks onto Indian soil. In no case have the needs of the local population been taken into account, and as in many parts of Africa, the designated wildlands are managed primarily for the benefit of rich tourists. Until very recently, wildlands preservation has been identified with environmentalism by the state and

the conservation elite; in consequence, environmental problems that impinge far more directly on the lives of the poor—for example, fuel, fodder, water shortages, soil erosion, and air and water pollution—have not been adequately addressed.[5]

Deep Ecology provides, perhaps unwittingly, a justification for the continuation of such narrow and inequitable conservation practices under a newly acquired radical guise. Increasingly, the international conservation elite is using the philosophical, moral, and scientific arguments used by deep ecologists in advancing their wilderness crusade. A striking but by no means atypical example is the recent plea by a prominent American biologist for the takeover of large portions of the globe by the author and his scientific colleagues. Writing in a prestigious scientific forum, the *Annual Review of Ecology and Systematics*, Daniel Janzen argues that only biologists have the competence to decide how the tropical landscape should be used. As "the representatives of the natural world," biologists are "in charge of the future of tropical ecology," and only they have the expertise and mandate to "determine whether the tropical agroscape is to be populated only by humans, their mutualist, commensals, and parasites, or whether it will also contain some islands of the greater nature—the nature that spawned humans, yet has been vanquished by them." Janzen exhorts his colleagues to advance their territorial claims on the tropical world more forcefully, warning that the very existence of these areas is at stake: "If biologists want a tropics in which to biologize, they are going to have to buy it with care, energy, effort, strategy, tactics, time, and cash."[6]

This frankly imperialist manifesto highlights the multiple dangers of the preoccupation with wilderness preservation that is characteristic of Deep Ecology. As I have suggested, it seriously compounds the neglect by the American movement of far more pressing environmental problems within the Third World. But perhaps more importantly, and in a more insidious fashion, it also provides an impetus to the imperialist yearning of Western biologists and their financial sponsors, organizations such as the WWF and IUCN. The wholesale transfer of a movement culturally rooted in American conservation history can only result in the social uprooting of human populations in other parts of the globe.

3. I come now to the persistent invocation of Eastern philosophies as antecedent in point of time but convergent in their structure with Deep Ecology. Complex and internally differentiated religious traditions—Hinduism, Budhism, and Taoism—are lumped together as holding a view of nature believed to be quintessentially biocentric. Individual philosophers such as the Taoist Lao Tzu are identified as being forerunners of Deep Ecology. Even an intensely political, pragmatic, and Christian-influenced thinker such as Gandhi has been accorded a wholly undeserved place in the deep ecological pantheon. Thus the Zen teacher Robert Aitken Roshi makes the strange claim that Gandhi's thought was not human-centered and that he practiced

an embryonic form of Deep Ecology which is "traditionally Eastern and is found with differing emphasis in Hinduism, Taoism and in Theravada and Mahayana Buddhism."[7] Moving away from the realm of high philosophy and scriptural religion, deep ecologists make the further claim that at the level of material and spiritual practice "primal" peoples subordinated themselves to the integrity of the biotic universe they inhabited.

I have indicated that this appropriation of Eastern traditions is in part dictated by the need to construct an authentic lineage and in part a desire to present Deep Ecology as a universalistic philosophy. Indeed, in his substantial and quixotic biography of John Muir, Michael Cohen goes so far as to suggest that Muir was the "Taoist of the [American] West."[8] This reading of Eastern traditions is selective and does not bother to differentiate between alternate (and changing) religious and cultural traditions; as it stands, it does considerable violence to the historical record. Throughout most of recorded history the characteristic form of human activity in the "East" has been a finely tuned but nonetheless conscious and dynamic manipulation of nature. Although mystics such as Lao Tzu did reflect on the spiritual essence of human relations with nature, it must be recognized that such ascetics and their reflections were supported by a society of cultivators whose relationship with nature was a far more *active* one. Many agricultural communities do have a sophisticated knowledge of the natural environment that may equal (and sometimes surpass) codified "scientific" knowledge; yet, the elaboration of such traditional ecological knowledge (in both material and spiritual contexts) can hardly be said to rest on a mystical affinity with nature of a deep ecological kind. Nor is such knowledge infallible; as the archaeological record powerfully suggests, modern Western man has no monopoly on ecological disasters.

In a brilliant article, the Chicago historian Ronald Inden points out that this romantic and essentially positive view of the East is a mirror image of the scientific and essentially pejorative view normally upheld by Western scholars of the Orient. In either case, the East constitutes the Other, a body wholly separate and alien from the West; it is defined by a uniquely spiritual and nonrational "essence," even if this essence is valorized quite differently by the two schools. Eastern man exhibits a spiritual dependence with respect to nature—on the one hand, this is symptomatic of his prescientific and backward self, on the other, of his ecological wisdom and deep ecological consciousness. Both views are monolithic, simplistic, and have the characteristic effect—intended in one case, perhaps unintended in the other—of denying agency and reason to the East and making it the privileged orbit of Western thinkers.

The two apparently opposed perspectives have then a common underlying structure of discourse in which the East merely serves as a vehicle for Western projections. Varying images of the East are raw material for political and cultural battles being played out in the West; they tell us far more about the Western commentator and his desires than about the "East." Inden's remarks apply not merely to Western scholarship on India, but to Orientalist constructions of China and Japan as well:

> Although these two views appear to be strongly opposed, they often combine together. Both have a similar interest in sustaining the Otherness of India. The holders of the dominant view, best exemplified in the past in imperial administrative discourse (and today probably by that of "development economics"), would place a traditional, superstition-ridden India in a position of perpetual tutelage to a modern, rational West. The adherents of the romantic view, best exemplified academically in the discourses of Christian liberalism and analytic psychology, concede the realm of the public and impersonal to the positivist. Taking their succour not from governments and big business, but from a plethora of religious foundations and self-help institutes, and from allies in the "consciousness industry," not to mention the important industry of tourism, the romantics insist that India embodies a private realm of the imagination and the religious which modern, western man lacks but needs. They, therefore, like the positivists, but for just the opposite reason, have a vested interest in seeing that the Orientalist view of India as "spiritual," "mysterious," and "exotic" is perpetuated.[9]

4. How radical, finally, are the deep ecologists? Notwithstanding their self-image and strident rhetoric (in which the label "shallow ecology" has an opprobrium similar to that reserved for "social democratic" by Marxist-Leninists), even within the American context their radicalism is limited and it manifests itself quite differently elsewhere.

To my mind, Deep Ecology is best viewed as a radical trend within the wilderness preservation movement. Although advancing philosophical rather than aesthetic arguments and encouraging political militancy rather than negotiation, its practical emphasis—namely, preservation of unspoiled nature—is virtually identical. For the mainstream movement, the function of wilderness is to provide a temporary antidote to modern civilization. As a special institution within an industrialized society, the national park "provides an opportunity for respite, contrast, contemplation, and affirmation of values for those who live most of their lives in the workaday world."[10] Indeed, the rapid increase in visitations to the national parks in postwar America is a direct consequence of economic expansion. The emergence of a popular interest in wilderness sites, the historian Samuel Hays points out, was "not a throwback to the primitive, but an integral part of the modern standard of living as people sought to add new 'amenity' and 'aesthetic' goals and desires to their earlier preoccupation with necessities and conveniences."[11]

Here, the enjoyment of nature is an integral part of the consumer society. The private automobile (and the life-style it has spawned) is in many respects the ultimate ecological villain, and an untouched wilderness the prototype of ecological harmony; yet, for most Americans it is perfectly consistent to drive a thousand miles to spend a holiday in a national park. They possess a vast, beautiful, and sparsely populated continent and are also able to draw upon the natural resources of large portions of the globe by virtue of their economic and political dominance. In consequence, America can simultaneously enjoy the material benefits of an expanding economy and the aesthetic benefits of unspoilt nature. The two poles of "wilderness" and "civilization" mutually coexist in an internally coherent whole, and philosophers of both poles are assigned a prominent place in this culture. Paradoxical as it may seem, it is no accident that Star Wars technology and Deep Ecology both find their fullest expression in that leading sector of Western civilization, California. . . .

The roots of global ecological problems lie in the disproportionate share of resources consumed by the industrialized countries as a whole *and* the urban elite within the Third World. Since it is impossible to reproduce an industrial monoculture worldwide, the ecological movement in the West must begin by cleaning up its own act. The greens advocate the creation of a "no-growth" economy, to be achieved by scaling down current (and clearly unsustainable) consumption levels.[12] This radical shift in consumption and production patterns requires the creation of alternate economic and political structures—smaller in scale and more amenable to social participation—but it rests equally on a shift in cultural values. The expansionist character of modern Western man will have to give way to an ethic of renunciation and self-limitation, in which spiritual and communal values play an increasing role in sustaining social life. This revolution in cultural values, however, has as its point of departure an understanding of environmental processes quite different from Deep Ecology.

Notes

1. Kirkpatrick Sale, "The Forest for the Trees: Can Today's Environmentalists Tell the Difference," *Mother Jones* 11, no. 8 (Nov. 1986): 26.
2. One of the major criticisms I make in this essay concerns Deep Ecology's lack of concern with inequalities *within* human society. In the article in which he coined the term "Deep Ecology," Naess himself expresses concerns about inequalities between and within nations. However, his concern with social cleavages and their impact on resource utilization patterns and ecological destruction is not very visible in the later writings of deep ecologists. See Arne Naess, "The Shallow and the Deep, Long-Range Ecology Movement: A Summary," *Inquiry* 16 (1973): 96 (I am grateful to Tom Birch for this reference).
3. Gary Snyder, quoted in Sale, "Forest for the Trees," 32. See also Dave Foreman "A Modest Proposal for a Wilderness System," *Whole Earth Review* 53 (Winter 1986–87): 42–45.
4. See, for example, Donald Worster, *Nature's Economy: The Roots of Ecology* (San Francisco: Sierra Club Books, 1977).
5. See Centre for Science and Environment, *India: The State of the Environment*

1982: *A Citizens Report* (New Delhi: Centre for Science and Environment, 1982); R. Sukumar, "Elephant-Man Conflict in Karnataka," in *The State of Karnataka's Environment* ed. Cecil Saldanha (Bangalore: Centre for Taxonomic Studies, 1985). For Africa, see the brilliant analysis by Helge Kjekshus, *Ecology Control and Economic Development in East African History* (Berkeley and Los Angeles: University of California Press, 1977).

6. Daniel Janzen, "The Future of Tropical Ecology," *Annual Review of Ecology and Systematics* 17 (1986): 305–6; emphasis added.

7. Robert Aitken Roshi, "Gandhi, Dogen, and Deep Ecology," reprinted as appendix C in Bill Devall and George Sessions, *Deep Ecology: Living as if Nature Mattered* (Salt Lake City: Peregrine Smith Books, 1985). For Gandhi's own views on social reconstruction, see the excellent three-volume collection edited by Raghavan Iyer, *The Moral and Political Writings of Mahatma Gandhi* (Oxford: Clarendon Press, 1986–87).

8. Michael Cohen, *The Pathless Way* (Madison: University of Wisconsin Press, 1984), 120.

9. Ronald Inden, "Orientalist Constructions of India," *Modern Asian Studies* 20 (1986): 442. Inden draws inspiration from Edward Said's forceful polemic, *Orientalism* (New York: Basic Books, 1980). It must be noted, however, that there is a salient difference between Western perceptions of Middle Eastern and Far Eastern cultures, respectively. Due perhaps to the long history of Christian conflict with Islam, Middle Eastern cultures (as Said documents) are consistently presented in pejorative terms. The juxtaposition of hostile and worshiping attitudes that Inden talks of applies only to Western attitudes toward Buddhist and Hindu societies.

10. Joseph Sax, *Mountains Without Handrails: Reflections on the National Parks* (Ann Arbor: University of Michigan Press, 1980), 42. Cf. also Peter Schmitt, *Back to Nature: The Arcadian Myth in Urban America* (New York: Oxford University Press, 1969), and Alfred Runte, *National Parks: The American Experience* (Lincoln: University of Nebraska Press, 1979).

11. Samuel Hays, "From Conservation to Environment: Environmental Politics in the United States since World War Two," *Environmental Review* 6 (1982): 21. See also the same author's book entitled *Beauty, Health and Permanence: Environmental Politics in the United States, 1955–85* (New York: Cambridge University Press, 1987).

12. From time to time, American scholars have themselves criticized these imbalances in consumption patterns. In the 1950s, William Vogt made the charge that the United States, with one-sixteenth of the world's population, was utilizing one-third of the globe's resources (Vogt, cited in E. F. Murphy, *Nature, Bureaucracy and the Rule of Property* [Amsterdam: North Holland, 1977, 29]). More recently, Zero Population Growth has estimated that each American consumes thirty-nine times as many resources as an Indian. See *Christian Science Monitor*, 2 Mar. 1987.

PART VI

Spiritual Ecology

Toward a Healing of Self and World

JOANNA MACY

A NEW PARADIGM IS emerging in our time. Through its lens we see reality structured in such a way that all life forms affect and sustain each other in a web of radical interdependence. This organic interconnectedness is what we call our Deep Ecology.

"Deep Ecology" is a term coined by Norwegian philosopher Arne Naess, to contrast with "shallow environmentalism," a Band-Aid approach applying piecemeal technological fixes for short-term goals. Deep Ecology teaches us that we humans are neither the rulers nor the center of the universe, but are embedded in a vast living matrix and subject to its laws of reciprocity. Deep Ecology represents a basic shift in ways of seeing and valuing, a shift beyond anthropocentrism:

> Anthropocentrism means human chauvinism. Similar to sexism, but substitute *human race* for man and *all other species* for woman. [It's about the human race being oppressive to other species and the environment.]
> When humans investigate and see through their layers of anthropocentric self-cherishing, a most profound change in consciousness begins to take place. Alienation subsides. The human is no longer an outsider, apart . . .
> What a relief then! The thousands of years of imagined separation are over and we begin to recall our true nature. That is, the change is a spiritual one . . . sometimes referred to as deep ecology (John Seed).

There are, of course, manifold ways of evoking or provoking this change in perspective. Methods of inspiring the experience of Deep Ecology range from prayer to poetry, from wilderness vision quests to the induction of

From: "Deep Ecology Work: Toward the Healing of Self and World," *Human Potential Magazine* 17, no. 1 (Spring 1992): 10–13, 29–31.

altered states of consciousness. The most reliable is direct action in defense of earth—and it is spreading today in many forms.

THE GREENING OF THE SELF

Something important is happening in our world that is not reported in the newspapers. I consider it the most fascinating and hopeful development of our time, and it is one of the reasons I am so glad to be alive today. It has to do with what is occurring to the notion of the self.

The self is the hypothetical piece of turf on which we construct our strategies for survival, the notion around which we focus our instincts for self-preservation, or need for self-approval, and the boundaries for our self-interest. The conventional notion of the self with which we have been raised and to which we have been conditioned by mainstream culture is being undermined. What Alan Watts called "the skin-encapsulated ego" and Gregory Bateson referred to as "the epistemological error of Occidental civilization" is being unhinged, peeled off. It is being replaced by concepts of identity and self-interest which are much wider than the conventional ego . . . by what you might call the ecological self, co-extensive with other beings and the life of our planet.

At a recent lecture on a college campus, I gave the students examples of activities which are currently being undertaken in defense of life on earth — actions in which people risk their comfort and even their lives to protect other species. In the Chipko, or tree-hugging, movement in northern India, for example, villagers fight the deforestation of their remaining woodlands. On the open seas, Greenpeace activists are intervening to protect marine mammals from slaughter. After that talk, I received a letter from a student I'll call Michael. He wrote:

> I think of the tree-huggers hugging my trunk, blocking the chainsaws with their bodies. I feel their fingers digging into my bark to stop the steel and let me breathe. I hear the bodhisattvas [note: a Buddhist term for an enlightened, compassionate being] in their rubber boats as they put themselves between the harpoons and me, so I can escape to the depths of the sea. I give thanks for your life and mine, and for life itself. I give thanks for realizing that I too have the powers of the tree-huggers and the bodhisattvas.

What is striking about Michael's words is the shift in identification. Michael is able to extend his sense of self to encompass the self of the tree and of the whale. Tree and whale are no longer removed, separate, disposable objects pertaining to a world "out there" (outside of humans and inside the environment); they are intrinsic to his own vitality. Through the power of his caring, his experience of self is expanded far beyond the skin-encapsulated ego. I quote Michael's words not because they are unusual, but, to the contrary, because they express a desire and a capacity that is being released from the prison cell of old constructs of self. This desire and capacity are arising in more and more people today as, out of deep concern for what is happening to our world, they begin to speak and act on its behalf.

Among those who are shedding these old constructs of self, like old skin or a confining shell, is John Seed, director of the *Rainforest Information Centre* in Australia. One day we were walking through the rain forest in New South Wales, where he has his office, and I asked him, "You talk about the struggle against the lumbering interests and politicians to save the remaining rain forest in Australia. How do you deal with the despair?" He replied, "I try to remember that it's not me, John Seed, trying to protect the rain forest. Rather I'm part of the rain forest protecting myself. I am that part of the rain forest recently emerged into human thinking." This is what I mean by the greening of the self. It involves a combining of the mystical with the practical and the pragmatic, transcending separateness, alienation, and fragmentation. It is a shift that Seed himself calls "a spiritual change," generating a sense of profound interconnectedness with all life.

This is hardly new to our species. In the past, poets and mystics have been speaking and writing about these ideas, but not people on the barricades agitating for social change. Now the sense of an encompassing self, the deep identity with the wider reaches of life, is a motivation for action. It is a source of courage that helps up stand up to the powers that are still, through force of inertia, destroying the fabric of life. I am convinced that this expanded sense of self is the only basis for adequate and effective action.

Three developments converge in our time to call forth the ecological self. They are: (1) the psychological and spiritual pressure exerted by current dangers of mass annihilation, (2) the emergence in science of the *systems* view of the world, and (3) a renaissance of nondualistic forms of spirituality.

CURRENT DANGERS OF MASS ANNIHILATION: PAIN FOR THE WORLD

The move to a wider ecological sense of self is in large part a function of the dangers that are threatening to overwhelm us. We are confronted by social breakdown, wars, nuclear proliferation, and the progressive destruction of our biosphere. Polls show that people today are aware that the world, as they know it, may come to an end. This loss of certainty that there will be a future is the pivotal psychological reality of our time.

Over the past twelve years my colleagues and I have worked with tens of thousands of people in North America, Europe, Asia, and Australia, helping them confront and explore what they know and feel about what is happening to their world. The purpose of this work, which was first known as *"Despair and Empowerment Work,"* is to overcome the numbing and powerlessness that result from suppression of painful responses to massively painful realities.

As their grief and fear for the world is allowed to be expressed without apology or argument and validated as a wholesome, life-preserving response, people break through their avoidance mechanisms, break through their sense of futility and isolation. Generally what they break through into is a larger sense of identity. It is as if the pressure of their acknowledged awareness of the suffering of our world stretches or collapses the culturally defined boundaries of the self.

It becomes clear, for example, that the grief and fear experienced for our world and our common future are categorically different from similar sentiments relating to one's personal welfare. This pain cannot be equated with dread of one's own individual demise. Its source lies less in concerns for personal survival than in apprehensions of collective suffering—of what looms for human life and other species and unborn generations to come. Its nature is akin to the original meaning of compassion— "suffering with." It is the distress we feel on behalf of the larger whole of which we are a part. And, when it is so defined, it serves as a trigger or getaway to a more encompassing sense of identity, inseparable from the web of life in which we are as intricately interconnected as cells in a larger body.

This shift in consciousness is an appropriate, adaptive response. For the crisis that threatens our planet, be it seen in its military, ecological, or social aspects, derives from a dysfunctional and pathogenic notion of the self. It is a mistake about our place in the order of things. It is the delusion that the self is so separate and fragile that we must delineate and defend its boundaries, that it is so small and needy that we must endlessly acquire and endlessly consume, that it is so aloof that we can—as individuals, corporations, nation-states, or as a species—be immune to what we do to other beings.

This view of human nature is not new, of course. Many have felt the imperative to extend self-interest to embrace the whole. What is notable in our situation is that this extension of identity can come not through an effort to be noble or good or altruistic, but simply to be present and own our pain. That is why this shift in the sense of self is credible to people. As the poet Theodore Roethke said, "I believe my pain."

SCIENCE AND THE SYSTEMS VIEW: CYBERNETICS OF THE SELF

The findings of twentieth-century science undermine the notion of a separate self, distinct from the world it observes and acts upon. As Einstein showed, the self's perceptions are shaped by its changing position in relation to other phenomena. And these phenomena are affected not only by location but, as Heisenberg demonstrated, by the very act of observation. Now contemporary systems science and systems cybernetics go yet further in challenging old assumptions about a distinct, separate, continuous self.

We are open, self-organizing systems; our very breathing, acting, and thinking arise in interaction with our shared world through the currents of matter, energy, and information that flow through us. In the web of relationships that sustain these activities, there are no clear lines demarcating a separate self. As systems theorists aver, there is no categorical "I" set over and against a categorical "you" or "it."

One of the clearer expositions of this is offered by Gregory Bateson, whom I earlier quoted as saying that the abstraction of a separate "I" is "the epistemological fallacy of Western civilization." He says that the process that decides and acts cannot be neatly identified with the isolated subjectivity of the individual or located within the confines of the skin. He contends that "the total self-corrective unit that processes information is a system whose boundaries do not at all coincide with t : boundaries either of the body or what is popularly called 'self' or 'consciousness.'" He goes on to say, "The self is ordinarily understood as only a small part of a much larger trial-and-error system which does the thinking, acting, and deciding."

Bateson uses the example of a woodcutter, about to fell a tree. His hands grip the handle of the axe. Whump, he makes a cut, and then whump, another cut. What is the feedback circuit, where is the information that is guiding that cutting down of the tree? That is the self-correcting unit, that is what is doing the chopping down of the tree. In another illustration, a blind person with a cane is walking along the sidewalk. Tap, tap, whoops, there's a fire hydrant, there's a curb. What is doing the walking? Where is the self then of the blind person? What is doing the perceiving and deciding? That self-corrective feedback circuit is the arm, the hand, the cane, the curb, the ear. At that moment that is the self that is walking. Bateson's point is that the self as we usually define it is an improperly delimited part of a much larger field of interlocking processes. And he maintains that

this false reification of the self is basic to the planetary ecological crisis in which we find ourselves. We have imagined that we are a unit of survival and we have to see to our own survival, and we imagine that the unit of survival is the separate individual or a separate species, whereas in reality through the history of evolution, it is the individual plus the environment, the species plus the environment, for they are essentially symbiotic.

The self is a metaphor. We can decide to limit it to our skin, our person, our family, our organization, or our species. We can select its boundaries in objective reality. As the systems theorists see it, our consciousness illuminates a small arc in the wider currents and loops of knowing that interconnect us. It is just as plausible to conceive of mind as coexistent with these larger circuits, the entire "pattern that connects" as Bateson said. Do not think that to broaden the construct of self this way involves an eclipse of one's distinctiveness. Do not think that you will lose your identity like a drop in the ocean merging into the oneness of Brahman. From the systems perspective, this interaction, creating larger wholes and patterns, fosters and even requires diversity. You become more yourself. Integration and differentiation go hand in hand.

NONDUALISTIC SPIRITUALITY: THE BOUNDLESS HEART OF THE BODHISATTVA

A third factor that nourishes deep ecological consciousness in our world today is the resurgence of nondualistic spirituality. We find it in many realms—in Sufism in Islam, Creation Spirituality in Christianity, and in Buddhism's historic coming to the West. Buddhism is distinctive in its clarity and sophistication about the dynamics of self. In much the same way as systems theory does, Buddhism undermines categorical distinctions between self and other. It then goes further than systems theory in showing the pathogenic character of any reifications of the self, and in offering methods for transcending these difficulties and healing this suffering. What the Buddha woke up to under the Bodhi tree was *paticca samuppada*, the dependent co-arising of phenomena, in which you cannot isolate a separate, continuous self.

We think, "What do we do with the self, this clamorous 'I,' always wanting attention, always wanting its goodies? Do we crucify it, sacrifice it, mortify it, punish it, or do we make it noble?" Upon awaking we realize, "Its just a convention!" When you take it too seriously, when you suppose that it is something enduring which you have to defend and promote, it becomes the foundation of delusion, the motive behind our attachments and our aversions. Consider the Tibetan portrayal of the wheel of life, that mythically depicts all the realms of being. At the very center of that wheel of suffering are three figures: the pig, the rooster, and the snake—they represent delusion, greed, and hatred—and they just chase one another around and around. The linchpin of all the pain is the notion of our self, the notion that we have to protect that self or conquer on its behalf—or do something with it.

The point of Buddhism, and, I think, of Deep Ecology too, is that we do not need to be doomed to the perpetual rat-race. The vicious circle can be broken. It can be broken by wisdom, meditation, and morality—that is, when we pay attention to our experience and our actions and discover that they do not have to be in bondage to a separate self. The sense of interconnectedness that can the arise is imaged—one of the most beautiful images coming out of the Mahayana—as the jewelled net of Indra. It is a vision of reality structured very much like the holographic view of the universe, so that each being is a jewel at each node of the net, and each jewel reflects all the other, reflecting back and catching the reflection, just as systems theory sees that the part contains the whole.

The awakening to our true self is the awakening to that entirety, breaking out of the prison-self of separate ego. The one who perceives this is the bodhisattva—and we are all bodhisattvas because we are all capable of experiencing that—it is our true nature. We are profoundly interconnected and therefore we are all able to recognize and act upon our deep, intricate, and intimate inter-existence with one another and all beings. That true nature of ours is already present in our pain for the world. When we turn our eyes away from that homeless figure, are we indifferent, or is the pain of seeing him or her too great? Do not be easily duped by the apparent indifference of those around you. What looks like apathy is really the fear of suffering. But the bodhisattva knows that to experience the pain of all beings it is necessary to experience their joy. It says in the Lotus Sutra that the bodhisattva hears the music of the spheres, and understands the language of the birds, while hearing the cries in the deepest levels of hell.

One of the things I like best about the ecological self that is arising in our time is that it is making moral exhortation irrelevant. Sermonizing is both boring and ineffective. As Arne Naess says, "The extensive moralizing within the ecological movement has given the public the false impression that they are being asked to make a sacrifice to show more responsibility, more concern, and a nicer moral standard. But all of that would flow naturally and easily if the self were widened and deepened so that the protection of nature was felt and perceived as protection of our very selves." Please note this important point: virtue is not required for the greening of the self or the emergence of the ecological self. The shift in identification at this point in our history is required precisely because moral exhortation doesn't work, and because sermons seldom hinder us from following our self-interest as we conceive it.

The obvious choice, then, is to extend our notions of self-interest. For example, it would not occur to me to plead with you, "Oh, don't saw off your leg. That would be an act of violence." It wouldn't occur to me, or you, because your leg is part of your body. Well, so are the trees in the Amazon rain basin. They are our external lungs. And we are beginning to realize that the world is our body. This ecological self, like any notion of selfhood, is a metaphoric construct and a dynamic one. It involves choice: choices can be made to identify at different moments with different dimensions or aspects of our systemically interrelated existence, be they hunted whales or homeless humans or the planet itself. In doing this, the extended self brings into play wider resources—courage, endurance, ingenuity—like a nerve cell in a neural net opening to the charge of the other neurons.

There is the sense of being acted through and sustained by those very beings on whose behalf one acts. This is very close to the religious concept of grace. In systems language we can talk about it as a synergy. But with this extension, this greening of the self, we can find a sense of buoyancy and resilience that comes from letting flow through us strengths and resources that come to us with continuous surprise and sense of blessing.

The Spiritual Dimension of Green Politics

CHARLENE SPRETNAK

HOW SHALL WE RELATE TO OUR CONTEXT, THE ENVIRONMENT?

IN 1967 LYNN WHITE, a professor of history at UCLA, published in *Science* "The Historical Roots of our Ecologic Crisis," a critical analysis of the attitudes Western religion has encouraged toward our environment. Since then ecologists often point to the injunctions in Genesis that humans should attempt to "subdue" the earth and have "dominion" over all the creatures of the earth as being bad advice with disastrous results. (Many of those critiques, however, have lacked a full sense of the Hebrew words.) Bill Devall, coauthor of *Deep Ecology*, spoke for many activists when he declared in August 1984, "Unless major changes occur in churches, ecologists and all those working in ecology movements will feel very uncomfortable sitting in the pews of most American churches."

The disparity between Judeo-Christian religion and ecological wisdom is illustrated by the experience of a friend of mine who once lived in a seminary overlooking Lake Erie and says he spent two years contemplating the sufferings of Christ without ever noticing that Lake Erie was dying.[1] Even when Catholic clergy speak today of St. Francis of Assisi, whom Lynn White nominated as the patron saint of ecologists, they often take pains to insist that he was not some "nature mystic,"[2] which, of course, would taint him with "paganism."[3] Religion that sets itself in opposition to nature and vehemently resists the resacralizing of the natural world on the grounds that it would be "pagan" to do so is not sustainable over time.

The cultural historian Thomas Berry has declared that we are entering a

---

From: *The Spiritual Dimension of Green Politics* (Santa Fe, N.M.: Bear and Co., 1986), 52–69.

new era of human history, the Ecological Age.[4] How could our religion reflect ecological wisdom and aid the desperately needed transformation of culture? First, I suggest that Judaism and Christianity should stop being ashamed of their "pagan" inheritance, *which is substantial*, and should proudly proclaim their many inherent ties to nature. How many of us realize that the church sets Easter on the first Sunday after the first *full moon* after the *vernal equinox* and that most of the Jewish holy days are determined by a lunar calendar?[5] Numerous symbols, rituals, and names in Jewish and Christian holy days have roots directly in the nature-revering Old Religion. The list is a long one and should be cause for self-congratulation and celebration among Christians and Jews.

Second, I hope the stewardship movement, which is gaining momentum in Christian and Jewish circles, will continue to deepen its analyses and its field of action. Those people are performing a valuable service by reinterpreting the overall biblical teachings about the natural world and finding ecological wisdom that balances or outweighs the "dominance" message. Virtually all spokespersons for the stewardship movement emphasize that nature is to be honored as God's creation.[6] In fact, that position is firmly rooted in the work of several noted theologians whose orientation is known as "creation spirituality." They emphasize the interrelatedness of all creation, the understanding that humans do not occupy the central position in the cosmic creation but have a responsible role to play, and the transformation of society in directions that will further the continuation of life. Hence peace is a central issue for creation theologians, as is justice. Nearly all of them give greater importance to the female dimension of creation than do other theologians. Among the Catholic, Protestant, and Jewish theologians of creation spirituality are Bernhard W. Anderson, Thomas Berry, Walter Bruggemann, Martin Buber, Marie-Dominique Chenu, Matthew Fox, Abraham Heschel, Jurgen Moltmann, Paul Santmire, Edward Schillebeeckx, Odil Hannes Steck, Pierre Teilhard de Chardin, and Samuel Terrien.[7]

The experience of knowing the Divine through communication with nature has been a recurrent theme in art. Recently Alice Walker described a theologically sophisticated, elementally spiritual experience in her Pulitzer Prize–winning novel *The Color Purple* when one black woman in rural Georgia explains to another that God "ain't a he or a she, but a It":

> It ain't a picture show. It ain't something you can look at apart from anything else, including yourself. I believe God is everything, say Shug. Everything that is or ever was or ever will be. And when you can feel that, and be happy to feel that, you've found it. . . . My first step away from the old white man was trees. Then air. Then birds. Then other people. But one day when I was sitting quiet and feeling like a motherless child, which I was, it come to me: that feeling of being part of everything, not separate at all. I knew that if I cut a tree, my arm would bleed. And I laughed and I cried and I run all around the house. I knew just what it

was. In fact, when it happen, you can't miss it. It sort of like you know what, she say, grinning and rubbing high up on my thigh.[8]

I am encouraged that a religion-based respect for nature is showing up in numerous articles and books, especially books like *The Spirit of the Earth* (1984), in which John Hart urges study of and respect for Native American religious perspectives on nature *because that is the indigenous tradition of our land* and suggests compatibility between their religion and the Judeo-Christian tradition. Yet why is it that attention to loving and caring for nature rarely makes it into the liturgy today? . . . Harold Gilliam in the *San Francisco Chronicle* describ[ed] a magnificent ecological service that spanned twenty-four hours, beginning at sunrise on the autumnal equinox, and took place in the gothic cathedral on Nob Hill in San Francisco, Grace Cathedral. At the sound of a bell and a conch shell, the Episcopal Bishop of California opened the service:

> We are gathered here at sunrise to express our love and concern for the living waters of the Central Valley of California and for the burrowing owls, white-tailed kites, great blue herons, migratory waterfowl, willow trees, cord grass, water lilies, beaver, possum, striped bass, anchovies, and women, children, and men of the Great Family who derive their life and spiritual sustenance from these waters. Today we offer our concerns and prayers for the ascending health and spirit of these phenomena of life and their interwoven habitats and rights. . . .

Poets, spiritual teachers, musicians, and ecologists all participated in the service, which included whale and wolf calls emanating from various corners of the cathedral's sound system, as well as the projection of nature photography onto the walls and pillars. Gary Snyder and his family read his "Prayer for the Great Family," which is based on a Mohawk prayer.[9] The celebrants poured water from all the rivers of California into the baptismal font. They committed themselves to changing our society and our environment into "a truly Great Family," and they assigned to each U.S. senator a totemic animal or plant from his or her region in order to accentuate the rights of our nonhuman family members. I read the account with awe and then noticed with sadness that it was dated 17 October 1971. (No subsequent ecological services took place in that church because a few influential members of the congregation pronounced it paganism.) How many species have been lost since then, how many tons of topsoil washed away, how many aquifers polluted—while we have failed to include nature in our religion?

Knowledge of nature must precede respect and love for it. We could urge that ecological wisdom regarding God's creation be incorporated in Sunday school as well as in sermons and prayer. We could suggest practices such as the planting of trees on certain holy days. We could mention in the church bulletin ecological issues that are crucial to our community.[10] There is no end to what we *could* do to focus spiritually based awareness and action on saving the great web of life.

HOW SHALL WE RELATE TO OTHER PEOPLE?

This last basic question has two parts: distinction by gender and then by other groups. Our lives are shaped to a great extent not by the differences between the sexes, but by the cultural response to those differences. There is no need to belabor the point that in patriarchal cultures the male is considered the norm and the female is considered "the Other." For our purposes here, however, it is relevant to note that Judeo-Christian religion has played a central role in constructing the subordinate role for women in Western culture. Suffice it to say that the eminent mythologist Joseph Campbell once remarked that in all his decades of studying religious texts worldwide he had never encountered a more relentlessly misogynist book than the Old Testament. Numerous Christian saints and theologians have continued the tradition.

The results for traditional society of denying women education and opportunity have been an inestimable loss of talent, intelligence, and creativity. For women it has meant both structural and direct violence. Of the former, Virginia Woolf observed that women under patriarchy are uncomfortable with themselves because they know society holds them in low esteem. The structural violence of forced dependency sometimes provides the conditions for physical violence, that is, battering. Finally, patriarchal culture usurps control over a woman's body from the woman herself, often inflicting torturous pain. It has been reported that in China today women are forced to undergo abortions even in the third trimester under the government's one-child-only policy. (The women who must undergo forced abortion are those who have incurred shame and the wrath of their husbands and in-laws by previously giving birth to a girl and later try desperately to carry a boy baby to term unnoticed by the government. Sometimes the women in that patriarchal culture simply drown themselves immediately after giving birth to a daughter.)

Some people accuse the greens of being hypocritical in calling themselves a "party of life" and adopting a "pro-choice" stance on abortion. In spring 1984, the European Greens, a coalition of green parties throughout Western Europe endorsed, after much debate, a position *against* social and political sanctions that force birthing and *for* free choice. The quality of the debate in green parties over abortion has more integrity than that currently being waged in American politics precisely because all aspects of the issue are considered. In our country half of the debate often seems to be missing: women's suffering. The issue is obviously complex, and there are people of good conscience on both sides. I offer my views merely as personal ones, not official positions of American greens.

Church leaders of many varieties are demanding an end to all legal (that is, medically safe) abortion. I suspect they can maintain a position demanding the criminalization of abortion only because they have never witnessed a woman going through pregnancy, labor, and delivery—or else they believe the biblical injunction that woman is *supposed* to suffer. Sometimes birth is textbook simple, but usually it is not. Some men say they remember their wife's screams for months. Many men say the birth experience made them "pro-choice" on the abortion issue because they would never want to force any woman to go through such an ordeal against her will.

As one of the most popular right-wing Christian preachers, Pat Robertson, likes to tell his TV audience (16.3 million households per month), "We are offering up 1½ million babies per year upon the altar of sensuality and selfishness."[11] Is that what it's all about—millions of sex-crazed, hedonistic women? Where is the compassion for the lonely teenager from an unnurturing family situation who tried to find affection and love where she could? Where is the compassion for the innocent victims of rape, including incestual rape and the increasing frequency of the "date rape" and "acquaintance rape"? Where is the compassion for *any* woman who discovers that she is "in trouble"? The number of abortions needed in this country would plummet if the problemmatic conditions were addressed effectively: disintegrated families, widespread pornography depicting violence against women, culturally approved hyper-macho behavior on dates, what has been called "patriarchy's dirty little secret" (the shocking statistics on sexual abuse by male relatives), selfishness and lack of spiritual grounding on the part of men who emotionally coerce their girlfriends, and lack of self-confidence and spiritual grounding in their own being among young women.

There comes a time at the end of many lives when life is not viable without machinery, and most people say they would like the machinery turned off if it came to that. Similarly, there is a time at the beginning of life when a fertilized egg and then a fetus is not viable life *unless* the woman is willing to give over her body and accept the suffering. To force a woman either to give birth or to abort is violence against the person. Most men and women know this in their hearts. They also know that countless women do not have the financial and other resources for the twenty-year task of raising a child. That is why a Gallup poll in June 1983 found that only 19 percent of American Catholics and 16 percent of the total American population want abortion to be illegal in all circumstances.[12] The Gallup organization released a poll on 20 February 1986 showing that the American public is evenly divided (45 percent to 45 percent) on the 1973 U.S. Supreme Court ruling that a woman may go to a doctor to have an abortion during the first three months of pregnancy. Interestingly, women's views were found to be statistically the same as in 1983, but men's support for the court ruling had fallen by 11 percent. Male fears of women's controlling their own sexuality are deeply rooted in patriarchal culture; during the Renaissance, for example, peasant healers were burned as "witches" for providing women with contraception and abortion. So the debate we are embroiled in is a very old one, and the campaign to "save the embryo; damn the woman" has been mounted many times before.

Men, too, suffer under patriarchal culture. Because woman is regarded as the denigrated Other, men are pressured to react and continually prove themselves very unlike the female. This dynamic result in what some men have called "the male machine." It has also skewed much of our behavioral and cognitive science since thousands of careers and volumes of commentary on "sex differences" have been funded but no recognized field of "sex similarities" exists.[13] That would be too unnerving. The most serious effect of men under patriarchy needing to prove themselves *very different* from women is the function of military combat as an initiation into true manhood and full citizenship. This deeply rooted belief surfaced as an unexpected element in the struggle to pass the Equal Rights Amendment, for instance. Feminist lobbyists in state legislatures throughout the 1970s were repeatedly informed, "When you ladies are ready to fight in a *war*, we'll be ready to discuss equal rights!"[14] Such an orientation is not sustainable in the nuclear age.

What role could religion play in removing the cultural insistence on women as Other and men as godlike and hence inherently superior? How could religion further the green principle of postpatriarchal consciousness? We know the answers because they are already being tried: women must have equal participation in ritual (as ministers, rabbis, and priests), language in sermons and translations must be inclusive, and the godhead must be considered female as well as male. These solutions are not new, but neither are they very effective, because so many people do not take either the need or the means seriously. Instead, they resent these efforts and feel silly and somewhat embarrassed with the notion of a female God. Being forced to say "God the Mother" once in a while is pointless if people have in mind Yahweh-with-a-skirt. We must first understand who She is: She is not in the sky; She is earth. Here is Her manifestation in the oldest creation story in Western culture:

The Myth of Gaia

Free of birth or destruction, of time or space, of form or condition, is the Void. From the eternal Void, Gaia danced forth and rolled Herself into a spinning ball. She molded mountains along Her spine, valleys in the hollows of Her flesh. A rhythm of hills and stretching plains followed Her contours. From Her warm moisture She bore a flow of gentle rain that fed Her surface and brought life. Wriggling creatures spawned in tidal pools, while tiny green shoots pushed upward through Her pores. She filled oceans and ponds and set rivers flowing through deep furrows. Gaia watched Her plants and animals grow. In time She brought forth from Her womb six women and six men. . . .

Unceasingly the Earth-Mother manifested gifts on Her surface and accepted the dead into her body. In return She was revered by all mortals. Offerings to Gaia of honey and barley cake were left in a small hole in the earth before plants were gathered. Many of Her temples were built near deep chasms where yearly the mortals offered sweet cakes into her womb. From within the darkness of Her secrets, Gaia received their gifts.[15]

Having addressed the self, nature, and gender, we now come to the last half of the last basic question, *How shall we relate to groups and other individuals?* There are, of course, a multiplicity of groups in society at the levels of family, community, region, state, nation, and planet. The following are merely some general considerations.

We must first analyze how our own mode of living affects others in the great family: Does the nature of our existence impose suffering on others— or does it support and assist those who are less privileged than we? Here we can enjoy the convergence of spiritual growth and political responsibility in the spiritual practice of cultivating moment-to-moment awareness, being fully "awake" and focused on our actions—a simple-sounding yet demanding task. There is a story in Zen of a student who studied very hard to master certain religious texts and then went before his spiritual teacher to be questioned. The *roshi* asked simply, "On which side of the umbrella did you place your shoes?" The student was defeated; he had lost awareness (or "spaced out," as we might say).

We can begin our day by focusing mindfulness on our every act. Turning on the water in the bathroom. Where does it come from? Is our town recklessly pumping water from the receding water table instead of calling for conservation measures? Where does our wastewater go when it leaves the sink? What happens after it is treated? Later we are in the kitchen, making breakfast. Where does our coffee come from? A worker-owned cooperative in the Third World or an exploitative multinational corporation? Obviously, it is exhausting to continue this practice very long unless one is adept. (It *is* difficult—so much so that a friend of mine has added an amendment to a popular spiritual saying: "Be here now—or now and then.") But everyone can practice *some* mindfulness.

If we analyze our own situation, we may discover that we are benefiting from the suffering of others—and that we ourselves are uncomfortable with the structural systems in which we work. When one thinks of religious people working for economic or social change, the "liberation theology" movement probably comes to mind because of its size in Latin America and its coverage in the press lately. In that movement, grass-roots Catholic groups (base communities) meet frequently to discuss the teachings in the Gospels and applications of Marxist analysis.

But there is another way: a religion-based movement for social change is beginning to flourish that is completely in keeping with green principles of private ownership and cooperative economics, decentralization, grass-roots democracy, nonviolence, social responsibility, global awareness—and the spiritual truth of oneness. This type of call for economic and social change is gaining momentum in Catholic, Protestant, and Jewish communities. We see it, for example, in the statement issued by the Catholic Bishops of Appalachia, *This Land Is My Home: A Pastoral Letter on Powerlessness in Appalachia*, which calls for worker-owned businesses and community-based economics. We see it in *Strangers and Guests: Toward Community in the Heartland* by the Catholic bishops of the heartland (Midwest) and in *The Land: God's Giving, Our Caring* by the American Lutheran Church, a statement which was then echoed by the Presbyterian Church. Both of these statements address ecological use of the land, and *Strangers and Guests* calls for small-is-beautiful *land reform* as the only sustainable course for rural America. Developing the applications of such principles as "the land should be distributed equitably" and "the land's workers should be able to become the land's owners," the heartland bishops discuss elimination of capital-gains tax laws which favor "wealthy investors and speculators" and disfavor "small and low-income farm families," taxation of agricultural land "according to its productive value rather than its speculative value," "taxing land progressively at a higher rate according to increases in size and quality of holdings" (a proposal in the Jeffersonian tradition), and low-interest loans to aspiring farmers as well as tax incentives for farmers with large holdings to sell land to them.

We see green-oriented economic and social change now promoted in the Jewish periodical *Menorah* and by the Protestant multidenominational association, Joint Strategy and Action Committee. The lead article in a 1984 issue of the JSAC newsletter began: "If you want to know what eco-justice is, read the Psalms. The dual theme of justice in the social order and integrity in the natural order is pervasive and prominent. The Book is, in large part, a celebration of interrelationships, the interaction, the mutuality, the organic oneness and wholeness of it all that is, that is to say, the Creator and the creation, human and nonhuman."

The green-oriented Jewish and Protestant leaders seek to locate justice *and* ecological wisdom in the Old Testament. Green-oriented Catholics usually turn to the papal encyclicals, especially Pope Pius XI's 1931 encyclical *Quadragesimo anno [Forty Years After]*,[16] which established three cardinal principles: *personalism* (the goal of society is to develop and enrich the individual human person), *subsidiarity* (no organization should be bigger than necessary and nothing should be done by a large and higher social unit than can be done effectively by a lower and smaller unit), and *pluralism* (that a healthy society is characterized by a wide variety of intermediate groups freely flourishing between the individual and the state).[17] Sounds like a lot of green party platforms I've read recently! Andrew Greeley argues in *No Bigger Than Necessary* that Catholic social theory is firmly rooted in the communitarian, decentralist tradition and that Catholics who drifted into Marxism in recent decades are simply unaware that their own tradition contains a better solution. Joe Holland, a Catholic activist with the Center of Concern in Washington, D.C., argues, however, that Left-oriented Catholics have never embraced "scientific Marxism" and the model of a machinelike centralized government and economy. They are attracted, rather, by communitarian ideals and are uncomfortable with the modernity of many socialist assumptions.[18] Hence, we may assume, and I believe Joe Holland would agree, that many of these lukewarm leftists in Catholic circles would readily become green.

The possibilities for locating and working with green-oriented activists in mainline religions have never been better. For example, a task force of the Presbyterian Church in Pennsylvania, Ohio, and West Virginia has been instrumental in introducing Rodale Press's Regeneration Project in economically depressed communities. The project's goals are to stimulate local economic vitality and to improve the overall quality of life.[19] Within our own green political organizations, however, the question remains of how much religious content is proper in pluralistic meetings and publications. I myself am uncertain about how much overt spirituality the "market will bear" in green conferences and statements, and I am often dissatisfied afterward because I and other greens have held back too much on spirituality so as not to exclude anyone in the group. I am not sure what the solutions may be, but I am certain I shall be influenced by learning recently of that Gallup statistic that *only 6 percent* of Americans do not believe in God "or a universal spirit." What a vocal minority! Perhaps I should simply avoid Manhattan and university towns.

Surely, no green, whatever his or her spiritual orientation, could object to our structuring our groups according to the Deep Ecology principles of diversity, interdependence, openness, and adaptability—as well as the spiritual principles of cultivating wisdom and compassion. These can be our guidelines as we evolve the everchanging forms of green politics.

Notes

1. Paul Ryan, "Relationships," *Talking Wood* 1, no. 4 (1980).
2. One example, although by no means the only one, is Murray Bodo, O.F.M., *The Way of St. Francis* (Garden City, N.Y.: Doubleday, 1984).
3. "Pagan" is from the Latin word for "country people," *pagani*. It has nothing to do with Satan worship.
4. Collections of Thomas Berry's papers, such as *The Riverdale Papers on the Earth Community*, are available from the Riverdale Center for Religious Research, 5801 Palisade Avenue, Riverdale, N.Y. 10471. His work is also presented by the physicist Brian Swimme in *The Universe is a Green Dragon* (Santa Fe, N.M.: Bear and Co., 1984).
5. Arthur Waskow, *Seasons of Our Joy: A Celebration of Modern Jewish Renewal* (New York: Bantam Books, 1982).
6. See, for example, Mary Evelyn Jegen and Bruno V. Manno, eds., *The Earth Is the Lord's: Essays on Stewardship* (New York: Paulist Press, 1978); Wesley Granberg-Michaelson, *A Worldly Spirituality: The Call to Take Care of the Earth* (New York: Harper and Row, 1984); John Hart, *The Spirit of the Earth: A*

Theology of the Land (New York: Paulist Press, 1984); John Carmody, *Ecology and Religion: Toward a New Christian Theology of Nature* (New York: Paulist Press, 1983); and Ian G. Barbour, ed., *Earth Might Be Fair: Reflections on Ethics, Religion, and Ecology* (Englewood Cliffs, N.J.: Prentice-Hall, 1972).

7. In addition to the scores of books by the creation theologians cited in the text, there is a relevant anthology, Philip N. Joranson and Ken Butigan, eds. *Cry of the Environment: Rebuilding the Christian Creation Tradition* (Santa Fe, N.M.: Bear and Co., 1984).

   A partial "family tree of creation-centered spirituality" may be found in Matthew Fox, *Original Blessing* (Santa Fe, N.M.: Bear and Co., 1983).

8. Alice Walker, *The Color Purple* (New York: Harcourt Brace Jovanovich, 1982), 167.

9. "Prayer for the Great Family" may be found in Gary Snyder's Pulitzer Prize-winning volume of poetry, *Turtle Island* (New York: New Directions Books, 1974).

10. See Byron Kennard, "Mixing Religion and Politics," Ecopinion, *Audubon*, 86, no. 2 (Mar. 1984): 14–19.

11. "Power, Glory—and Politics," *Time*, 17 Feb. 1986, 65.

12. Center for Religion Research, Gallup Organization, Princeton, N.J. (Information given via telephone to the author on 16 Oct. 1984.)

13. Ruth Bleier, *Science and Gender* (New York: Pergamon Press, 1984).

14. See Charlene Spretnak, "Naming the Cultural Forces That Push Us toward War," *Journal of Humanistic Psychology* 23, no. 1 (Summer, 1983), 104–14; also in *Nuclear Strategy and the Code of the Warrior: Face of Mars and Shiva in the Crisis of Human Survival*, ed. Richard Grossinger and Lindy Hough (Berkeley, Calif.: Atlantic Books, 1984). Also see the chapters on "The Soldier" and "War" in Mark Gerzon, *A Choice of Heroes* (Boston: Houghton Mifflin, 1982).

15. Charlene Spretnak, *Lost Goddesses of Early Greece: A Collection of Pre-Hellenic Myths* (Boston: Beacon Press, 1981). Also see Virginia Ramey Mollenkott, *The Divine Female: The Biblical Imagery of God as Female* (New York: Crossroad, 1984) for some useful compromise positions on the Great Mother.

16. *Quadragesimo anno* was a commemoration and expansion of Pope Leo XIII's 1891 encyclical *Rerum Novarum (Of the New Situation of the Working Class)*.

17. Andrew M. Greeley, *No Bigger than Necessary* (New York: New American Library, 1977), 10.

18. Joe Holland, *The Postmodern Paradigm Implicit in the Church's Shift to the Left* (Washington, D.C.: Center of Concern, 1984).

19. Regeneration Project, Rodale Press, 33 East Minor St., Emmaus, Penn. 18049.

Why Women Need the Goddess

CAROL CHRIST

AT THE CLOSE OF Ntosake Shange's stupendously successful Broadway play "For Colored Girls Who Have Considered Suicide When the Rainbow Is Enuf," a tall beautiful black woman rises from despair to cry out, "I found God in myself and I loved her fiercely."[1] Her discovery is echoed by women around the country who meet spontaneously in small groups on full moons, solstices, and equinoxes to celebrate the Goddess as symbol of life and death powers and waxing and waning energies in the universe and in themselves.[2]

> It is the night of the full moon. Nine women stand in a circle, on a rocky hill above the city. The western sky is rosy with the setting sun; in the east the moon's face begins to peer above the horizon. . . . The woman pours out a cup of wine onto the earth, refills it and raises it high. "Hail, Tana, Mother of mothers!" she cries. "Awaken from your long sleep, and return to your children again!"[3]

What are the political and psychological effects of this fierce new love of the divine in themselves for women whose spiritual experience has been focused by the male God of Judaism and Christianity? Is the spiritual dimension of feminism a passing diversion, an escape from difficult but necessary political work? Or does the emergence of the symbol of Goddess among women have significant political and psychological ramifications for the feminist movement?

To answer this question, we must first understand the importance of religious symbols and rituals in human life and consider the effect of male symbolism

From: *Womanspirit Rising: A Feminist Reader in Religion* by Carol P. Christ and Judith Plaskow (San Francisco: Harper and Row, 1979), 273–89. Also printed in Carol P. Christ, *Laughter of Aphrodite: Reflections on a Journey to the Goddess* (San Francisco: Harper and Row, 1987), 117–32. First published in *Heresies No. 5: The Great Goddess*, 1978.

of God on women. According to anthropologist Clifford Geertz, religious symbols shape a cultural ethos, defining the deepest values of a society and the persons in it. "Religion," Geertz writes "is a system of symbols which act to produce powerful, pervasive, and long-lasting moods and motivations"[4] in the people of a given culture. A "mood for Geertz is a psychological attitude such as awe, trust, and respect, while a "motivation" is the *social* and *political* trajectory created by a mood that transforms mythos into ethos, symbol system into social and political reality. Symbols have both psychological and political effects, because they create the inner conditions (deep-seated attitudes and feelings) that lead people to feel comfortable with or to accept social and political arrangements that correspond to the symbol system.

Because religion has such a compelling hold on the deep psyches of so many people, feminists cannot afford to leave it in the hands of the fathers. Even people who no longer "believe in God" or participate in the institutional structure of patriarchal religion still may not be free of the power of the symbolism of God the Father. A symbol's effect does not depend on rational assent, for a symbol also functions on levels of the psyche other than the rational. Religion fulfills deep psychic needs by providing symbols and rituals that enable people to cope with limit situations[5] in human life (death, evil, suffering) and to pass through life's important transitions (birth, sexuality, death). Even people who consider themselves completely secularized will often find themselves sitting in a church or synagogue when a friend or relative gets married, or when a parent or friend has died. The symbols associated with these important rituals cannot fail to affect the deep or unconscious structures of the mind of even a person who has rejected these symbolisms on a conscious level—especially if the person is under stress. The reason for the continuing effect of religious symbols is that the mind abhors a vacuum. Symbol systems cannot simply be rejected, they must be replaced. Where there is not any replacement, the mind will revert to familiar structures at times of crisis, bafflement, or defeat.

Religions centered on the worship of a male God create "moods" and "motivations" that keep women in a state of psychological dependence on men and male authority, while at the same legitimating the *political* and *social* authority of fathers and sons in the institutions of society.

Religious symbol systems focused around exclusively male images of divinity create the impression that female power can never be fully legitimate or wholly beneficent. This message need never be explicitly stated (as, for example, it is in the story of Eve) for its effect to be felt. A woman completely ignorant of the myths of female evil in biblical religion nonetheless acknowledges the anomaly of female power when she prays exclusively to a male God. She may see herself as like God (created in the image of God) only by denying her own sexual identity and affirming God's transcendence of sexual identity. But she can never have the experience that is freely available to every man and boy in her culture, of having her full sexual identity affirmed as being in the image and likeness of God. In Geertz's terms, her "mood" is one of trust in male power as salvific and distrust of female power in herself and other women as inferior or dangerous. Such a powerful, pervasive, and longlasting "mood" cannot fail to become a "motivation" that translates into social and political reality.

In *Beyond God the Father*, feminist theologian Mary Daly detailed the psychological and political ramifications of father religion for women. "If God in 'his' heaven is a father ruling his people," she wrote, "then it is the 'nature' of things and according to divine plan and the order of the universe that society be male dominated. Within this context, a *mystification of roles* takes place: The husband dominating his wife represents God 'himself.' The images and values of a given society have been projected into the realm of dogmas and 'Articles of Faith,' and these in turn justify the social structures which have given rise to them and which sustain their plausibility."[6]

Philosopher Simone de Beauvoir was well aware of the function of patriarchal religion as legitimater of male power. As she wrote, "Man enjoys the great advantage of having a god endorse the code he writes; and since man exercises a sovereign authority over women it is especially fortunate that this authority has been vested in him by the Supreme Being. For the Jew, Mohammedans, and Christians, among others, man is Master by divine right; the fear of God will therefore repress any impulse to revolt in the downtrodden female."[7]

This brief discussion of the psychological and political effects of God religion puts us in an excellent position to begin to understand the significance of the symbol of Goddess for women. In discussing the meaning of the Goddess, my method will first be phenomenological. I will isolate a meaning of the symbol of the Goddess as it has emerged in the lives of contemporary women. I will then discuss its psychological and political significance by contrasting the "moods" and "motivations" engendered by Goddess symbols with those engendered by Christian symbolism. I will also correlate Goddess symbolism with themes that have emerged in the women's movement, in order to show how Goddess symbolism undergirds and legitimates the concerns of the women's movement, much as God symbolism in Christianity undergirded the interests of men in patriarchy. I will discuss four aspects of Goddess symbolism here: the Goddess as affirmation of female power, the female body, the female will, and women's bonds and heritage. There are, of course, many other meanings of the Goddess that I will not discuss here.

The sources for the symbol of the Goddess in contemporary spirituality are traditions of Goddess worship and modern women's experience. The ancient Mediterranean, pre-Christian European, Native American, Mesoamerican, Hindu, African, and other traditions are rich sources for Goddess symbolism. But these traditions are filtered through modern women's experiences. Traditions of goddesses, subordination to gods, for example, are ignored. Ancient traditions are tapped selectively and eclecticly, but they are not considered authoritative for modern consciousness. The Goddess symbol has emerged spontaneously in the dreams, fantasies, and thoughts of many women around the country in the past several years. Kirsten Grimstad and Susan Rennie reported that they were surprised to discover widespread interest in spirituality, including the Goddess, among feminists around the country in the summer of 1974.[8] *WomanSpirit* magazine, which published its first issue in 1974 and has contributors from across the United States, has expressed the grass-roots nature of the women's spirituality movement. In 1976, a journal, *Lady Unique*, devoted to the Goddess emerged. In 1975, the first women's spirituality conference was held in Boston and attended by eighteen hundred women. In 1978, a University of California Santa Cruz course on the Goddess drew over five hundred people. Sources for this essay are these manifestations of the Goddess in modern women's experiences as reported in *WomanSpirit*, *Lady Unique*, and elsewhere, and as expressed in conversations I have had with women who have been thinking about the Goddess and women's spirituality.

The simplest and most basic meaning of the symbol of Goddess is the acknowledgment of the legitimacy of female power as a beneficient and independent power. A woman who echoes Ntosake Shange's dramatic statement, "I found God in myself and I loved her fiercely," is saying "Female power is strong and creative." She is saying that the divine principle, the saving and sustaining power, is in herself, that she will no longer look to men or male figures as saviors. The strength and independence of female power can be intuited by contemplating ancient and modern images of the Goddess. This meaning of the symbol of Goddess is simple and obvious, and yet it is difficult for many to comprehend. It stands in sharp contrast to the paradigms of female dependence on males that have been predominant in Western religion and culture. The internationally acclaimed novelist Monique Wittig captured the novelty and flavor of the affirmation of female power when she wrote, in her mythic work *Les Guerilleres*,

> There was a time when you were not a slave, remember that. You walked alone, full of laughter, you bathed bare-bellied. You say you have lost all recollection of it, remember . . . you say there are no words to describe it, you say it does not exist. But remember. Make an effort to remember. Or, failing that, invent.[9]

While Wittig does not speak directly of the Goddess here, she captures the "mood" of joyous celebration of female freedom and independence that is created in women who define their identities through the symbol of Goddess. Artist Mary Beth Edelson expressed the political "motivations" inspired by the Goddess when she wrote,

> The ascending archetypal symbols of the feminine unfold today in the psyche of modern Everywoman. They encompass the multiple forms of the Great Goddess. Reaching across the centuries we take the hands of

our Ancient Sisters. The Great Goddess alive and well is rising to announce to the patriarchs that their 5,000 years are up—Hallelujah! Here we come.[10]

The affirmation of female power contained in the Goddess symbol has both psychological and political consequences. Psychologically, it means the defeat of the view engendered by patriarchy that women's power is inferior and dangerous. This new "mood" of affirmation of female power also leads to new "motivations"; it supports and undergirds women's trust in their own power and the power of other women in family and society.

If the simplest meaning of the Goddess symbol is an affirmation of the legitimacy and beneficience of female power, then a question immediately arises: Is the Goddess simply female power writ large, and if so, why bother with the symbol of Goddess at all? Or does the symbol refer to a Goddess "out there" who is not reducible to a human potential? The many women who have rediscovered the power of Goddess would give three answers to this question: (1) the Goddess is divine female, a personification who can be invoked in prayer and ritual; (2) the Goddess is symbol of the life, death, and rebirth energy in nature and culture, in personal and communal life; and (3) the Goddess is symbol of the affirmation of the legitimacy and beauty of female power (made possible by the new becoming of women in the women's liberation movement). If one were to ask these women which answer is the "correct" one, different responses would be given. Some would assert that the Goddess definitely is *not* "out there," that the symbol of a divinity "out there" is part of the legacy of patriarchal oppression, which brings with it the authoritarianism, hierarchicalism, and dogmatic rigidity associated with biblical monotheistic religions. They might assert that the Goddess symbol reflects the sacred power within women and nature, suggesting the connectedness between women's cycles of menstruation, birth, and menopause, and the life and death cycles of the universe. Others seem quite comfortable with the notion of Goddess as a divine female protector and creator and would find their experience of Goddess limited by the assertion that she is not *also* out there as well as within themselves and in all natural processes. When asked what the symbol of Goddess means, feminist priestess Starhawk replied, "It all depends on how I feel. When I feel weak, she is someone who can help and protect me. When I feel strong, she is the symbol of my own power. At other times I feel her as the natural energy in my body and the world."[11] How are we to evaluate such a statement? Theologians might call these the words of a sloppy thinker. But my deepest intuition tells me they contain a wisdom that Western theological thought has lost.

To theologians, these differing views of the "meaning" of the symbol of Goddess might seem to threaten a replay of the trinitarian controversies. Is there, perhaps, a way of doing theology, which would not lead immediately into dogmatic controversy, which would not require theologians to say definitively that one understanding is true and the others are false? Could

people's relation to a common symbol be made primary and varying interpretations be acknowledged? The diversity of explications of the meaning
of the Goddess symbol suggests that symbols have a richer significance than
any explications of their meaning can express, a point literary critics have
long insisted on. This phenomenological fact suggests that theologians may
need to give more than lip service to a theory of symbol in which the
symbol is viewed as the primary fact and the meanings are viewed as secondary.
It also suggests that a *thealogy*[12] of the Goddess would be very different
from the *theology* we have known in the West. But to spell out this notion
of the primacy of *symbol* in thealogy in contrast to the primacy of the
explanation in theology would be the topic of another paper. Let me simply
state that women, who have been deprived of a female religious symbol
system for centuries, are therefore in an excellent position to recognize the
power and primacy of symbols. I believe women must develop a theory of
symbol and thealogy congruent with their experience at the same time as
they "remember and invent" new symbol systems.

A second important implication of the Goddess symbol for women is the
affirmation of the female body and the life cycle expressed in it. Because of
women's unique position as menstruants, birthgivers, and those who have
traditionally cared for the young and the dying, women's connection to the
body, nature, and this world has been obvious. Women were denigrated
because they seemed more carnal, fleshy, and earthy than the culture-creating
males.[13] The misogynist anti*body* tradition in Western thought is symbolized
in the myth of Eve, who is traditionally viewed as a sexual temptress, the
epitome of women's carnal nature. This tradition reaches its nadir in the
Malleus Maleficarum (*The Hammer of Evil-Doing Women*), which states, "All
witchcraft stems from carnal lust, which in women is insatiable."[14] The Virgin
Mary, the positive female image in Christianity, does not contradict Christian
denigration of the female body and its powers. The Virgin Mary is revered
because she, in her perpetual virginity, transcends the carnal sexuality attributed
to most women.

The denigration of the female body is expressed in cultural and religious
taboos surrounding menstruation, childbirth, and menopause in women. While
menstruation taboos may have originated in a perception of the awesome
powers of the female body,[15] they degenerated into a simple perception that
there is something "wrong" with female bodily functions. Menstruating women
were forbidden to enter the sanctuary in ancient Hebrew and premodern
Christian communities. Although only Orthodox Jews still enforce religious
taboos against menstruant women, few women in our culture grow up affirming
their menstruation as a connection to sacred power. Most women learn
that menstruation is a curse and grow up believing that the bloody facts of
menstruation are best hidden away. Feminists challenge this attitude toward
the female body. Judy Chicago's art piece "Menstruation Bathroom" broke
these menstrual taboos. In a sterile white bathroom, she exhibited boxes of

Tampax and Kotex on an open shelf, and the wastepaper basket was overflowing with bloody tampons and sanitary napkins.[16] Many women who viewed the piece felt relieved to have their "dirty secret" out in the open.

The denigration of the female body and its powers is further expressed in Western culture's attitudes toward childbirth.[17] Religious iconography does not celebrate the birthgiver, and there is no theology or ritual that enables a woman to celebrate the process of birth as a spiritual experience. Indeed, Jewish and Christian traditions also had blood taboos concerning the woman who had recently given birth. While these religious taboos are rarely enforced today (again, only by Orthodox Jews), they have secular equivalents. Giving birth is treated as a disease requiring hospitalization, and the woman is viewed as a passive object, anesthetized to ensure her acquiescence to the will of the doctor. The women's liberation movement has challenged these cultural attitudes, and many feminists have joined with advocates of natural childbirth and home birth in emphasizing the need for women to control and take pride in their bodies, including the birth process.

Western culture also gives little dignity to the postmenopausal or aging women. It is no secret that our culture is based on a denial of aging and death, and that women suffer more severely from this denial than men. Women are placed on a pedestal and considered powerful when they are young and beautiful, but they are said to lose this power as they age. As feminists have pointed out, the "power" of the young woman is illusory, since beauty standards are defined by men, and since few women are considered (or consider themselves) beautiful for more than a few years of their lives. Some men are viewed as wise and authoritative in age, but old women are pitied and shunned. Religious iconography supports this cultural attitude toward aging women. The purity and virginity of Mary and the female saints is often expressed in the iconographic convention of perpetual youth. Moreover, religious mythology associates aging women with evil in the symbol of the wicked old witch. Feminists have challenged cultural myths of aging women and have urged women to reject patriarchal beauty standards and to celebrate the distinctive beauty of women of all ages.

The symbol of Goddess aids the process of naming and reclaiming the female body and its cycles and processes. In the ancient world and among modern women, the Goddess symbol represents the birth, death, and rebirth processes of the natural and human worlds. The female body is viewed as the direct incarnation of waxing and waning, life and death, cycles in the universe. This is sometimes expressed through the symbolic connection between the twenty-eight-day cycles of menstruation and the twenty-eight-day cycles of the moon. Moreover, the Goddess is celebrated in the triple aspect of youth, maturity, and age, or maiden, mother, and crone. The potentiality of the young girl is celebrated in the nymph or maiden aspect of the Goddess. The Goddess as mother is sometimes depicted giving birth, and giving birth is viewed as a symbol for all the creative, life-giving powers of the universe.[18]

The life-giving powers of the Goddess in her creative aspect are not limited
to physical birth, for the Goddess is also seen as the creator of all the arts
of civilization, including healing, writing, and the giving of just law. Women
in the middle of life who are not physical mothers may give birth to poems,
songs, and books, or nurture other women, men, and children. They too
are incarnations of the Goddess in her creative, life-giving aspect. At the
end of life, women incarnate the crone aspect of the Goddess. The wise old
woman, the woman who knows from experience what life is about, the
woman whose closeness to her own death gives her a distance and perspective
on the problems of life, is celebrated as the third aspect of the Goddess.
Thus, women learn to value youth, creativity, and wisdom in themselves
and other women.

The possibilities of reclaiming the female body and its cycles have been
expressed in a number of Goddess-centered rituals. Hallie Mountainwing
and Barby My Own created a summer solstice ritual to celebrate menstruation
and birth. The women simulated a birth canal and birthed each other into
their circle. They raised power by placing their hands on each other's bellies
and chanting together. Finally they marked each other's faces with rich,
dark menstrual blood saying, "This is the blood that promises renewal. This
is the blood that promises sustenance. This is the blood that promises life."[19]
From hidden dirty secret to symbol of the life power of the Goddess, women's
blood has come full circle. Other women have created rituals that celebrate
the crone aspect of the Goddess. Z. Budapest believes that the crone aspect
of the Goddess is predominant in the fall, especially at Halloween, an ancient
holiday. On this day, the wisdom of the old woman is celebrated, and it is
also recognized that the old must die so that the new can be born.

The "mood" created by the symbol of the Goddess in triple aspect is one
of positive, joyful affirmation of the female body and its cycles and acceptance
of aging and death as well as life. The "motivations" are to overcome menstrual
taboos, to return the birth process to the hands of women, and to change
cultural attitudes about age and death. Changing cultural attitudes toward
the female body could go a long way toward overcoming the spirit-flesh,
mind-body dualisms of Western culture, since, as Ruether has pointed out,
the denigration of the female body is at the heart of these dualisms. The
Goddess as symbol of the revaluation of the body and nature thus also
undergirds the human potential and ecology movements. The "mood" is
one of affirmation, awe, and respect for the body and nature, and the
"motivation" is to respect the teachings of the body and the rights of all
living beings.

A third important implication of the Goddess symbol for women is the
positive valuation of will in a Goddess-centered ritual, especially in Goddess-
centered ritual magic and spellcasting in woman-spirit and feminist witchcraft
circles. The basic notion behind ritual magic and spellcasting is energy as
power. Here the Goddess is a center or focus of power and energy; she is

the personification of the energy that flows between beings in the natural and human worlds. In Goddess circles, energy is raised by chanting or dancing. According to Starhawk, "Witches conceive of psychic energy as having form and substance that can be perceived and directed by those with a trained awareness. The power generated within the circle is built into a cone form, and at its peak is released—to the Goddess, to reenergize the members of the coven, or to do a specific work such as healing."[20] In ritual magic, the energy raised is directed by willpower. Women who celebrate in Goddess circles believe they can achieve their wills in the world.

The emphasis on the will is important for women, because women traditionally have been taught to devalue their wills, to believe that they cannot achieve their will through their own power, and even to suspect that the assertion of will is evil. Faith Wildung's poem "Waiting," from which I will quote only a short segment, sums up women's sense that their lives are defined not by their own will, but by waiting for others to take the initiative.

> Waiting for my breasts to develop
> Waiting to wear a bra
> Waiting to menstruate
>
> . . .
>
> Waiting for life to begin, Waiting—
> Waiting to be somebody
>
> . . .
>
> Waiting to get married
> Waiting for my wedding day
> Waiting for my wedding night
>
> . . .
>
> Waiting for the end of the day
> Waiting for sleep. Waiting . . .[21]

Patriarchal religion has enforced the view that female initiative and will are evil through the juxtaposition of Eve and Mary. Eve caused the fall by asserting her will against the command of God, while Mary began the new age with her response to God's initiative, "Let it be done to me according to thy word" (Luke 1:38). Even for men, patriarchal religion values the passive will subordinate to divine initiative. The classical doctrines of sin and grace view sin as the prideful assertion of will and grace as the obedient subordination of the human will to the divine initiative or order. While this view of will might be questioned from a human perspective, Valerie Saiving has argued that it has particularly deleterious consequences for women in Western culture. According to Saiving, Western culture encourages males in the assertion of will, and thus it may make some sense to view the male form of sin as an excess of will. But since culture discourages females in the assertion of will, the traditional doctrines of sin and grace encourage women

to remain in their form of sin, which is self-negation or insufficient assertion of will.[22] One possible reason the will is denigrated in a patriarchal
religious framework is that both human and divine will are often pictured
as arbitrary, self-initiated, and exercised without regard for other wills.

In a Goddess-centered context, in contrast, the will is valued. *A woman
is encouraged to know her will, to believe that her will is valid, and to believe
that her will can be achieved in the world,* three powers traditionally denied to
her in patriarchy. In a Goddess-centered framework, a woman's will is not
subordinated to the Lord God as king and ruler, nor to men as his representatives. Thus a woman is not reduced to waiting and acquiescing in the
wills of others as she is in patriarchy. But neither does she adopt the egocentric
form of will that pursues self-interest without regard for the interests of others.

The Goddess-centered context provides a different understanding of the
will than that available in the traditional patriarchal religious framework.
In the Goddess framework, will can be achieved only when it is exercised
in harmony with the energies and wills of other beings. Wise women, for
example, raise a cone of healing energy at the full moon or solstice when
the lunar or solar energies are at their high points with respect to the earth.
This discipline encourages them to recognize that not all times are propi-
tious for the achieving of every will. Similarly, they know that spring is a
time for new beginnings in work and love, summer a time for producing
external manifestations of inner potentialities, and fall or winter times for
stripping down to the inner core and extending roots. Such awareness of
waxing and waning processes in the universe discourages arbitrary ego-cen-
tered assertion of will, while at the same time encouraging the assertion of
individual will in cooperation with natural energies and the energies created by the wills of others. Wise women also have a tradition that whatever is sent out will be returned and this reminds them to assert their wills
in cooperative and healing, rather than egocentric and destructive, ways.
This view of will allows women to begin to recognize, claim, and assert
their wills without adopting the worst characteristics of the patriarchal
understanding and use of will. In the Goddess-centered framework, the "mood"
is one of positive affirmation of personal will in the context of the energies
of other wills or beings. The "motivation" is for women to know and assert
their wills in cooperation with other wills and energies. This of course does
not mean that women always assert their wills in positive and life-affirming
ways. Women's capacity for evil is, of course, as great as men's. My purpose
is simply to contrast the differing attitudes toward the exercise of will per
se, and the female will in particular, in Goddess-centered religion and in
the Christian God-centered religion.

The fourth and final aspect of Goddess symbolism that I will discuss
here is the significance of the Goddess for a revaluation of woman's bonds
and heritage. As Virginia Woolf has said, "Chloe liked Olivia," a statement
about a woman's relation to another woman, is a sentence that rarely oc-

curs in fiction. Men have written the stories, and they have written about women almost exclusively in their relations to men.[23] The celebrations of women's bonds to each other, as mothers and daughters, as colleagues and co-workers, as sisters, friends, and lovers, is beginning to occur in the new literature and culture created by women in the women's movement. While I believe that the revaluing of each of these bonds is important, I will focus on the mother-daughter bond, in part because I believe it may be the key to the others.

Adrienne Rich has pointed out that the mother-daughter bond, perhaps the most important of woman's bonds, "resonant with charges ... the flow of energy between two biologically alike bodies, one of which has lain in amniotic bliss inside the other, one of which has labored to give birth to the other,"[24] is rarely celebrated in patriarchal religion and culture. Christianity celebrates the father's relation to the son and the mother's relation to the son, but the story of mother and daughter is missing. So, too, in patriarchal literature and psychology, the mothers and the daughters rarely exist. Volumes have been written about the Oedipal complex, but little has been written about the girl's relation to her mother. Moreover, as de Beauvoir has noted, the mother-daughter relation is distorted in patriarchy because the mother must give her daughter over to men in a male-defined culture in which women are viewed as inferior. The mother must socialize her daughter to become subordinate to men, and if her daughter challenges patriarchal norms, the mother is likely to defend the patriarchal structures against her own daughter.[25]

These patterns are changing in the new culture created by women in which the bonds of women to women are beginning to be celebrated. Holly Near has written several songs that celebrate women's bonds and women's heritage. In one of her finest songs she writes of an "old-time woman" who is "waiting to die." A young woman feels for the life that has passed the old woman by and begins to cry, but the old woman looks her in the eye and says, "If I had not suffered, you wouldn't be wearing those jeans / Being an old-time woman ain't as bad as it seems."[26] This song, which Near has said was inspired by her grandmother, expresses and celebrates a bond and a heritage passed down from one woman to another. In another of Near's songs, she sings of "a hiking-boot mother who's seeing the world / For the first time with her own little girl." In this song, the mother tells the drifter who has been traveling with her to pack up and travel alone if he thinks "traveling three is a drag" because "I've got a little one who loves me as much as you need me / And darling, that's loving enough."[27] This song is significant because the mother places her relationship to her daughter above her relationship to a man, something women rarely do in patriarchy.[28]

Almost the only story of mothers and daughters that has been transmitted in Western culture is the myth of Demeter and Persephone that was the basis of religious rites celebrated by women only, the Thesmophoria, and later formed the basis of the Eleusian mysteries, which were open to all who spoke Greek. In this story, the daughter, Persephone, is raped away from her mother, Demeter, by the God of the underworld. Unwilling to accept this state of affairs, Demeter rages and withholds fertility from the earth until her daughter is returned to her. What is important for women in this story is that a mother fights for her daughter and for her relation to her daughter. This is completely different from the mother's relation to her daughter in patriarchy. The "mood" created by the story of Demeter and Persephone is one of celebration of the mother-daughter bond, and the "motivation" is for mothers and daughters to affirm the heritage passed on from mother to daughter and to reject the patriarchal pattern where the primary loyalties of mother and daughter must be to men.

The symbol of Goddess has much to offer women who are struggling to be rid of the "powerful, pervasive, and long-lasting moods and motivations" of devaluation of female power, denigration of the female body, distrust of female will, and denial of the women's bonds and heritage that have been engendered by patriarchal religion. As women struggle to create a new culture in which women's power, bodies, will, and bonds are celebrated, it seems natural that the Goddess would reemerge as symbol of the newfound beauty, strength, and power of women.

Notes

1. From the original cast album, Buddah Records, 1976.
2. See Susan Rennie and Kristen Grimstad, "Spiritual Explorations Cross-Country," *Quest* 1, no. 4 (1975): 49–51; and *WomanSpirit* magazine.
3. See Starhawk, "Witchcraft and Women's Culture," in *Womanspirit Rising: A Feminist Reader in Religion*, ed. Carol P. Christ and Judith Plaskow (San Francisco: Harper and Row, 1979), 259–68.
4. Clifford Geertz, "Religion as a Cultural System," in *Reader in Comparative Religion*, ed. William L. Lessa and Evon V. Vogt, 2d ed. (New York: Harper and Row, 1972), 206.
5. Geertz, "Religion as a Cultural System," 210.
6. Mary Daly, *Beyond God the Father*, (Boston: Beacon Press, 1974), 13; emphasis added.
7. Simone de Beauvoir, *The Second Sex*, trans. H. M. Parshleys (New York: Alfred A. Knopf, 1953).
8. See Grimstad and Rennie, "Spiritual Explorations Cross-Country."
9. Monique Wittig, *Les Guerilleres* trans. David LeVay (New York: Avon Books, 1971), 89. Also quoted in Morgan MacFarland, "Witchcraft: The Art of Remembering," *Quest* 1, no. 4 (1975): 41.
10. Mary Beth Edelson, "Speaking for Myself," *Lady Unique* 1 (1976): 56.
11. Personal communication.
12. A term coined by Naomi Goldenberg to refer to reflection on the meaning of the symbol of goddess.
13. This theory of the origins of the Western dualism is stated by Rosemary Ruether in *New Woman: New Earth* (New York: Seabury Press, 1975), and elsewhere.

14. Heinrich Kramer and Jacob Sprenger (New York: Dover, 1971), 47.
15. See Rita M. Gross, "Menstruation and Childbirth as Ritual and Religious Experience in the Religion of the Australian Aborigines," in *Journal of the American Academy of Religion*, 45, no. 4 (1977): supplement 1147–81.
16. Judy Chicago, *Through the Flower* (New York: Doubleday, 1975), plate 4, pp. 106–7.
17. See Adrienne Rich, *Of Woman Born* (New York: Bantam Books, 1977), chaps. 6 and 7.
18. See James Mellaart, *Earliest Civilizations of the Near East* (New York: McGraw-Hill, 1965), 92.
19. Barbry My Own, "Ursa Maior: Menstrual Moon Celebration," in *Moon, Moon* ed. Anne Kent Rush, (Berkeley, Calif: Moon Books, Random House, 1976), 374–87.
20. Starhawk, "Witchcraft and Women's Culture."
21. In Judy Chicago, *Through the Flower*, 213–17.
22. Valerie Saiving, "The Human Situation: A Feminine View," *Journal of Religion*, 40 (1960): 100, and reprinted in this volume.
23. Virginia Woolf, *A Room of One's Own* (New York: Harcourt Brace Jovanovich, 1928), 86.
24. Rich, *Of Woman Born*, 226.
25. De Beauvior, *Second Sex*, 448–49.
26. "Old Time Woman," lyrics by Jeffrey Langley and Holly Near, from *Holly Near: A Live Album*, Redwood Records, 1974.
27. "Started Out Fine," by Holly Near, from *Holly Near: A Live Album*.
28. Rich, *Of Woman Born*, 223.

Ecology and Process Theology

JOHN COBB, JR.

DURING THE PAST DECADE, people have become aware of the dangers to the human future resulting from exploitation of the environment. This exploitation has been consistent with both the dominant economic theories and the dominant theologies of the nineteenth and twentieth centuries. Ideally these theories called for treatment of all human beings as ends rather than as means, but the power of the dominant theories has been such that their objectifying categories are readily extended to human beings. People, too, become resources, and the term human resources has become prevalent. In practice powerless human beings and powerless societies have been treated as resources for exploitation by those who have the economic and political power to establish goals and to pursue them. In response to this situation, the task cannot be simply to improve practice in light of existing theory. It must be to change the theory. And because our theory, both in economics and theology, has both shaped and expressed our dominant perceptions and sensibility, it is necessary to change our vision of reality as well.

Process thought has been protesting for some time against some of the features of our dominant practice, theory, and sensibility which are now more widely recognized as damaging. In particular, process thought has offered an alternative to the dominant dualisms of soul and body, spirit and nature, mind and matter, self and other. It should be able to speak with some relevance to the contemporary situation. The theology which has appropriated these contributions of process thought is often called "process theology."

The dominant thinker behind the distinctive approach of process theology is the English mathematician-philosopher Alfred North Whitehead (1861–

From: "Process Theology and An Ecological Model," *Pacific Theological Review* 15, no. 2 (Winter 1982): 24–27, 28; original title: "Process Theology and an Ecological Model."

1947). He taught at Harvard University during his later years and has had his largest following in the United States. In his most important book, his Gifford Lectures of 1927–28, entitled *Process and Reality*, he proposed a "philosophy of organism." By this he meant that the actual entities of which the world is ultimately made up are better thought of as organisms than as material or mental substances.

The main significance of the idea of organism here is that each entity exists only in its relation to its environment. One cannot think first of an entity and then, incidentally, of its relations to the rest of the world. On the contrary, every entity is relational in its most fundamental nature. It is constituted by its relations. Even in thought it cannot be abstracted from them.

Professor L. Charles Birch, an Australian biologist active in the World Council of Churches, and I have authored a book in which we develop this point. We call ours an ecological model of living things, and we show that this model fits the evidence of biological sciences better than the substantialist, the materialist, and mechanist models that have dominated their development. The ecological model is also more appropriate to field, relativity, and quantum theory in physics. There are a number of converging trends in the natural sciences that support a Whiteheadian, ecological understanding of nature.

If process thought accepted a dualism between humanity and nature, a changed view of nature would not seem very important theologically. But those who follow Whitehead understand human beings as part of nature. The ecological model applies to us as well. Human experience, too, is constituted by its relations with the body with other people, and with nonhuman creatures.

Three further clarifications of the meaning of these assertions for theology, as well as for the natural sciences, are needed. First, the ecological model depicts the actual entities as events rather than as objects that exist through time. An event occurs in its fullness and is over. Whereas our previous models have usually explained events by the motions of atoms and particles, the ecological model explains atoms and particles as well as tables and mountains, in terms of events. The explanation is finally sought at the level of the ultimate, indivisible unit events into which larger events can be analyzed. These include subatomic events but also momentary human experiences. Whitehead called these unit events "actual occasions" or "occasions of experience." The world is a vast field of actual occasions within which there emerge relatively enduring patterns of many varieties. The atom, the subatomic particle, the table, and the mountain are all societies of actual occasions with relatively enduring patterns. A human person is too.

Second, our usual ideas of such dualities as mental and physical or subjective and objective are derived from reflection about societies of actual occasions, such as atoms and mountains, wrongly supposed to be substances. These ideas cannot be applied to events or occasions. Mind and matter, as conceived according to these usual ideas, do not exist. But understood more loosely, "mental" and "physical" aspects can be found in every occasion whatsoever. Every occasion is a synthesis of features derived from its environing world. In this sense it is physical. Because this synthesis is not merely the mechanical product of the forces that impinge upon it, because it includes an element of selectivity and self-determination, we can also say that the occasion has a mental aspect. There are no purely mental occasions and no purely physical ones. There are only occasions in which the element of self-origination is relatively important and others in which it is negligible.

Similarly, when we think of subjects and objects in the usual way, we cannot say that an event or occasion is a subject or that it is an object. As it takes place it is the locus of subjectivity, that is, it receives the activity of the past, and it, in its turn, acts. It is subject to the action of other events, and it is an agent determining its response to these influences and, thereby, its effects upon the world. It actively takes account of its world. When it has taken place, it becomes part of the objective world in which other occasions occur. All occasions are "subjects" in the moment of their occurrence and "objects" for other occasions as soon as they are over.

Third, every occasion in some measure transcends its world. It transcends it, first, simply by being a new occasion. No occasion can recur. However similar an occasion may be to antecedent ones, it is a different entity. In addition, no occasion is qualitatively identical with any other. It cannot be, because no two occasions have the same spatio-temporal locus, and every locus defines a different environment or world. Since this world enters constitutively into the occasion, the occasion must be qualitatively unique. Further, as already noted, no occasion is the mere product of its world. However much the world of the occasion determines its character, the exact form of the occasion is decided only when it happens. Every occasion transcends its world by determining just how it constitutes itself out of that world. And, finally, many occasions, indeed, all living occasions, constitute themselves not only out of the world of past occasions but also out of the world of possibility or the Not Yet. They transcend their world by incorporating possibilities derived, not from it, but from God. This need not involve conscious choice, but it can, and in the human case it often does. This is the context in which we can speak meaningfully of freedom.

Those who are accustomed to contrast material or merely phenomenal nature with human existence as spiritual may be shocked by the insistence of process thought that human beings are part of an inclusive ecosystem. But it is well to recall that the dualism of the human and the natural is a modern one. For the Bible there is the one world of creatures of which human beings are a part. The welfare of humanity and the welfare of other creatures are interdependent. Process theologians rejoice that Christians are beginning to recover the sense of the unity of creation against the modern philosophical dualism of spirit and nature.

Neither for the Bible nor for process theologians does the insistence that human beings are part of the world of creatures mean that we are not distinctive. Human experiences are the most valuable of all the events on this planet, but they are not the only loci of value. In Genesis God declares all other creatures good even before and apart from the creation of human beings, and process theologians share this biblical conviction. But human beings are created in the image of God, and process theologians agree that we share in God's power of creation in quite unique ways. As Genesis indicates, we exercise a unique power over all the other creatures. Hence we are responsible for the welfare of the whole created order on this planet as no other creature could be. That we are exercising that power so extensively for the degradation of the biosphere as well as the oppression of our powerless sisters and brothers expresses the depth of our betrayal of God's trust.

The doctrine of God to which this brings us is at the center of process theology. It is above all Charles Hartshorne who has developed this aspect of process thought. Process theology has been very critical of some features of classical theism. That theism often pictures God as an immutable substance, whereas process theologians see God as the most perfect exemplification of the ecological model. Divine perfection does not consist in being totally self-contained and unaffected by creaturely joy and suffering but in being totally open, totally receptive, and perfectly responsive. God, too, is constituted by relations to all things. In the case of God, these relations are complete and express the perfection of love. Through this perfection of relations, what is perpetually perishing in the world attains everlastingness in the divine life. The threat to meaning that is inherent in transience is thereby overcome.

Too often, also, classical theism has spoken of God's power in a way that suggests the control of a tyrant or a dictator. It is thought that everything that happens in the world, even an earthquake or a war, must express the will of God. Many have turned against this God, and process theologians do so, too. But process theologians do not go to the extreme of asserting that God is powerless and is to be found only in suffering. God is certainly found there, but God is found also in a child's enjoyment of play, in the joy of a happy marriage, and in the creative ecstasy of an artist. Process theology teaches that God's power is perfect and that perfect power is not coercive. Following the biblical image of parental power, we see that coercion expresses the failure of power, not its perfection. God's power lies in the gift to every occasion of some measure of transcendence or freedom and in the call to employ that freedom in love. It is this gift and call which have brought into being all that is good in our world; indeed, they have created our world. But creatures, and especially human creatures, resist the gift of freedom and abuse it. God does not will the evil which we thereby bring about.

The language of process theology about God often leads readers to suppose that process theologians are optimistic, and it is true that they are hopeful. They hope because the giver of freedom and direction is always surprising us. The future cannot be foreseen. It is not just the unfolding of the past. It is the place where the new can be received if we will accept it. However bleak the projections of past trends into the future, we hope for a new heaven and a new earth. . . .

Process theology developed chiefly at the Divinity School of the University of Chicago. During the first three decades of this century that school was committed to socio-historical research in the service of the social gospel. During the 1930s it became more philosophical, and the influence of Alfred North Whitehead grew. The concerns for political freedom, peace, and justice did not disappear, but they ceased to control the theological program. The dominant emphasis among the theologians was an empirical, rational, and speculative approach to understanding God and God's relationship to human beings. The label "process theology" came to be used during the sixties to emphasize the rejection of static and substantial modes of thought.

When the 1960s brought renewed concern for human liberation as the focus of Christian theology, most process theologians recognized that their work had become too theoretical and abstract. During the [1970s] they made efforts to display the relevance of their cosmological vision to the practical needs of the church. Thus far success has been greatest in relation to interfaith dialogue, ecology, feminism, and pastoral work. Much remains to be done, especially in relation to the world of politics, although a beginning has been made there as well.

The Woman I Love Is a Planet

PAULA GUNN ALLEN

OUR PHYSICALITY—WHICH ALWAYS and everywhere includes our spirituality, mentality, emotionality, social institutions, and processes—is a microform of all physicality. Each of us reflect, in our attitudes toward our body and the bodies of other planetary creatures and plants, our inner attitude toward the planet. And, as we believe, so we act. A society that believes that the body is somehow diseased, painful, sinful, or wrong, a people that spends its time trying to deny the body's needs, aims, goals, and processes—whether these be called health or disease—is going to misunderstand the nature of its existence and of the planet's and is going to create social institutions out of those body-denying attitudes that wreak destruction not only on human, plant, and other creaturely bodies but on the body of the earth herself.

The planet, our mother, grandmother earth, is *physical* and therefore a spiritual, mental, and emotional being. Planets are alive, as are all their byproducts or expressions, such as animals, vegetables, minerals, and climatic and meteorological phenomena.

Believing that our mother, the beloved earth, is inert matter is destructive to yourself. (There's little you can do to her, believe it or not.) Such beliefs point to a dangerously diseased physicality.

Being good, holy, and/or politically responsible means being able to accept whatever life brings—and that includes just about everything you usually think of as unacceptable, like disease, death, and violence. Walking in balance, in harmony, and in a sacred manner requires staying in your body, accepting its discomforts, decayings, witherings, and blossomings and respecting them. Your body is also a planet, replete with creatures that live in and on it. Walking in balance requires knowing that living and dying

From: Irene Diamond and Gloria Orenstein, eds., *Reweaving the World: The Emergence of Ecofeminism* (San Francisco: Sierra Club Books, 1990), 52–57.

are twin beings, gifts of our mother, the earth, and honoring her ways does not mean cheating her of your flesh, your pain, your joy, your sensuality, your desires, your frustrations, your unmet and met needs, your emotions, your life. In the end you can't cheat her successfully, but in the attempt to do so, you can do great harm to the delicate and subtle balance of the vital processes of planetary being.

A society based on body hate destroys itself and causes harm to all of grandmother's grandchildren.

In the United States, where milk and honey cost little enough, where private serenity is prized above all things by the wealthy, privileged, and well washed, where tension, intensity, passion, and the concomitant loss of self-possession are detested, the idea that your attitudes and behaviors vis-à-vis your body are your politics and your spirituality may seem strange. Moreover, when I suggest that passion—whether it be emotional, muscular, sexual, or intellectual—is spirituality, the idea might seem even stranger. In the United States of the privileged, going to ashrams and centers to meditate on how to be in one's immediate experience, on how to be successful at serenity when the entire planet is overwrought, tense, far indeed from serene, the idea that connected spirituality consists in accepting overwroughtness, tension, yes, and violence, may seem not only strange but downright dangerous. The patriarchs have long taught the Western peoples that violence is sin, that tension is the opposite of spiritual life, that the overwrought are denied enlightenment. But we must remember that those who preached and taught serenity and peacefulness were teaching the oppressed how to act—docile slaves who deeply accept their place and do not recognize that in their anguish lies also their redemption, their liberation, are not likely to disturb the tranquillity of the ruling class. Members of the ruling class are, of course, utterly tranquil. Why not? As long as those upon whose labor and pain their serenity rests don't upset the apple cart, as long as they can make the rules for human behavior—in its inner as well as its outer dimensions—they can be tranquil indeed and can focus their attention on reaching nirvanic bliss, transcendence, or divine peace and love.

And yet, the time for tranquillity, if there ever was time for it, is not now. Now we have only to look, to listen, to our beloved planet to see that tranquillity is not the best word to describe her condition. Her volcanic passions, her hurricane storms of temper, her tremblings and shakings, her thrashings and lashings indicate that something other than serenity is going on. And after careful consideration, it must occur to the sensitive observer that congruence with self, which must be congruence with spirit, which must therefore be congruence with the planet, requires something more active than serenity, tranquillity, or inner peace.

Our planet, my beloved, is in crisis; this, of course, we all know. We, many of us, think that her crisis is caused by men, or white people, or capitalism, or industrialism, or loss of spiritual vision, or social turmoil, or war, or psychic disease. For the most part, we do not recognize that the reason for her state is that she is entering upon a great initiation—she is becoming someone else. Our planet, my darling, is gone coyote, *heyoka*, and it is our great honor to attend her passage rites. She is giving birth to her new consciousness of herself and her relationship to the other vast intelligences, other holy beings in her universe. Her travail is not easy, and it occasions her intensity, her conflict, her turmoil—the turmoil, conflict, and intensity that human and other creaturely life mirror. And as she moves, growing and learning ever closer to the sacred moment of her realization, her turmoil, intensity, agony, and conflict increase.

We are each and all a part of her, an expression of her essential being. We are each a small fragment that is not the whole but that, perforce, reflects in our inner self, our outer behavior, our expressions and relationships and institutions, her self, her behaviors, her expressions and relationships, her forms and structures. We humans and our relatives the other creatures are integral expressions of her thought and being. We are not her, but we take our being from her, and in her being we have being, as in her life we have life. As she is, so are we.

In this time of her emergence as one of the sacred planets in the grandmother galaxy, we necessarily experience, each of us in our own specific way, our share or form of her experience, her form. As the initiation nears completion we are caught in the throes of her wailings and contractions, her muscular, circulatory, and neurologic destabilization. We should recognize that her longing for the culmination of the initiatory process is at present nearly as intense as her longing to remain as she was before the initiation ceremony began, and our longing for a new world that the completion of the great ceremony will bring, almost as great as our longing to remain in the systems familiar to us for a very long time, correspond. Her longing for completion is great, as is ours; our longing to remain as we have been, our fear that we will not survive the transition, that we will fail to enter the new age, our terror at ourselves becoming transformed, mutated, unrecognizable to ourselves and all we have known correspond to her longing to remain as she has been, her fear that she will fail the tests as they arise for her, her terror at becoming new, unrecognizable to herself and to all she has known.

What can we do in times such as these? We can rejoice that she will soon be counted among the blessed. That we, her feathers, talons, beak, eyes, have come crying and singing, lamenting and laughing, to this vast climacteric.

I am speaking of all womankind, of all mankind. And of more. I am speaking of all our relatives, the four-leggeds, the wingeds, the crawlers; of the plants and seasons, the winds, thunders, and rains, the rivers, lakes, and streams, the pebbles, rocks, and mountains, the spirits, the holy people, and the gods and goddesses—of all the intelligences, all the beings. I am speaking even of the tiniest, those no one can see; and of the vastest, the planets and stars. Together you and I and they and she are moving with increasing rapidity and under ever increasing pressure toward transformation.

Now, now is the time when mother becomes grandmother, when daughter becomes mother, when the living dead are released from entombment, when the dead live again and walk once again in her ways. Together we all can rejoice, take up the tasks of attending, take up the joy of giving birth and of being born, of transforming in recognition of the awfulness of what is entailed, in recognition of what it is we together can and must and will do. I have said that this is the time of her initiation, of her new birth. I could also say it is the time of mutation, for transformation means to change form; I could also say it is the climacteric, when the beloved planet goes through menopause and takes her place among the wise women planets that dance among the stars.

At a time such as this, what indeed can we do? We can sing *Heya-hey* in honoring all that has come to pass, all that is passing. Sing, honoring, *Heya-hey* to all the beings gathering on all the planes to witness this great event. From every quadrant of the universe they are coming. They are standing gathered around, waiting for the emergence, the piercing moment when she is counted among those who are counted among the wise. We can sing *Heya-hey* to the familiar and the estranged, to the recognized and the disowned, to each shrub and tree, to each flower and vine, to each pebble and stone, to each mountain and hill. We can sing *Heya-hey* honoring the stars and the clouds, the winds and the rains, the seasons and the temperature. We can think with our hearts, as the old ones do, and put our brains and muscles in the service of the heart, our mother and grandmother earth, who is coming into being in another way. We can sing *Heya-hey*, honoring.

What can we do, rejoicing and honoring, to show our respect? We can heal. We can cherish our bodies and honor them, sing *Heya-hey* to our flesh. We can cherish our being—our petulances and rages, our anguishes and griefs, our disabilities and strengths, our desires and passions, our pleasures and delights. We can, willingly and recognizing the fullness of her abundance, which includes scarcity and muchness, enter inside ourselves to seek and find her, who is our own dear body, our own dear flesh. For the body is not the dwelling place of the spirit—it is the spirit. It is not a tomb, it is life itself. And even as it withers and dies, it is born; even as it is renewed and reborn, it dies.

Think: How many times each day do you habitually deny and deprive her in your flesh, in your physicality? How often do you willfully prevent her from moving or resting, from eating or drinking what she requests, from eliminating wastes or taking breath? How many times do you order your body to produce enzymes and hormones to further your social image, your "identity," your emotional comfort, regardless of your actual situation and hers? How many of her gifts do you spurn, how much of her abundance do you deny? How often do you interpret disease as wrong, suffering as abnormal, physical imperatives as troublesome, cravings as failures, deprivation and denial of appetite as the right thing to do? In how many ways do you refuse to experience your vulnerability, your frailty, your mortality? How often do you refuse these expressions of the life force of the mother in your lovers, your friends, your society? How often do you find yourself interpreting sickness, weakness, aging, fatness, physical differences as pitiful, contemptible, avoidable, a violation of social norm and spiritual accomplishment? How much of your life is devoted to avoiding any and/or all of these? How much of her life is devoted to avoiding any and all of these?

The mortal body is a tree; it is holy in whatever condition; it is truth and myth because it has so many potential conditions; because of its possibilities, it is sacred and profane; most of all, it is your most precious talisman, your own connection to her. Healing the self means honoring and recognizing the body, accepting rather than denying all the turmoil its existence brings, welcoming the woes and anguish flesh is subject to, cherishing its multitudinous forms and seasons, its unfailing ability to know and be, to grow and wither, to live and die, to mutate, to change. Healing the self means commiting ourselves to a wholehearted willingness to be what and how we are—beings frail and fragile, strong and passionate, neurotic and balanced, diseased and whole, partial and complete, stingy and generous, safe and dangerous, twisted and straight, storm-tossed and quiescent, bound and free.

What can we do to be politically useful, spiritually mature attendants in this great transformation we are privileged to participate in? Find out by asking as many trees as you meet how to be a tree. Our mother, in her form known as Sophia, was long ago said to be a tree, the great tree of life: Listen to what they wrote down from the song she gave them:

> I have grown tall as a cedar on Lebanon,
> as a cypress on Mount Hermon;
> I have grown tall as a palm in Engedi,
> as the rose bushes of Jericho;
> as a fine olive on the plain,
> as a plane tree I have grown tall.
> I have exhaled perfume like cinnamon and acacia;
> I have breathed out a scent like choice myrrh,
> like galbanum, onzcha and stacte,
> like the smoke of incense in the tabernacle.
> I have spread my branches like a terebinth,
> and my branches are glorious and graceful.
> I am like a vine putting out graceful shoots,
> my blossoms bear the fruit of glory and wealth.
> Approach me, you who desire me,
> and take your fill of my fruits.[1]

Note

1. Quoted by Susan Cady, Marian Ronan, and Hal Taussig, *SOPHIA: The Future of Feminist Spirituality* (San Francisco: Harper & Row, 1986), 29.

PART VII

Postmodern Science

Systems Theory and the New Paradigm

Fritjof Capra

Crisis and Transformation in Science and Society

THE DRAMATIC CHANGE IN concepts and ideas that happened in physics during the first three decades of this century has been widely discussed by physicists and philosophers for more than fifty years. It led Thomas Kuhn to the notion of a scientific paradigm, a constellation of achievements—concepts, values, techniques, and so on—shared by a scientific community and used by that community to define legitimate problems and solutions. Changes of paradigms, according to Kuhn, occur in discontinuous, revolutionary breaks called paradigm shifts.[1]

Today, [three decades] after Kuhn's analysis, we recognize paradigm shifts in physics as an integral part of a much larger cultural transformation.[2] The intellectual crisis of quantum physicists in the 1920s is mirrored today by a similar but much broader cultural crisis. The major problems of our time—the growing threat of nuclear war, the devastation of our natural environment, our inability to deal with poverty and starvation around the world, to name just the most urgent ones—are all different facets of one single crisis, which is essentially a crisis of perception. Like the crisis in quantum physics, it derives from the fact that most of us, and especially our large social institutions, subscribe to the concepts of an outdated world view, inadequate for dealing with the problems of our overpopulated, globally interconnected world. At the same time, researchers in several scientific disciplines, various social movements, and numerous alternative organiza-

From: "Physics and the Current Change of Paradigms," in *The World View of Contemporary Physics: Does It Need a New Metaphysics?*, ed. Richard F. Kitchener (Albany: State University of New York Press, 1988), 144–52.

tions and networks are developing a new vision of reality that will form the basis of our future technologies, economic systems, and social institutions.

What we are seeing today is a shift of paradigms not only within science but also in the larger social arena. To analyze that cultural transformation, I have generalized Kuhn's account of a scientific paradigm to that of a *social paradigm*, which I define as "a constellation of concepts, values, perceptions, and practices shared by a community, which form a particular vision of reality that is the basis of the way the community organizes itself."[3]

The social paradigm now receding has dominated our culture for several hundred years, during which it has shaped our modern Western society and has significantly influenced the rest of the world. This paradigm consists of a number of ideas and values, among them the view of the universe as a mechanical system composed of elementary building blocks, the view of the human body as a machine, the view of life in a society as a competitive struggle for existence, the belief in unlimited material progress to be achieved through economic and technological growth and—last but not least—the belief that a society, in which the female is everywhere subsumed under the male, is one that follows from some basic law of nature. During recent decades, all of these assumptions have been found severely limited and in need of radical revision.

Indeed, such a revision is now taking place. The emerging new paradigm may be called a holistic, or an ecological, world view, using the term *ecological* here in a much broader and deeper sense than it is commonly used. Ecological awareness, in that deep sense, recognizes the fundamental interdependence of all phenomena and the embeddedness of individuals and societies in the cyclical processes of nature.

Ultimately, deep ecological awareness is spiritual or religious awareness. When the concept of the human spirit is understood as the mode of consciousness in which the individual feels connected to the cosmos as a whole, which is the root meaning of the word *religion* (from the Latin *religare*, meaning "to bind strongly"), it becomes clear that ecological awareness is spiritual in its deepest essence. It is, therefore, not surprising that the emerging new vision of reality, based on deep ecological awareness, is consistent with the "perennial philosophy" of spiritual traditions, for example, that of Eastern spiritual traditions, the spirituality of Christian mystics, or with the philosophy and cosmology underlying the Native American traditions.[4]

The Systems Approach

In science, the language of systems theory, and especially the theory of living systems, seems to provide the most appropriate formulation of the new ecological paradigm.[5] Since living systems cover such a wide range of phenomena—individual organisms, social systems, and ecosystems—the theory provides a common framework and language for biology, psychology, medicine, economics, ecology, and many other sciences, a framework in which the so urgently needed ecological perspective is explicitly manifest.

The conceptual framework of contemporary physics, and especially those aspects [suggesting a new metaphysics is needed], may be seen as a special case of the systems approach, dealing with nonliving systems and exploring the interface between nonliving and living systems. It is important to recognize, I believe, that in the new paradigm physics is no longer the model and source of metaphors for the other sciences. Even though the paradigm shift in physics is still of special interest, since it was the first to occur in modern science, physics has now lost its role as the science providing the most fundamental description of reality.

I would now like to specify what I mean by the systems approach. To do so, I shall identify five criteria of systems thinking that, I claim, hold for all the sciences—the natural sciences, the humanities, and the social sciences. I shall formulate each criterion in terms of the shift from the old to the new paradigm, and I will illustrate the five criteria with examples from contemporary physics. However, since the criteria hold for all the sciences, I could equally well illustrate them with examples from biology, psychology, or economics.

1. *Shift from the part to the whole.* In the old paradigm, it is believed that in any complex system the dynamics of the whole can be understood from the properties of the parts. The parts themselves cannot be analyzed any further, except by reducing them to still smaller parts. Indeed, physics has been progressing in that way, and at each step there has been a level of fundamental constituents that could not be analyzed any further.

In the new paradigm, the relationship between the parts and the whole is reversed. The properties of the parts can be understood only from the dynamics of the whole. In fact, ultimately there are no parts at all. What we call a part is merely a pattern in an inseparable web of relationships. The shift from the part to the whole was the central aspect of the conceptual revolution of quantum physics in the 1920s. Werner Heisenberg was so impressed by this aspect that he entitled his autobiography *Der Teil und das Ganze* (*The Part and the Whole*).[6] More recently, the view of physical reality as a web of relationships has been emphasized by Henry Stapp, who showed how this view is embodied in S-matrix theory.[7]

2. *Shift from structure to process.* In the old paradigm, there are fundamental structures, and then there are forces and mechanisms through which these interact, thus giving rise to processes. In the new paradigm, every structure is seen as the manifestation of an underlying process. The entire web of relationships is intrinsically dynamic. The shift from structure to process is evident, for example, when we remember that mass in contemporary physics is no longer seen as measuring a fundamental substance but rather as a form of energy, that is, as measuring activity or processes. The

shift from structure to process is also apparent in the work of Ilya Prigogine, who entitled his classic book *From Being to Becoming*.[8]

3. *Shift from objective to "epistemic" science.* In the old paradigm, scientific descriptions are believed to be objective, that is, independent of the human observer and the process of knowing. In the new paradigm, it is believed that epistemology—the understanding of the process of knowledge—has to be included explicitly in the description of natural phenomena. This recognition entered into physics with Heisenberg and is closely related to the view of physical reality as a web of relationships. Whenever we isolate a pattern in this network and define it as a part, or an object, we do so by cutting through some of its connections to the rest of the network, and this may be done in different ways. As Heisenberg put it, "What we observe is not nature itself, but nature exposed to our method of questioning."[9]

This method of questioning, in other words epistemology, inevitably becomes part of the theory. At present, there is no consensus about what is the proper epistemology, but there is an emerging consensus that epistemology will have to be an integral part of every scientific theory.

4. *Shift from "building" to "network" as metaphor of knowledge.* The metaphor of knowledge as a building has been used in Western science and philosophy for thousands of years. There are *fundamental* laws, *fundamental* principles, basic building blocks, and so on. The edifice of science must be built on firm foundations. During periods of paradigm shift, it was always felt that the foundations of knowledge were shifting, or even crumbling, and that feeling induced great anxiety. Einstein (1949), for example, wrote in his autobiography about the early days of quantum mechanics: "All my attempts to adapt the theoretical foundations of physics to this (new type) of knowledge failed completely. It was as if the ground had been pulled out from under one, with no firm foundation to be seen anywhere, upon which one could have built."[10]

In the new paradigm, the metaphor of knowledge as a building is being replaced by that of the network. Since we perceive reality as a network of relationships, our descriptions, too, form an interconnected network of concepts and models in which there are no foundations. For most scientists this metaphor of knowledge as a network with no firm foundations is extremely uncomfortable. It is explicitly expressed in physics in Geoffrey Chew's bootstrap theory of particles.[11] According to Chew, nature cannot be reduced to any fundamental entities, but has to be understood entirely through self-consistency. There are no fundamental equations or fundamental symmetries in the bootstrap theory. Physical reality is seen as a dynamic web of interrelated events. Things exist by virtue of their mutually consistent relationships, and all of physics has to follow uniquely from the requirement that its components be consistent with one another and with themselves. This approach is so foreign to our traditional scientific ways of thinking that it is pursued today only by a small minority of physicists.

When the notion of scientific knowledge as a network of concepts and models, in which no part is any more fundamental than the others, is applied to science as a whole, it implies that physics can no longer be seen as the most fundamental level of science. Since there are no foundations in the network, the phenomena described by physics are not any more fundamental than those described, for example, by biology or psychology. They belong to different systems levels, but none of those levels is any more fundamental than the others.

5. *Shift from truth to approximate descriptions.* The four criteria of systems thinking presented so far are all interdependent. Nature is seen as an interconnected, dynamic web of relationships, in which the identification of specific patterns as "objects" depends on the human observer and the process of knowledge. This web of relationships is described in terms of a corresponding network of concepts and models, none of which is any more fundamental than the others.

This new approach immediately raises an important question: If everything is connected to everything else, how can you ever hope to understand anything? Since all natural phenomena are ultimately interconnected, in order to explain any one of them we need to understand all the others, which is obviously impossible.

What makes it possible to turn the systems approach into a scientific theory is the fact that there is such a thing as approximate knowledge. This insight is crucial to all of modern science. The old paradigm is based on the Cartesian belief in the certainty of scientific knowledge. In the new paradigm, it is recognized that all scientific concepts and theories are limited and approximate. Science can never provide any complete and definitive understanding. Scientists do not deal with truth in the sense of a precise correspondence between the description and the described phenomena. They deal with limited and approximate descriptions of reality. Heisenberg often pointed out that important fact. For example, he wrote in *Physics and Philosophy*, "The often discussed lesson that has been learned from modern physics [is] that every word or concept, clear as it may seem to be, has only a limited range of applicability."[12]

SELF-ORGANIZING SYSTEMS

The broadest implications of the systems approach are found today in a new theory of living systems, which originated in cybernetics in the 1940s and emerged in its main outlines over the last twenty years.[13] As I mentioned before, living systems include individual organisms, social systems, and ecosystems, and thus the new theory can provide a common framework and language for a wide range of disciplines—biology, psychology, medicine, economics, ecology, and many others.

The central concept of the new theory is that of self-organization. A living system is defined as a self-organizing system, which means that its order is not imposed by the environment but is established by the system itself. In other words, self-organizing systems exhibit a certain degree of autonomy. This does not mean that living systems are isolated from their environment; on the contrary, they interact with it continually, but this interaction does not determine their organization.

In this essay, I can give only a brief sketch of the theory of self-organizing systems. To do so, let me distinguish three aspects of self-organization:

1. *Pattern of organization*: the totality of relationships that define the system as an integrated whole
2. *Structure*: the physical realization of the pattern of organization in space and time
3. *Organizing activity*: the activity involved in realizing the pattern of organization

For self-organizing systems, the pattern of organization is characterized by a mutual dependency of the system's parts, which is necessary and sufficient to understand the parts. This is quite similar to the pattern of relationships between subatomic particles in Chew's bootstrap theory. However, the pattern of self-organization has the additional property that gives the whole system an individual identity.

The pattern of self-organization has been studied extensively and described precisely by Humberto Maturana and Francisco Varela, who have called it *autopoiesis*, which means literally self-production. Sometimes it is also called operational closure.[14]

An important aspect of the theory is the fact that the description of the pattern of self-organization does not use any physical parameters, such as energy or entropy, nor does it use the concepts of space and time. It is an abstract mathematical description of a pattern of relationships. This pattern can be realized in space and time in different physical structures, which are then described in terms of the concepts of physics and chemistry. But such a description alone will fail to capture the biological phenomenon of self-organization. In other words, physics and chemistry are not enough to understand life; we also need to understand the pattern of self-organization, which is independent of physical and chemical parameters.

The structure of self-organizing systems has been studied extensively by Ilya Prigogine, who has called it a dissipative structure.[15] The two main characteristics of a dissipative structure are (1) that it is an open system, maintaining its pattern of organization through continuous exchange of energy and matter with its environment; and (2) that it operates far from thermodynamic equilibrium and thus cannot be described in terms of classical thermodynamics. One of Prigogine's greatest contributions has been to create a new thermodynamics to describe living systems.

The organizing activity of living, self-organizing systems, finally, is cognition, or mental activity. This implies a radically new concept of mind, which was first proposed by Gregory Bateson.[16] Mental process is defined as the organizing activity of life. This means that all interactions of a living system with its environment are cognitive, or mental interactions. With this new concept of mind, life and cognition become inseparably connected. Mind, or more accurately, mental process is seen as being immanent in matter at all levels of life.

I have taken some time to outline the emerging theory of self-organizing systems because it is today the broadest scientific formulation of the ecological paradigm with the most wide-ranging implications. The world view of contemporary physics, in my view, will have to be understood within that broader framework. In particular, any speculation about human consciousness and its relation to the phenomena described by physics will have to take into account the notion of mental process as the self-organizing activity of life.

Science and Ethics

A further reason why I find the theory of self-organizing systems so important is that it seems to provide the ideal scientific framework for an ecologically oriented ethics.[17] Such a system of ethics is urgently needed, since most of what scientists are doing today is not life-furthering and life-preserving but life-destroying. With physicists designing nuclear weapons that threaten to wipe out all life on the planet, with chemists contaminating our environment, with biologists releasing new and unknown types of microorganisms into the environment without really knowing what the consequences are, with psychologists and other scientists torturing animals in the name of scientific progress, with all these activities occurring, it seems that it is most urgent to introduce ethical standards into modern science.

It is generally not recognized in our culture that values are not peripheral to science and technology but constitute their very basis and driving force. During the scientific revolution in the seventeenth century, values were separated from facts, and since that time we have tended to believe that scientific facts are independent of what we do and, therefore, independent of our values. In reality, scientific facts emerge out of an entire constellation of human perceptions, values, and actions—in a word, out of a paradigm—from which they cannot be separated. Although much of the detailed research may not depend explicitly on the scientist's value system, the larger paradigm within which this research is pursued will never be value-free. Scientists, therefore, are responsible for their research not only intellectually but also morally.

One of the most important insights of the new systems theory of life is that life and cognition are inseparable. The process of knowledge is also the process of self-organization, that is, the process of life. Our conventional model of knowledge is one of a representation or an image of independently existing facts, which is the model derived from classical physics. From the new systems point of view, knowledge is part of the process of life, of a dialogue between object and subject.

Knowledge and life, then, are inseparable, and, therefore, facts are inseparable from values. Thus, the fundamental split that made it impossible to include ethical considerations in our scientific world view has now been healed. At present, nobody has yet established a system of ethics that expresses the same ecological awareness on which the systems view of life is based, but I believe that this is now possible. I also believe that it is one of the most important tasks for scientists and philosophers today.

Notes

1. Thomas Kuhn, *The Structure of Scientific Revolutions* (Chicago: University of Chicago Press, 1970). The definition quoted is my own synthesis of several definitions given by Kuhn.
2. See Fritjof Capra, *The Turning Point* (New York: Bantam Books, 1983).
3. Fritjof Capra, "The Concept of Paradigm and Paradigm Shift" and "New Paradigm Thinking in Science," *ReVision* 9, no. 1 (Summer/Fall 1986): 11.
4. Fritjof Capra, *The Tao of Physics*, 2d ed. (New York: Bantam Books, 1984).
5. Capra, *Turning Point*, chap. 9.
6. Werner Heisenberg, *Der Teil und das Ganze* (Piper, 1970); published in the United States as *Physics and Beyond* (New York: Harper and Row, 1971).
7. Henry P. Stapp, "S-Matrix Interpretation of Quantum Theory," *Physical Review D, Particles and Fields* 3, no. 6 (15 Mar. 1971): 1303–20; Henry P. Stapp, "The Copenhagen Interpretation," *American Journal of Physics* 40, no. 7 (July 1972): 1098–1116.
8. Ilya Prigogine, *From Being to Becoming* (San Francisco: W. H. Freeman, 1980).
9. Werner Heisenberg, *Physics and Philosophy* (New York: Harper and Row, 1971), 58.
10. Albert Einstein in P. A. Schilpp, ed., *Albert Einstein: Philosopher-Scientist* (Evanston, Ill.: Library of Living Philosophers, 1949), 45.
11. See Fritjof Capra, "Bootstrap Physics: A Conversation with Geoffrey Chew," in *A Passion for Physics: Essays in Honor of Geoffrey Chew*, including an interview with Chew, ed. C. De Tar, J. Finkelstein, and Chung-I Tang (Philadelphia: World Scientific, 1985), 247–86.
12. Heisenberg, *Physics and Philosophy*, p. 125.
13. Capra, *Turning Point*, chap. 9.
14. Humberto R. Maturana and Francisco J. Varela, *Autopoiesis and Cognition: The Realization of the Living* (Boston: Reidel, 1980).
15. Prigogine, *From Being to Becoming*.
16. Gregory Bateson, *Mind and Nature* (New York: Dutton, 1979).
17. Fritjof Capra, ed., "Science and Ethics," Elmwood Discussion Transcript No. 1, Elmwood Institute, Berkeley, Calif.

Postmodern Science and a Postmodern World

DAVID BOHM

MODERN PHYSICS AND THE MODERN WORLD

WITH THE COMING OF the modern era, human beings' view of their world and themselves underwent a fundamental change. The earlier, basically religious approach to life was replaced by a secular approach. This approach has assumed that nature could be thoroughly understood and eventually brought under control by means of the systematic development of scientific knowledge through observation, experiment, and rational thought. This idea became powerful in the seventeenth and eighteenth centuries. In fact, the great seal of the United States has as part of its motto "the new secular order," showing the way the founders of the country were thinking. The main focus of attention was on discerning the order of the universe as it manifests itself in the laws of nature. The principal path to human happiness was to be in the discovery of these laws, in complying with them, in utilizing them wherever possible for the benefit of humankind.

So great is the change in the whole context of thought thereby brought about that Huston Smith and some others have described it as the onset of the modern mind.[1] This mind is in contrast with the mind of the medieval period, in which it was generally supposed that the order of nature was beyond human comprehension and in which human happiness consisted in being aware of the revealed knowledge of God and carrying out the divine commandments. A total revolution occurred in the way people were aiming to live.

The modern mind went from one triumph to another for several centuries through science, technology, industry, and it seemed to be solidly based

From: David Ray Griffin, ed., *The Reenchantment of Science: Postmodern Proposals* (Albany: State University of New York Press, 1988), 57–58, 60–66, 68.

for all time. But in the early twentieth century, it began to have its foundations questioned. The challenge coming from physics was especially serious, because it was in this science that the modern mind was thought to have its firmest foundation. In particular, relativity theory, to a certain extent, and quantum theory, to a much greater extent, led to questioning the assumption of an intuitively imaginable and knowable order in the universe. The nature of the world began to fade out into something almost indescribable. For the most part, physicists began to give up the attempt to grasp the world as an intuitively comprehensible whole; they instead restricted their work mostly to developing a mathematical formalism with rules to apply in the laboratory and eventually in technology. Of course, a great deal of unity has emerged in this work, but it is almost entirely in the mathematical formalism. It has little or no imaginative or intuitive expression (whereas Newton's ideas were quite easily understandable by any reasonably educated person). . . .

The possibility of a postmodern physics, extended also to postmodern science in general, may be of crucial significance. A postmodern science should not separate matter and consciousness and should therefore not separate facts, meaning, and value. Science would then be inseparable from a kind of intrinsic morality, and truth and virtue would not be kept apart as they currently are in science. This separation is part of the reason we are in our present desperate situation.

Of course, this proposal runs entirely contrary to the prevailing view of what science should be, which is a morally neutral way of manipulating nature, either for good or for evil, according to the choices of the people who apply it. I hope in this essay to indicate how a very different approach to science is possible, one that is consistent and plausible and that fits better the actual development of modern physics than does the current approach.

MECHANISTIC PHYSICS

I begin by outlining briefly the mechanistic view in physics, which was characteristic of the modern view and which reached its highest point toward the end of the nineteenth century. This view remains the basis of the approach of most physicists and other scientists today. Although the more recent physics has dissolved the mechanistic view, not very many scientists and even fewer members of the general public are aware of this fact; therefore, the mechanistic view is still the dominant view as far as effectiveness is concerned. In discussing this mechanistic view, I start by listing the principal characteristics of mechanism in physics. To clarify this view, I contrast it with that of ancient times, which was organic rather than mechanistic.

The first point about mechanism is that the world is reduced as far as possible to a set of basic elements. Typically, these elements take the form of particles. They can be called atoms or sometimes these are broken into electrons, protons, and neutrons; now the most elementary particles are called quarks, maybe there will be a subquark. Whatever they may be called, the assumption is that a basic element exists which we either have or hope to have. To these elementary particles, various continuous fields, such as electromagnetic and gravitational fields, must be added.

Second, these elements are basically external to each other; not only are they separate in space, but even more important, the fundamental nature of each is independent of that of the other. Each particle just has its own nature; it may be somewhat affected by being pushed around by the others, but that is all. The elements do not grow organically as parts of a whole, but are rather more like parts of a machine whose forms are determined externally to the structure of the machine in which they are working. By contrast, organic parts, the parts of an organism, all grow together with the organism.

Third, because the elements only interact mechanically by sort of pushing each other around, the forces of interaction do not affect their inner natures. In an organism or a society, by contrast, the very nature of each part is profoundly affected by changes in the other parts, so that the parts are internally related. If a man comes into a group, the consciousness of the whole group may change, depending on what he does. He does not push people's consciousnesses around as if they were parts of a machine. In the mechanistic view, this sort of organismic behavior is admitted, but it is explained eventually by analyzing everything into still smaller particles out of which the organs of the body are made, such as DNA molecules, ordinary molecules, atoms, and so on. This view says that eventually everything is reducible to something mechanical.

The mechanistic program has been very successful and is still successful in certain areas, for example, in genetic engineering to control heredity by treating the molecules on which heredity depends. Advocates do admit that the program still has much to achieve, but this mechanistic reductionistic program assumes that there is nothing that cannot eventually be treated in this way—that if we just keep on going this way we will deal with anything that may arise.

The adherence to this program has been so successful as to threaten our very existence as well as to produce all sorts of other dangers, but, of course, such success does not prove its truth. To a certain extent the reductionistic picture is still an article of faith, and faith in the mechanistic reductionistic program still provides the motivation of most of the scientific enterprise, the faith that this approach can deal with everything. This is a counterpart of the religious faith that people had earlier which allowed them to do great things.

How far can this faith in mechanism be justified? People try endlessly to justify faith in their religions through theology, and much similar work has gone into justifying faith in mechanism through the philosophy of science. Of course, that the mechanism works in a very important domain is given, thereby bringing about a revolution in our life.

During the nineteenth century, the Newtonian world view seemed so certain and complete that no serious scientist was able to doubt it. In fact, we may refer to Lord Kelvin, one of the leading theoretical physicists at the time. He expressed the opinion that physics was more or less finished, advising young people not to go into the field because further work was only a matter of minor refinements. He did point, however, to two small clouds on the horizon. One was the negative results of the Michelson-Morley experiment and the other was the difficulty in understanding black-body radiation. Now he certainly chose his clouds well: the first one led to the theory of relativity and the second to quantum theory. Those little clouds became tremendous storms; but the sky is not even as clear today as it was then—plenty of clouds are still around. The fact that relativity and quantum together overturned the Newtonian physics shows the danger of complacency about world view. It shows that we constantly must look at our world views as provisional, as exploratory, and to inquire. We must have a world view, but we must not make it an absolute thing that leaves no room for inquiry and change. We must avoid dogmatism.

THE BEGINNING OF NONMECHANISTIC PHYSICS: RELATIVITY THEORY

Relativity theory was the first important step away from the mechanistic vision. It introduced new concepts of space, time, and matter. Instead of having separate little particles as the constituents of matter, Einstein thought of a field spread through all space, which would have strong and weak regions. Some strong regions, which are stable, represent particles. If you watch a whirlpool or a vortex, you see the water going around and you see that the movement gets weaker the farther away it is from the center, but it never ends. Now the vortex does not actually exist; there is only the moving water. The vortex is a pattern and a form your mind abstracts from the sensations you have of moving water. If two vortices are put together, they will affect each other; a changing pattern will exist where they modify each other, but it will still be only one pattern. You can say that two exist, but this is only a convenient way of thinking. As they become even closer together, they may merge. When you have flowing water with patterns in them, none of those patterns actually has a separate existence. They are appearances or forms in the flowing movement, which the mind abstracts momentarily for the sake of convenience. The flowing pattern is the ultimate reality, at least at that level. Of course, all the nineteenth-century physicists knew this perfectly well, but they said that *really* water is made of little atoms, that neither the vortices nor the water are the reality: the reality is the little atoms out of which it is all made. So the problem did not bother them.

But with the theory of relativity, Einstein gave arguments showing that thinking of these separate atoms as existent would not be consistent. His solution was to think of a field not so different from the flowing water, a field that spreads through all space and time and in which every particle is a stable form of movement, just as the vortex or whirlpool is a temporarily stable form that can be thought of as an entity which can be given a name. We speak of a whirlpool, but one does not exist. In the same way, we can speak of a particle, but one does not exist: "particle" is a name for a certain form in the field of movement. If you bring two particles together, they will gradually modify each other and eventually become one. Consequently, this approach contradicted the assumption of separate, elementary, mechanical constituents of the universe. In doing so, it brought in a view which I call "unbroken wholeness or flowing wholeness": it has also been called "seamless wholeness." The universe is one seamless, unbroken whole, and all the forms we see in it are abstracted by our way of looking and thinking, which is convenient at times, helping us with our technology, for example.

Nonetheless, relativity theory retains certain essential features of mechanism, in that the fields at different points in space were thought to exist separately and not to be internally related. The separate existence of these basic elements was emphasized by the idea that they were only locally connected, that the field at one point could affect a field only infinitesimally nearby. There was no direct effect of a field here on something far away. This notion is now being called "locality" by physicists; it is the notion of no long-distance connection. This notion is essential to the kind of mechanistic materialism developing throughout the science of the modern era, the notion of separate elements not internally related and not connected to things far away. The animistic view of earlier times was that spirits were behind everything and that these spirits were not located anywhere. Therefore, things far away would tend to be related. This view was taken to be most natural by astrologers and alchemists. But that view had been turned completely around in the modern period, and the modern view seemed so fruitful and so powerful that there arose the utter conviction of its truth.

MORE FULLY NONMECHANISTIC PHYSICS: QUANTUM THEORY

With quantum theory, a much bigger change occurred. The main point is that all action or all motion is found in a discrete indivisible unit called a "quantum." In the early form of the theory, electrons had to jump from one orbit to the other without passing in between. The whole idea of the continuous motion of particles, an idea at the heart of mechanism, was thereby being questioned. The ordinary visible movement, like my hand moving, was thought to comprise a vast number of quantum movements, just as, if enough fine grains of sand are in the hourglass, the flow seems continuous. All movements were said to comprise very tiny, discrete movements that do not, as it were, go from one place to another by passing through the space in between. This was a very mysterious idea.

Second, matter and energy had a dual nature; they manifest either like a wave or like a particle, according to how they were treated in an experiment. An electron is ordinarily a particle, but it can also behave like waves, and light which ordinarily behaves like waves can also behave like particles; their behavior depends on the context in which they are treated. That is, the quality of the thing depends on the context. This idea is utterly opposed to mechanism, because in mechanism the particle is just what it is, no matter what the context. Of course, with complex things, this is a familiar fact; it is clear, for example, that organs depend very much on context, that the brain depends on the context, that the mind functions differently in a different context. The new suggestion of quantum theory is that this context-dependence is true of the ultimate units of nature. They hence begin to look more like something organic than like something mechanical.

A third point of quantum theory was the property of nonlocal connection. In certain areas, things could apparently be connected with other things any distance away without any apparent force to carry the connection. This "nonlocality" was very opposed to what Einstein wanted and very opposed to mechanism.

A fourth new feature of quantum physics, which was against mechanism, was that the whole organizes the parts, even in ordinary matter. One can see it doing so in living matter, in organisms, where the state of the whole organizes the various parts in the organism. But something a bit similar happens in electrons, too, in various phenomena such as superconductivity. The whole of chemistry, in fact, depends on this idea.

In summary, according to quantum physics, ultimately no continuous motion exists; an internal relationship between the parts and the whole, among the various parts, and a context-dependence, which is very much a part of the same thing, all do exist. An indivisible connection between elements also exists which cannot be further analyzed. All of that adds up to the notion that the world is one unbroken whole. Quantum physics thereby says what relativity theory said, but in a very different way.

These phenomena are evident only with highly refined modes of observation. At the ordinary order of refinement, which was available during the nineteenth century, there was no evidence that any of this was occurring. People formed the mechanistic philosophy on the basis of fairly crude observations, which demonstrates the danger of deciding a final philosophy on the basis of any particular observations; even our present observations may be too crude for something still deeper.

Now one may ask: if there has been such a disproof of mechanism, why is it that most scientists are still mechanistic? The first reason is that this disproof takes place only in a very esoteric part of modern physics, called "quantum mechanical field theories," which only a few people understand, and most of those only deal with it mathematically, being committed to the idea they could never understand it beyond that level. Second, most other physicists have only the vaguest idea of what quantum mechanical field theorists are doing, and scientists in other fields have still less knowledge about it. Science has become so specialized that people in one branch can apply another branch without really understanding what it means. In a way this is humorous, but it has some very serious consequences.

Unbroken Wholeness and Postmechanistic Physics

I propose a view that I have called "unbroken wholeness." Relativity and quantum physics agree in suggesting unbroken wholeness, although they disagree on everything else. That is, relativity requires strict continuity, strict determinism, and strict locality, while quantum mechanics requires just the opposite—discontinuity, indeterminism, and nonlocality. The two basic theories of physics have entirely contradictory concepts which have not been brought together; this is one of the problems that remains. They both agree, however, on the unbroken wholeness of the universe, although in different ways. So it has seemed to me that we could use this unbroken wholeness as our starting point for understanding the new situation.

The question is then how to understand this wholeness. The entire language of physics is now analytic. If we use this language, we are committed to analyzing into parts, even though our intention may be quite the opposite. Therefore, the task is quite difficult.

What I want to suggest is that one of the most important problems is that of *order*. World views have always had views of order. The ancient Greeks had the view of the earth as the center of the universe and of various spheres in order of increasing perfection. In Newtonian physics, the order is that of the particles and the way they move. That is a mechanical order, and coordinates are used mathematically to express that order. What kind of order will enable us to consider unbroken wholeness?

What *is* order? That is a very deep question, because everything we say presupposes order. A few examples: There is the order of the numbers, the order of the words here, the order of the walls, the order in which the body works, the order in which thought works, the order in which language works. We cannot really define order, but we nevertheless understand order somewhat, because we cannot think, talk, or do anything without beginning from some kind of order.

The order physics has been using is the order of separation. Here the lens is the basic idea. If one takes a photograph, one point on the object corresponds to one point on the image. This fact has affected us very greatly, suggesting that everything is made of points. The camera was thereby a very important instrument for helping to strengthen the mechanistic philosophy. It gives an experience that allows everybody to see what is meant by the idea that the universe is nothing but separate parts.

Another instrument, the holograph, can also illustrate this point. The Greek word *holo* means "whole," and *graph* means "to write"; consequently, a holograph writes the whole. With the aid of a laser, which produces highly ordered light, the waves of light from everywhere can be brought to one spot, and just the waves, rather than the image of the object, can be photographed. What is remarkable is that in the resulting picture, each part of it can produce an image of the whole object. Unlike the picture produced by a camera, no point-to-point correspondence with the object obtains. Information about each object is enfolded in each part; an image is produced when this enfolded information is unfolded. The holograph hence suggests a new kind of knowledge and a new understanding of the universe in which information about the whole is enfolded in each part and in which the various objects of the world result from the unfolding of this information.

In my proposal of unbroken wholeness, I turn the mechanistic picture upside down. Whereas the mechanistic picture regarded discrete objects as the primary reality, and the enfolding and unfolding of organisms, including minds, as secondary phenomena, I suggest that the unbroken movements of enfolding and unfolding, which I call the "holomovement," is primary while the apparently discrete objects are secondary phenomena. They are related to the holomovement somewhat as the vortex, in the above example, is related to the unbroken flow of water. An essential part of this proposal is that the whole universe is *actively* enfolded to some degree in each of the parts. Because the whole is enfolded in each part, so are all the other parts, in some way and to some degree. Hence, the mechanistic picture, according to which the parts are only externally related to each other, is denied. That is, it is denied to be the primary truth; external relatedness is a secondary, derivative truth, applicable only to the secondary order of things, which I call the explicate or unfolded order. This is, of course, the order on which modern science has focused. The more fundamental truth is the truth of internal relatedness, because it is true of the more fundamental order, which I call the implicate order, because in this order the whole and hence all the other parts are enfolded in each part.

In my technical writings,[2] I have sought to show that the mathematical laws of quantum theory can be understood as describing the holomovement, in which the whole is enfolded in each region, and the region is unfolded into the whole. Whereas modern physics has tried to understand the whole reductively by beginning with the most elementary parts, I am proposing a postmodern physics which begins with the whole.

POSTMODERN SCIENCE AND QUESTIONS OF MEANING AND VALUE

We have seen that fragmentary thinking is giving rise to a reality that is constantly breaking up into disorderly, disharmonious, and destructive partial activities. Therefore, seriously exploring a mode of thinking that starts from the most encompassing possible whole and goes down to the parts (subwholes) in a way appropriate to the actual nature of things seems reasonable. This approach tends to bring about a different reality, one that is orderly, harmonious, and creative. For this actually to happen, however, a thoroughgoing end to fragmentation is necessary. . . .

Because we are enfolded inseparably in the world, with no ultimate division between matter and consciousness, *meaning and value are as much integral aspects of the world as they are of us.* If science is carried out with an amoral attitude, the world will ultimately respond to science in a destructive way. Postmodern science must therefore overcome the separation between truth and virtue, value and fact, ethics and practical necessity. To call for this nonseparation, is, of course, to ask for a tremendous revolution in our whole attitude to knowledge. But such a change is now necessary and indeed long overdue. Can humanity meet in time the challenge of what is required? The coming years will be crucial in revealing the answer to this question.

Notes

1. See Huston Smith, *Beyond the Post-Modern Mind* (New York: Crossroad, 1982), esp. chap. 8, "Beyond the Modern Western Mind-Set."
2. See David Bohm, *Wholeness and the Implicate Order* (London: Routledge and Kegan Paul, 1980) and other references given therein.

Gaia

JAMES LOVELOCK

THE QUESTION..., "IS the earth a living organism?" is a very unusual question, because most scientists, and that includes most biologists, would answer, without doubt, that it is not alive; so why should I, a fairly hard scientist, wish to contradict them? I have my reasons, and tonight I want to tell you of them, why I think it is alive, how I think it works, and what kind of life it is. Now in doing this, I am speaking as a radical. I have had to become one because science is usually reluctant, and rightly so, to accept new hypotheses as fact; and the Gaia hypothesis, that postulates the earth to be the largest living thing in the solar system, has only been around for about fifteen years.

We don't know if our very real feelings of loss about the destruction of the natural world is just some kind of hypochondria or if they are instincts that are telling us that there are mortal dangers lying ahead. If Gaia is alive, then it could die. The notion of a living planet came, as far as I was concerned, from space research. My part in the story began in 1965 when I was privileged to be called as an advisor to the Jet Propulsion Laboratory at Pasadena in California, from which so much of the glorious exploration of the planets of the solar system has taken place. Above all, it is a place that has taught us so much about the earth by looking at it from outside. My job there was to help design sensitive instruments to measure the soil of the moon and later Mars, to see if life, if it was there at all, even in abundance, could be detected. NASA, like all large organizations, when faced with a problem like "How do you find life on Mars?" goes out and seeks a group of experts on the topic. Well the problem was there weren't any experts on life on Mars because nobody had been there.

It occurred to us after a while that you could in fact find an entropy reduction on a planet by looking for a change in the composition of its

From: *Planet Earth* (Fall 1986): 3–22.

atmosphere, and the argument goes roughly like this. If there is life on the planet, it would be bound to use the mobile media that are available to it, like the oceans or the atmosphere, as a sort of conveyor belt for raw materials and waste products. Such a use of the atmosphere would be bound to change its chemical composition in a way that would make it recognizably different from the atmosphere of a dead planet.

Of course, Mars has an atmosphere in which there is only one mobile medium, so it seemed a good idea to just do a Mars atmospheric composition analysis to tell us if there was life there. In the late 1960s you didn't even have to go to Mars for this, because very competent astronomers with infrared telescopes had done some very complete analyses of both the Marsian and Venusian atmospheres. These showed that both planets are dominated by carbon dioxide and very close to the equilibrium state; so according to our theory, Mars was dead. But in order to prove it, we had to test the thesis on a planet that did have life, and of course, the only one we know is the earth. Now, if you were to put an infrared telescope, an imaginary one, somewhere on Mars, on Nix Olympus 3, for example, and cause it to look back at the earth, that telescope could tell you that there are present the gases oxygen, water vapor, carbon dioxide, methane, and nitrogen. And also it would tell you roughly how much of each of these was present. Now, any competent Marsian chemist could say straightaway from that evidence, and that alone, that the earth bears life in abundance and what kind of life it is. The way he would argue would be as follows: just take two of the gases—methane and oxygen; methane is present at one-and-a-half parts a million, a mere trace you may think. But to sustain it at a steady-state concentration in the presence of oxygen with which it reacts would mean that there would have to be at the earth a source of no less than a thousand million tons of methane every year. More than this, since the methane reaction with oxygen would use up the oxygen, there would also have to be a source at the earth's surface of no less than four thousand million tons of oxygen per year. There's just no way that lifeless chemistry can do it. There would have to be something there—presumably life—that was doing this job. You can use the same argument with the other gases. So, what we've done is to use our method to show that Mars is almost certainly lifeless and the earth is rich and abundant in life.

When I got back to England, in the late 1960s, I couldn't help but wonder how it was that the earth had such a strange and so reactive an atmosphere. You see, it's an atmosphere that's like the gases that go into the intake manifold of your car, a mixture of hydrocarbons and oxygen, not explosive, but that sort of mixture. Still more remarkable to me was the fact that the unstable and reactive mixture was just right for life. And then I began to wonder, could it be we've got it the wrong way around, that the atmosphere isn't an environment in which life swims but that it is a part of life, something that life makes, to sustain for itself a chosen and comfortable environment? In other words, the air was not alive, but as much a part of living things as is the fur of a cat or the shell of a snail.

At that time, the only scientists I could find to take the least interest in this were the chemist Sidney Epton and the eminent biologist Lynn Margulis, and Lynn and I have collaborated on this topic ever since. There was one other person who was encouraging and interested in this hypothesis, and that was the author William Golding, who happens to be a near neighbor in the village where I live in southern England. On a walk around the village discussing the idea with him, he said, "You ought to give it a good solid name. I think you should call it the 'Gaia hypothesis.'" He was a classicist, I wasn't, and I was actually glad and grateful to him, if only because Gaia is a nice, simple four-letter word. Perhaps that's a superficial reason, but nevertheless, I am very grateful to Bill for suggesting it.

Let's look briefly at the evidence there was, nearly thirty years ago now, that justified the postulation of the Gaia hypothesis. Figure 1 shows the atmospheric composition of the earth as it is now, and what it would be if life had never evolved on the earth at all, or alternatively, if we had managed to kill it off. What is so remarkable about our present atmosphere is this coexistence of these reactive gases. It's so remarkable that it shouts the kind of song of life in the infrared that can be heard far beyond the solar system by any roving space craft from anywhere that wanted to look at the earth in that particular radiation spectrum. It is the existence of this beautifully anomalous atmosphere that tells you that the earth bears life. The persistence of it at a steady state suggests strongly that there is some controlling mechanism on earth that's keeping it constant; and, of course, that is what I mean by Gaia.

If life ceased tomorrow, those reducing gases would be gone in a hundred years or so. In a million years, a mere moment in geological history, as George Wald has told us, the earth's atmosphere would fall to a dull state with carbon dioxide dominant, oxygen and nitrogen reduced to quite low levels, and none of the other interesting gases. Let me show you the same diagram expressed as fluxes (in figure 2), that is to say, the quantity of each of these gases that's flowing through the atmosphere just as much as is the most abundant gas, nitrogen. Again, on the dead earth, the fluxes are quite trival.

The other piece of evidence we had which seemed to suggest that we have a living planet was what the astronomers called the "problem of the cool sun." It is one of the less-unshakeable facts of astronomy that stars as they grow old warm up. There is no reason to suppose that our sun is any exception. That being so, when life started on earth, the sun was something like 30 percent cooler than it is now. You might well ask, if that was the case, why wasn't the earth frozen solid, and how did life start, or alternatively, if it was warm enough for life to start, how is it that we're not now boiling?

An intuition that life on the planetary scale might be more than just a catalog of species has persisted in science for a very long time indeed. More recently it led two pioneers, Vladimir Vernadsky in Russia and George Evelyn Hutchinson in this country, to lay the foundations of biological chemistry. It also led courageous ecologists like Eugene Odum and C. S. Holling to postulate that ecosystems might be self-regulating. It led Lewis Thomas to say, with the understanding and wisdom of a physician, that the earth was alive. Now, none of this, nor the scientific evidence that I have just given you, has cut much ice amongst establishment scientists generally. So before we go on to talk about Gaia, I think it is important to look at the criticisms that they have made.

In a splendid recent textbook on the evolution of the atmosphere and the oceans, the geochemist Dick Holland commented that Gaia is a charming idea, but it's not needed to explain the facts of the earth. He expressed the general consensus amongst earth scientists by saying, "This is the best of all worlds, but only for those who have adapted to it." As for the climatologists, Stephen Schneider, in a book published last year, similarly referred to Gaia as teleological, echoing the objection of many of his colleagues. The real fallacy of the earth scientists lies in their failure to recognize the implications of their own research. They have discovered, but they haven't yet digested, the fact that the air, the ocean, the rocks, are all massively modified or are the direct process of life. This means that life doesn't inhabit a dead world of geochemistry and try to adapt to it. It is, rather, adapting to and adapting a living world that it, itself, has made.

A much more serious and interesting criticism of Gaia came from the molecular biologist W. Ford Doolittle. He rejected Gaia on the grounds that planetary self-regulation would need foresight and planning by the biota. Clearly this was impossible, he said. He also observed that there was no way for the global scale to evolve by natural selection. Similarly, co-evolution always occurs between closely coupled entities, and the concept of Gaia would require the existence of some kind of giant nanny who has looked after the earth ever since life began. This kind of criticism is quite easy to make, but it's just as easy to dispose of. It arises because most scientists are quite unable to distinguish between the classical circular false argument and the circular logic that describes the operation both of self-regulating mechanisms like thermostats and automatic pilots, and of living things. If you doubt it, try explaining to yourself how the simple thermostat that operates in your kitchen oven works without using a circular argument.

This was, I think, the most formidable criticism, and it didn't come from dogma or from the conservatism of an entrenched position. For a while I wondered if Gaia was just another vague idea in the same category as Sheldrake's morphogenic field—entertaining, but unprovable. Then it occured to me that there was an answer to Doolittle's criticism. The problem was that at first it had seemed that the feedback loops that link life with the environment were so numerous and so intricate that there wasn't a dog's chance of quantifying, let alone understanding, them. As you probably know, in places like NCAR (National Center for Atmospheric Research) in Boulder, Colorado, there are some of the largest computers in the world; and they are stretched to their limits to predict even next week's weather. So to try to predict biology and geochemistry and everything else is quite beyond our capabilities at the moment. It was understandable, then, to think that you couldn't really hope to understand this system.

What we can do in science, just as we do in art, is to abstract the essence. What I thought to do is to make a line drawing of Gaia which would show perhaps a likeness. For this purpose I reduced the environment to a single variable—temperature. It didn't absolutely have to be temperature, it could have been acidity, or something else. I reduced the biology to a single species— daisies. It needn't have been daisies, it could have been bacteria or whales, but I chose daisies for reasons that will shortly become apparent.

What I want you to do is imagine a planet which is very like the earth but has a lot less ocean. It travels at the earth's orbit around a star of the same mass as our sun, and the planet spins just as the earth spins. Its atmosphere has few clouds and what greenhouse gases are present are always constant, so its climate is very uncomplicated. In such circumstances, it is easy to calculate what the mean surface temperature of the planet would be from the balance between the radiation it receives from its star and the heat it reradiates away into space.

If you know the color of the planet, the "albedo" as the astronomers call it, you know its mean temperature. There's enough water on our imaginary planet for everything to be well watered, but it's not an ocean planet like earth. Now figure 3 shows you what would happen to that planet's temperature as its star warmed up, as stars do. The line across the (A) diagram shows the rise of temperature, the vertical axis is in degrees Celsius and the horizontal axis is solar luminosity, with "1" being equivalent to the output of our own present sun. So, as you can see, the dead planet's temperature smoothly rises as the sun warms up. The (B) diagram next on the curve is what happens to much life on earth, how it grows at different temperatures. It starts growing when the temperature is plus five Celsius. It grows best around about twenty, and it dies off when the temperature reaches about anything above forty, when it's too hot for it. So the evolution of life on this imaginary world according to standard textbook biogeochemistry would be like that diagram. As the sun warms the planet up, life would start, it would come maximal when the temperature reached twenty degrees Celsius and would begin to die off as it is heated up further. There would be some adaptation, to be sure, but the general condition would be like this.

I think what actually happens is very different indeed. Consider what would happen to our imaginary world if on it we had a single colored species of daisies that happened to be lighter in color than the planet's surface.

When the planet's temperature reached five degrees Celsius, warm enough for life to start, the white daisies would start growing. As they grew and spread, they would alter the color of the planet and make it reflect more sunlight back into space, instead of warming up. As on the dotted line, the planet's temperature would follow the (A) curve. The growth of a single species of daisies as it spread across the planet would regulate the planet's temperature for a wide range of solar luminosities. Eventually it would get too hot, and the system would crash.

Now, as most of you who are ecologists know, a single-species culture is not a very good or stable ecosystem. So let's see what happens if you have two species, dark and light, both present on the planet. Here again, when the temperature reaches five degrees Celsius, the daisies start growing, but this time there are dark ones present. Dark daisies have a mean advantage over light ones because, being darker, they will be warmer. So they will spread rapidly, and as they spread they will warm not only themselves but the planet, and so their spread is explosive. A positive feedback takes place then, and the temperature rapidly rises, as does the population of black daisies to somewhere past the optimum temperature for life on the planet. It won't go on rising, because it would then get too hot for the black daisies. They would die, and the planet would cool off.

But now, of course, it would be warm enough for the white ones to grow as well. As the star evolves, so do the population of the two species, and together they regulate the planetary temperature, now much closer to an ideal. Two species, again, isn't much diversity, so I thought I would go whole hog and put ten species on earth. If any of you are theoretical ecologists, you would know that this is impossible. You try to model the competition of ten species, and the mathematics blows up on you; but what theoretical ecologists possibly don't know is that if you join the environment back as a feedback into the system, you can model as many species as you like, as you see from the (A) diagram. With ten species present, temperature regulation works much better over a much wider range. On diagram (B) you see the interplay of the different daisies' colors. They are all slightly different in color, uniformly spaced, and they grow when they are fit to grow and the environment is fit for them.

The top diagram (C) shows you the diversity which is an ecologist's parameter. We see in this diagram that a lot of diversity is a good thing because at the position of maximum stability in the middle of the temperature curve there are the most species—the height of the greatest diversity. There is the least diversity when the system is under maximum stress, and I think, again, this is rather contrary to standard ecological teachings. Now, you can go further: I have made a model world in which there are five species of daisies that are being eaten by rabbits, and the rabbits are being eaten by the foxes; the foxes in turn are being shot by men. This ecosystem is even more stable. You can perturb it at intervals and it will bounce back and regain its homeostasis. So daisy world is a fertile planet and it can accommodate very many strange ecologies, and nearly always there are unplanned adaptations and competition leading to homeostasis. The models have in common four essentials which I think are the four essentials of the sort of Gaia I am talking about, being the conditions for planetary homeostasis.

Well, of course, daisy world is just a model. It doesn't follow that the earth regulates its temperature by growing daisies. The purpose of the model was to provide insights into the mechanism of the real world and of Gaia. Daisy world showed that the alteration of the environment by the growth of daisies and the effect of the environment upon their growth served to bind life and its environment in a web of feedback. This means that at least as far as that model goes and perhaps as far as the world itself goes, the evolution of life and the evolution of the rocks are a single and indivisible process.

We also begin to see Gaia as a collective; it's a kind of "colligative property" of life. This is a term used by physicists about collections of things like molecules. If you have a lot of molecules, you have properties like temperature and pressure that an individual molecule does not possess, and I think one can regard life in that sort of way. Obviously, it's a colligative that is a lot more powerful than a termite's nest or a nation-state. Also by the same reckoning, it's very much less organized than the simplest of living cells. But it's much more alive than a virus, for that's just a piece of software.

A theory, of course, is judged by the accuracy of its predictions and by its practical utility. Interestingly, Gaia has performed fairly well under this test. I will just conclude the evidence by taking one single example, going back to the affair of the cool sun, and of how the climate has stayed constant since life has been on earth. When I say "constant," somebody is bound to say, "But there are ice ages, surely that's not constant." The answer there is that an ice age doesn't really seriously affect the whole earth as a living system. An ice age is the desired state of Gaia, and our present warm interglacial period is a real fever—and interestingly, of course, we are adding to it. The reason that the planet was warm enough for life to start three and half thousand million years ago was that at that time the earth's atmosphere was just like that of Mars or Venus. It was dominated by carbon dioxide. There was something like 30 percent carbon dioxide in the atmosphere, and that was more than enough to give a comfortable greenhouse that compensated for the cool sun back then.

But the intriguing question is, what happens next? As soon as life appears on the planet, it is bound to change the atmosphere by the same arguments as I gave earlier for the life-detection experiment. The only source of carbon available to life was the carbon dioxide in the atmosphere, and being greedy it would start eating. In other words, it was eating the blanket that was keeping it warm, and such is the capacity of life. Even the early life of this planet, which might have been much less efficient than the present life, could nevertheless have taken all of the carbon dioxide in as little a time as a million years. It didn't. We know that because there is no sign that the earth grows solid in any time of its history.

What I think happened was that a Gaian situation very like daisy world grew up. The first protozoan did take the carbon dioxide out and the temperature did fall, but from a warm state to one that was still not too cold. As this process went on, the dead bodies of photosynthesis that had died of old age accumulated on the bottom of whatever shallow seas there were then and this opened a gigantic niche for the methane gases. There are organisms that live in biogas generators and in swamps and the like. They came along and recycled the carbon back into the atmosphere as a mixture of methane and carbon dioxide and so restored the greenhouse. So you have a model of photosynthesis acting like the white daisies tending to cool the planet by removing carbon dioxide, and the methane gases acting like dark daisies restoring the warmth. This situation persisted for a thousand million years. Just dumb bugs ran a stable biosphere, a stable Gaia back then.

All living things are in the business of fixing carbon dioxide and putting it into the ground. When a tree grows, it puts tons of carbon underground because there's as much tree beneath the earth as there is above it. When that tree dies, tons of carbon dioxide are released in the soil right near the rocks that need weathering, so the whole of the biosphere is a kind of giant pump, pumping carbon dioxide out of the air to keep the planet cool. The same process goes on in the ocean with other creatures that live on the surface. And of course as usual, we are trying to swamp that process by adding carbon dioxide to the atmosphere.

The interesting point about this particular system (ours) is that it's very like the daisy world system before it crashed. We can't reduce carbon dioxide much below its present level without having too little for mainstream plants to grow on. During the last glaciation, it was down to 180 parts per million, and mainstream plants would find it very difficult to grow at much below that. We're very near the edge; and you can predict that if the process is to be used for climate regulation, it has about a hundred million years to go, which is a very short time as far as Gaia goes. What we are doing by perturbing it is singularly dangerous, because if you perturb a stabilizing system that is near the end of its capacity to regulate, it is very liable to jump to a new stable state. It will certainly be one that will not be anywhere near as comfortable for us as the present one. Gaia in that sense is no nanny looking after us.

Well, that then is some of the evidence and some of the criticisms of the Gaia hypothesis. If I had given the whole lot to you, it still would not prove her existence; but in science we're not really concerned with the unattainable perfection of truth. What we really need are good guesses, working hypotheses, and facts; and on this basis, I think that it would be a good idea to assume that the earth and Gaia are alive. Then you can enjoy being a part of things and feel the outrage over the destruction of some part which is part of yourself as something real and not as just a sentimentality.

I would want to stress that in no way is Gaia fragile. Gaia has withstood devastations far beyond our powers at least thirty times during the three and a half billion years of her life-span. Nothing that we can do threatens her. But, of course, if we transgress in our pollutions and our forest clearance, Gaia can move to a new stable state, and one that's no longer comfortable for us. So living with Gaia is not so different from a human relationship. It is an affair of the heart as well as the head; and if we are to do it lovingly, it is something that must be renewed on a daily basis if it is to succeed.

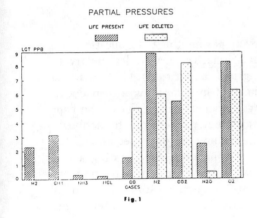

Fig. 1

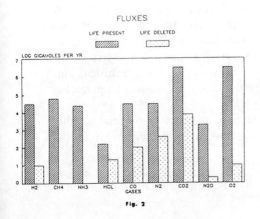

Fig. 2

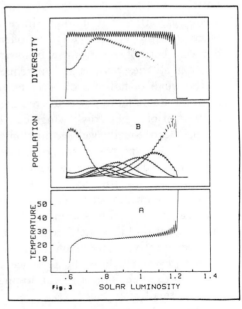

Fig. 3

Predictability: Does the Flap of a Butterfly's Wings in Brazil Set Off a Tornado in Texas?

EDWARD LORENZ

LEST I APPEAR FRIVOLOUS in even posing the title question, let alone suggesting that it might have an affirmative answer, let me try to place it in proper perspective by offering two propositions: (1) if a single flap of a butterfly's wings can be instrumental in generating a tornado, so also can all the previous and subsequent flaps of its wings, as can the flaps of the wings of millions of other butterflies, not to mention the activities of innumerable more powerful creatures, including our own species; and (2) if the flap of a butterfly's wings can be instrumental in generating a tornado, it can equally well be instrumental in preventing a tornado. More generally, I am proposing that over the years, minuscule disturbances neither increase nor decrease the frequency of occurrence of various weather events such as tornados; the most that they may do is to modify the sequence in which these events occur. The question which really interests us is whether they can do even this—whether, for example, two particular weather situations differing by as little as the immediate influence of a single butterfly will generally, after sufficient time, evolve into two situations differing by as much as the presence of a tornado. In more technical language: Is the behavior of the atmosphere *unstable* with respect to perturbations of small amplitude?

The connection between this question and our ability to predict the weather is evident. Since we do not know exactly how many butterflies there are,

Paper presented at the annual meeting of the American Association for the Advancement of Science, Section on Environmental Sciences: New Approaches to Global Weather: The Global Atmospheric Research Program; 29 December 1972.

nor where they are all located, let alone which ones are flapping their wings at any instant, we cannot, if the answer to our question is affirmative, accurately predict the occurrence of tornados at a sufficiently distant future time. More significantly, our general failure to detect systems even as large as thunderstorms when they slip between weather stations may impair our ability to predict the general weather pattern even in the near future.

How can we determine whether the atmosphere is unstable? The atmosphere is not a controlled laboratory experiment; if we disturb it and then observe what happens, we shall never know what would have happened if we had not disturbed it. Any claim that we can learn what would have happened by referring to the weather forecast would imply that the question whose answer we seek has already been answered in the negative.

The bulk of our conclusions are based upon computer simulation of the atmosphere. The equations to be solved represent our best attempts to approximate the equations actually governing the atmosphere by equations which are compatible with present computer capabilities. Generally two numerical solutions are compared. One of these is taken to simulate the actual weather, while the other simulates the weather which would have evolved from slightly different initial conditions, that is, the weather which would have been predicted with a perfect forecasting technique but imperfect observations. The difference between the solutions therefore simulates the error in forecasting. New simulations are continually being performed as more powerful computers and improved knowledge of atmospheric dynamics become available.

Although we cannot claim to have proven that the atmosphere is unstable, the evidence that it is so is overwhelming. The most significant results are the following:

1. Small errors in the coarser structure of the weather pattern—those features which are readily resolved by conventional observing networks—tend to double in about three days. As the errors become larger the growth rate subsides. This limitation alone would allow us to extend the range of acceptable prediction by three days every time we cut the observation error in half and would offer the hope of eventually making good forecasts several weeks in advance.

2. Small errors in the finer structure—for example, the positions of individual clouds—tend to grow much more rapidly, doubling in hours or less. This limitation alone would not seriously reduce our hopes for extended-range forecasting, since ordinarily we do not forecast the finer structure at all.

3. Errors in the finer structure, having attained appreciable size, tend to induce errors in the coarser structure. This result, which is less firmly established than the previous ones, implies that after a day or so there will be appreciable errors in the coarser structure, which will thereafter grow just as if they had been present initially. Cutting the observation error in the

finer structure in half—a formidable task—would extend the range of acceptable prediction of even the coarser structure only by hours or less. The hopes for predicting two weeks or more in advance are thus greatly diminished.

4. Certain special quantities, such as weekly average temperatures and weekly total rainfall, may be predictable at a range at which entire weather patterns are not.

Regardless of what any theoretical study may imply, conclusive proof that good day-to-day forecasts can be made at a range of two weeks or more would be afforded by any valid demonstration that any particular forecasting scheme generally yields good results at that range. To the best of our knowledge, no such demonstration has ever been offered. Of course, even pure guesses will be correct a certain percentage of the time.

Returning now to the question as originally posed, we notice some additional points not yet considered. First of all, the influence of a single butterfly is not only a fine detail—it is confined to a small volume. Some of the numerical methods which seem to be well adapted for examining the intensification of errors are not suitable for studying the dispersion of errors from restricted to unrestricted regions. One hypothesis, unconfirmed, is that the influence of a butterfly's wings will spread in turbulent air, but not in calm air.

A second point is that Brazil and Texas lie in opposite hemispheres. The dynamical properties of the tropical atmosphere differ considerably from those of the atmosphere in temperate and polar latitudes. It is almost as if the tropical atmosphere were a different fluid. It seems entirely possible that an error might be able to spread many thousands of miles within the temperate latitudes of either hemisphere, while yet being unable to cross the equator.

We must therefore leave our original question unanswered for a few more years, even while affirming our faith in the instability of the atmosphere. Meanwhile, today's errors in weather forecasting cannot be blamed entirely, or even primarily, upon the finer structure of weather patterns. They arise mainly from our failure to observe even the coarser structure with near completeness, our somewhat incomplete knowledge of the governing physical principles, and the inevitable approximations which must be introduced in formulating these principles as procedures which the human brain or the computer can carry out. These shortcomings cannot be entirely eliminated, but they can be greatly reduced by an expanded observing system and intensive research. It is to the ultimate purpose of making not exact forecasts but the best forecasts which the atmosphere is willing to have us make that the Global Atmospheric Research Program is dedicated.

Science in a World of Limited Predictability

ILYA PRIGOGINE

THE TWENTIETH CENTURY HAS been a remarkable century in physics. It started with completely new theories and conceptions—quantum mechanics and relativity. Then in the second third of this century came some absolutely unexpected discoveries, which nobody could have predicted, such as the discovery that matter is unstable, and that elementary particles can transform into each other. The second discovery was that our universe has a history. Classically, the idea was that there could be no history of the universe because the universe contains everything. And the third discovery, I would say, was the discovery that non-equilibrium irreversibility can be a source of organization. These changes severely radicalize our views of space and time. And in essence, what we are trying to do now is to incorporate these unexpected discoveries into a more consistent picture. I want to emphasize that from the point of view of classical physics, there was a dichotomy—on the one hand, physics had the view of the universe as a giant automaton, at some stage we were satisfied with time-reversible and deterministic laws. On the other hand, when we see our own internal spiritual life, we see the importance of creativity, the fact that time is irreversible, and the fact that we have at least the feeling that we see, in a sense, order coming out of disorder—new ideas from fragments coming together.

Now, the view of classical physics is not an accident; in fact, it was elaborated in a very famous discussion between Gottfried Wilhelm Leibniz and Samuel Clarke (speaking in the name of Sir Isaac Newton). This is still a very interesting discussion to read today, nearly two centuries after it took place. Leibniz attacks Newton by saying that Newton has a poor idea

From: "The Rediscovery of Time: Science in a World of Limited Predictability" (Paper read to the International Congress on "Geist & Natur," Hanover 21–27 May 1988), excerpts.

of God because he believes that God is inferior to a good watchmaker. In the view of Newton, God had to repair the universe from time to time. In other words, in more modern language, one would say that Newton curiously was in favor of an evolutionary view of the universe. Clarke (Newton) said the idea of Leibniz, that everything was created in a single step, made God a *rex autiosus*, a *roi infini*, because once the creation of the universe had taken place, God had no more role. What is so important is that there is a theological component in this discussion in the way in which people understood the role of the scientist. It is clear that as people imagined God in the seventeenth century, he, of course, knew everything. For him there could be no difference between past, present, and future. And Leibniz's view became more and more important until Laplace said that for a sufficiently informed being nothing can happen which is not already in the present. For a scientist, who would be the representative of God on earth, time would be an illusion, as Einstein said. In other words, one of the main components of classical science was to try to reach a knowledge of the universe which would be a knowledge we could imagine God would have. In physics, there have been a lot of demons—Maxwell's and Laplace's—the idea being that there may be a knowledge which would involve a kind of knowledge from above. One of the important things which we have learned in this century is that this is impossible. In essence, we have only a window on nature, and out of the elements we have, we have to extrapolate. We are ourselves involved in the nature we describe.

The conflict between time as perceived by an evolutionary theory or by our existential experience and time of fundamental classical physics came to the forefront after the work of Ludwig Boltzmann. Boltzmann tried to give a mechanical interpretation of the second law of thermodynamics, of the increase of entropy, when he obtained a very interesting description of irreversibility. However, people were not convinced. They said, How can you even hope to obtain a description of irreversibility when the basic laws of nature are deterministic and reversible? That is impossible. Finally, Boltzmann had to retreat and to accept that on the basic level there is no time.

I was astonished that nobody, or very few people, protested. Nobody, or very few people at least, said, but there must be something difficult there, something which still has to be revised. How can we accept this dichotomy between a phenomenological description in which time exists and a fundamental description in which time would disappear? One of the main points in the evolution of this century is that now we understand better how we can go beyond this dichotomy; now we can understand irreversibility on a fundamental basis.

I would like to quote a sentence by the British mathematician James Lighthill, who wrote two years ago: "I have to speak on behalf of the broad global fraternity of researchers on mechanics. We collectively wish to apologize for having misled the general educated public by spreading ideas of all the determinisms of systems satisfying Newton's law of motion that after 1960 were to be proved incorrect." This statement is quite amazing. It helps us, of course, that an individual scientist apologizes for mistakes he may have made. That someone in the name of all the people working in classical dynamics would apologize for a mistake made for three centuries is something unlikely in the history of science. In other words, he says that the whole interpretation which was the basis of all our philosophy of space and time—the way in which Immanuel Kant, Henri Bergson, and others have seen classical science—was wrong.

I would like to explain briefly how these new aspects entered classical physics. And I will first speak briefly about non-equilibrium processes— thermodynamics—and then about classical mechanics, because I believe that it is there that we see the greatest change. Then I would like to speak about the way in which these ideas begin to appear in modern physics, that is, in quantum mechanics. Finally, I will say a few words about recent cosmology and the role of time and irreversibility in the creation of our universe.

First, thermodynamics is based, as you know, on the second law. The second law says that entropy for an isolated system will increase. If it is not isolated in any way, entropy will be produced by irreversible processes. Now, people had always tried to minimize this statement. They have said it is rather trivial. If I have two boxes, with a hundred particles in the left box and twenty particles in the right, the boxes bearing the same volume, of course, then after some time I will have fifty-fifty. That is rather trivial. Therefore, they said, don't worry about the second law. However, we now understand that this is completely wrong. Irreversible processes always create—or nearly always create—both order and disorder at the same time. Take again the example of the two boxes. If I have two boxes of hydrogen and nitrogen and I heat one and cool the other, then there will be more hydrogen on one side and more nitrogen on the other. In other words, I produce order. It is true that I have to pay a price. I have to pay a price because I have to dissipate energy through the heat flow. That this price leads to order in this and many other examples is now well understood. In fact, we are changing our paradigm. The classical paradigm was that the crystal is an ordered subject, an ordered object. However, we now know that even a nice crystal is far from being so ordered. The particles are moving, and this motion is described by incoherent thermal waves. On the other hand, what we believe to be disorder, turbulence, for example, appears today as highly ordered. In order to have turbulence, particles have to be correlated for an enormous distance, distances much larger than intermolecular distances, distances which are macroscopic. The coherence has to be on a macroscopic scale. We see now that non-equilibrium processes contain much more order than most or even all equilibrium processes.

This completely changes our view of the relation between order and dis-order, on the one hand, and equilibrium and non-equilibrium on the other. And over the last few years, more and more of these kinds of phenomena have been discovered in hydrodynamics and chemistry. Not a month goes by without discovering another non-equilibrium type of structure. In es-sence, we see a kind of unexpected evolution of chemistry because chemis-try is a science of irreversible processes. It is a science of nonlinear processes because we have collisions and interactions between particles. What we see now is that irreversibility and nonlinearity lead to self-organization. And we can follow this in many fields: the science of materials and many other examples. Another aspect I want to emphasize is that irreversibility leads to an enhancement of the role of fluctuation. We give up a deterministic description because nonlinearity leads through bifurcations to the possibil-ity of different solutions. When you come to this type of bifurcation, the system can only be described in a probabilistic way. You cannot predict with certainty where the system will go. Quantum mechanics has become famous for introducing probabilistic concepts on the microscopic scale. Here we have a probabilistic concept on the macroscopic scale. So, as I said, non-equilibrium physics is in a period of exponential expansion. This is still phenomenology because it involves many particles. This involves sys-tems which I cannot solve exactly. It is therefore interesting to understand how this irreversibility can be related to dynamics. And here we come to the revolution in dynamics to which Lighthill alluded.

What type of new dynamical systems have been discovered? The great names are Henri Poincaré at the end of the last century and the beginning of this century, and Andrei Nikolaevich Kolmogoroff, perhaps the greatest mathematician of this century. What Kolmogoroff discovered is that there are many dynamical systems which are so unstable that whatever the dis-tance between two trajectories, the two trajectories will diverge exponen-tially in time. In other words, even if the distance between the two trajectories was initially very small, after some time it will become as large as you want. The way in which this expansion is growing is called the Lyapunov expo-nent, and this Lyapunov exponent determines what we can call the tempo-ral horizon of the dynamical system. After the temporal horizon, we can no longer speak about trajectories. The concept of a trajectory is lost after some time, whatever the initial condition may be, unless you have infinite knowledge and unless you know exactly the initial condition. But in phys-ics we never know exactly the initial condition. We may have a knowledge as precise as you want, but we never know a point; we know a small region. And what is important is that for these unstable systems, there is an essen-tial difference between knowing a small region and going to the limit of a point. The point behaves differently from the region. The finite knowledge corresponds to a type of behavior—and that is our situation—different from the limit of infinite knowledge, and that is true for unstable dynamical systems. For stable dynamical systems, there is no difficulty. You can go to the limit of infinite knowledge—it is the same type of behavior. For this reason, astronomers can predict the position of the earth around the sun in about five million years, which is already quite a long time. However, nobody can hope to predict the weather for better than, let's say, one to two weeks. In other words, it is not that we are more stupid in meteorology or climatology than in astronomy, but that the systems are unstable. The climatological system, the meteorological system, are unstable dynamical systems. They have Lyapunov exponents, and that makes the problem of prediction quite different.

Let me explain the meaning of these highly unstable systems by giving a simple example. I take a piece of money and I throw it and I suppose it is a game of chance. Therefore, I find the game to be fifty-fifty. Now, once I say fifty-fifty, it is a problem of probability. However, this piece of money is also a heavy object. It satisfies Newton's equation of motion. How then does probability enter? Well, probability enters because I do not know my initial conditions. Therefore, let me make my initial conditions more and more precise. Then two possibilities can happen. My initial condition can be made so precise that I can predict what will happen, and then, of course, it is a deterministic system. My ignorance was in looking at the system as a probabilistic system. Whatever I do, whatever the precision of my initial conditions, I always find the chances fifty-fifty. How is this possible? It is possible if each initial condition which leads to one side is so rounded by the initial condition which leads to the other side. Everybody remembers the nice law about number theory and everybody may remember that a rational number is rounded by irrationals and an irrational number is rounded by rationals. So you imagine briefly that rational numbers would give one result, irrational numbers would give the other result. But I never know if this is a rational or an irrational number. Even if I know a hundred digits, a billion digits, I still do not know if this is a rational number or an irrational number. And what is so extraordinary—and for me still a surprise—is that this type of unstable dynamic is prevalent. We find it everywhere: in liquids, in gases, in quantum mechanics, everywhere. In fact, the reason is (but I do not want to go into technicalities) that you find this type of instability every time you have strong resonances, every time you have an oscillator, which has the same frequency or frequencies which are very closely related and can transfer energy. Resonances lead to very chaotic behavior, and it is, finally, for this reason that we find so many systems which are unstable in nature.

Now, how do unstable systems behave? How do they behave in contrast, let us say, with planetary motion? The important element is that unstable systems are not controllable. The classical view of nature, the classical view of the laws of nature, of our relation with nature, was domination. That we can control everything. If we change our initial conditions, the trajectories change slightly. If we change our initial wave function, the final wave function is changed slightly. We can, therefore, control things. And this remains true for many situations. When we send a Sputnik into space, fortunately we can control its trajectory. But that is not the general situation. If, for example, I take a few hundred particles and I give them initially all kinds of positions and velocities, whatever I do, I cannot prevent the system from reaching thermal equilibrium and a Maxwell-Boltzmann distribution after some time. In other words, the system escapes my control and behaves—if I can say so—like it wants to behave independently of the instructions which I gave it. You could say again that a system with a few hundred particles, a liquid or a gas, is a very complex system and it is not so astonishing that I cannot control it. But recently some of my colleagues performed numerical analyses, computer experiments, and other analytical calculations for the simplest problem in electrodynamics, the emission of radiation. In the emission of radiation, if I have a charge which is oscillating, it radiates according to classical electrodynamics and sends electromagnetic waves, that is, light, into space. In quantum mechanics, we would say it sends photons into space. Now what we have shown is that we can prepare the initial electron of the initial charge in many ways and we can even prevent the system from radiating for some time. But whatever we do, after some time it behaves in agreement with the laws of electrodynamics. In other words, it is escaping from our control. It is dissipating its vibrational energy into radiation, whatever we do in the preparation of the system.

The main point which I want to emphasize is that we see in nature the appearance of spontaneous processes which we cannot control in the strict sense in which it was imagined to be possible in classical mechanics. This is not giving up scientific rationality. After all, we have not chosen the world in which we are living. We scientists have to describe the world in which we are living. And the world in which we are living is highly unstable. What I want to emphasize, however is that this knowledge of instability may lead to other types of strategies, may lead to other types of interacting systems.

Well, thermodynamics, non-equilibrium processes, and classical dynamics are, of course, parts of classical physics. And it is very remarkable that parts of classical physics which were supposed to be in a final form or a nearly final form are precisely some of the fields which are changing so much in the present evolution of science. But, of course, we can expect that this evolution of classical physics will have very deep consequences in the more modern fields of quantum mechanics and relativity, because all the other fields of physics are, in essence, born inside classical physics. And they have used some of the concepts of stable dynamical systems, which were believed for such a long time to be the only type of dynamical systems. . . .

Quantum mechanics has this strange dual structure. On the one hand, you have the deterministic Schrödinger equation. Once you know the wave function, you can predict what will happen in the future just as you could calculate what happened in the past. Irreversibility and stochasticity come from our measurement. This is not such an absurd idea, because if I look, for example, at planetary motion, I know that it is reversible motion. But when I look at it, I use for the chemical reaction an irreversible process. Therefore, my observation of a time-reversible process introduces irreversibility. But quantum mechanics made an enormous statement when it said that all of irreversibility is due to my observation—in other words, to my measurement—and that was exactly the situation of classical physics before the discoveries of Kolmogoroff and Poincaré. What is very difficult to believe now is that in quantum physics as in classical physics, we have all the spontaneous processes: disintegration, emission of photons which again have nothing to do with our measurement, which comes from the intrinsic instability of dynamical systems. So I believe that the future revolution of quantum mechanics will be dealing with a better understanding of the classification of dynamical quantum systems. . . .

It is quite remarkable to me that this change . . . has appeared at the very moment humanity is going through an age of transition, when instability, irreversibility, fluctuation, amplification are found in every human activity. The idea of chaos, of amplification, became very popular in the United States after 19 October 1987, the famous Black Monday of the crash of Wall Street. From this time on, everybody became interested in chaos and amplification. What is so interesting is that there is a kind of overall cultural atmosphere, be it in science, be it in human science, which is developing at the end of this century.

Conclusion

Principles of Environmental Justice

The First National People of Color Environmental
Leadership Summit
October 24–27, 1991
Washington, D.C.

PREAMBLE

WE, THE PEOPLE OF color, gathered together at this multinational People of Color Environmental Leadership Summit, to begin to build a national and international movement of all peoples of color to fight the destruction and taking of our lands and communities, do hereby re-establish our spiritual interdependence to the sacredness of our Mother Earth; to respect and celebrate each of our cultures, languages and beliefs about the natural world and our roles in healing ourselves; to insure environmental justice; to promote economic alternatives which would contribute to the development of environmentally safe livelihoods; and to secure our political, economic and cultural liberation that has been denied for over 500 years of colonization and oppression, resulting in the poisoning of our communities and land and the genocide of our peoples, do affirm and adopt these Principles of Environmental Justice:

1. Environmental justice affirms the sacredness of Mother Earth, ecological unity and the interdependence of all species, and the right to be free from ecological destruction.
2. Environmental justice demands that public policy be based on mutual respect and justice for all peoples, free from any form of discrimination or bias.
3. Environmental justice mandates the right to ethical, balanced and responsible uses of land and renewable resources in the interest of a sustainable planet for humans and other living things.
4. Environmental justice calls for universal protection from nuclear testing, extraction, production and disposal of toxic/hazardous wastes and poisons and nuclear testing that threaten the fundamental right to clean air, land, water, and food.
5. Environmental justice affirms the fundamental right to political, economic, cultural and environmental self-determination of all peoples.

6. Environmental justice demands the cessation of the production of all toxins, hazardous wastes, and radioactive materials, and that all past and current producers be held strictly accountable to the people for detoxification and the containment at the point of production.

7. Environmental justice demands the right to participate as equal partners at every level of decision-making including needs assessment, planning, implementation, enforcement and evaluation.

8. Environmental justice affirms the right of all workers to a safe and healthy work environment, without being forced to choose between an unsafe livelihood and unemployment. It also affirms the right of those who work at home to be free from environmental hazards.

9. Environmental justice protects the right of victims of environmental injustice to receive full compensation and reparations for damages as well as quality health care.

10. Environmental justice considers governmental acts of environmental injustice a violation of international law, the Universal Declaration On Human Rights, and the United Nations Convention on Genocide.

11. Environmental justice must recognize a special legal and natural relationship of Native Peoples to the U.S. government through treaties, agreements, compacts, and covenants affirming sovereignty and self-determination.

12. Environmental justice affirms the need for urban and rural ecological policies to clean up and rebuild our cities and rural areas in balance with nature, honoring the cultural integrity of all our communities, and providing fair access for all to the full range of resources.

13. Environmental justice calls for the strict enforcement of principles of informed consent, and a halt to the testing of experimental reproductive and medical procedures and vaccinations on people of color.

14. Environmental justice opposes the destructive operations of multinational corporations.

15. Environmental justice opposes military occupation, repression and exploitation of lands, peoples and cultures, and other life forms.

16. Environmental justice calls for the education of present and future generations which emphasizes social and environmental issues, based on our experience and an appreciation of our diverse cultural perspectives.

17. Environmental justice requires that we, as individuals, make personal and consumer choices to consume as little of Mother Earth's resources and to produce as little waste as possible; and make the conscious decision to challenge and reprioritize our life-styles to insure the health of the natural world for present and future generations.

Adopted, 27 October 1991
The First National People of Color Environmental Leadership Summit
Washington, D.C.

INDEX